Electrochemical Biosensors for Whole Blood Analysis

Fan Xia · Hui Li · Shaoguang Li · Xiaoding Lou
Editors

Electrochemical Biosensors for Whole Blood Analysis

Springer

Editors
Fan Xia
State Key Laboratory of Biogeology and Environmental Geology, Engineering Research Center of Nano-Geomaterials of Ministry of Education, Faculty of Materials Science and Chemistry
China University of Geosciences
Wuhan, China

Hui Li
State Key Laboratory of Biogeology and Environmental Geology, Engineering Research Center of Nano-Geomaterials of Ministry of Education, Faculty of Materials Science and Chemistry
China University of Geosciences
Wuhan, China

Shaoguang Li
State Key Laboratory of Biogeology and Environmental Geology, Engineering Research Center of Nano-Geomaterials of Ministry of Education, Faculty of Materials Science and Chemistry
China University of Geosciences
Wuhan, China

Xiaoding Lou
State Key Laboratory of Biogeology and Environmental Geology, Engineering Research Center of Nano-Geomaterials of Ministry of Education, Faculty of Materials Science and Chemistry
China University of Geosciences
Wuhan, China

ISBN 978-981-99-5643-2 ISBN 978-981-99-5644-9 (eBook)
https://doi.org/10.1007/978-981-99-5644-9

This Springer imprint is published by the registered company Springer Nature Singapore Pte Ltd.
The registered company address is: 152 Beach Road, #21-01/04 Gateway East, Singapore 189721, Singapore

Paper in this product is recyclable.

Foreword

Welcome to "Electrochemical Biosensors for Whole Blood Analysis," a comprehensive exploration of the fascinating field that lies at the intersection of electrochemistry, biology, and healthcare. This book examines the role of electrochemical biosensors in revolutionizing whole-blood analysis for diagnostic purposes. Biosensors have revolutionized diagnostics by seamlessly integrating sensing technologies with biological systems. Electrochemical biosensors, in particular, offer high sensitivity, specificity, and portability, enabling real-time monitoring of analytes in whole blood. Machine learning techniques can process electrical signal data with interference and overlap, thus making electrochemical biosensors intelligent.

This book introduces the principles of electrochemistry and the complexities of whole blood as an analyte matrix. It explores various transduction mechanisms, such as potentiometric and amperometric techniques, impedance spectroscopy, and field-effect devices, used in electrochemical biosensors. The book focuses on biological recognition elements, including enzymes, antibodies, nucleic acids, and aptamers, which are crucial for biosensor design. Immobilization strategies, stability enhancement techniques, and the impact of biofouling phenomena are discussed. Specific applications of electrochemical biosensors in whole blood analysis are examined, including the detection and quantification of biomarkers associated with infectious diseases, cardiovascular disorders, cancer, and metabolic disorders. Emerging trends like point-of-care diagnostics and wearable biosensors are also discussed. Expert contributions provide insights into current challenges, future directions, and the potential of electrochemical biosensors in personalized medicine and healthcare delivery.

"Electrochemical Biosensors for Whole Blood Analysis" serves as an essential academic reference for researchers, scientists, clinicians, and students. It aims to unlock the possibilities of electrochemical biosensors and drive innovation in healthcare for improved outcomes.

Chunhai Fan
Shanghai Jiao Tong University
Shanghai, China

Contents

1 Introduction 1
Le Jing, Qianqian Li, Shaoguang Li, Hui Li, and Fan Xia

2 Electrochemical Biosensors and the Signaling 17
Xuewei Du, Wanxue Zhang, Suyan Yi, Hui Li, Shaoguang Li, and Fan Xia

3 Electrochemical Biosensors for Ions Detection 39
Kai Zhang, Xiaojin Zhang, and Fan Xia

4 Electrochemical Biosensors for Neurotransmitters Detection 61
Qitao Zhou, Jing Pan, Hui Li, Shaoguang Li, and Fan Xia

5 Electrochemical Biosensors for Metabolites Detection 77
Jiafu Shen, Yu Dai, and Fan Xia

6 Electrochemical Biosensors for Drugs Detection 101
Ruiqi Ma, Rui Liu, and Fan Xia

7 Electrochemical Biosensors for Amino Acids Detection 119
Yanyan Wang, Li Dai, Cihui Luo, Yidan Tu, Xugang Wang, and Yu Huang

8 Electrochemical Biosensors for Nucleic Acids Detection 137
Wenjing Chu, Mengyu Yang, Meihua Lin, and Fan Xia

9 Electrochemical Biosensors for Hepatic and Cardiac Biomarkers Detection 155
Jing Pan, Qitao Zhou, Hui Li, Shaoguang Li, and Fan Xia

10 Electrochemical Biosensors for Inflammatory Biomarkers Detection 171
Zhijuan Duan, Danlong Chen, Hong Liu, Huiying Xue, Fujian Huang, and Fan Xia

11 Electrochemical Biosensors for Tumor Biomarkers Detection 197
Linfeng Chen and Yanan Li

12 Electrochemical Biosensors for Virus Detection 217
Zhi Zheng, Haiyang Li, Hui Li, and Fan Xia

13 Electrochemical Biosensors for Cells and Cell Exosomes Detection 235
Zhi Zheng, Haiyang Li, Hui Li, and Fan Xia

14 Electrochemical Biosensor with Machine Learning 253
Yunliang Chen, Xiaohui Huang, Ao Long, and Wei Han

Chapter 1
Introduction

Le Jing, Qianqian Li, Shaoguang Li, Hui Li, and Fan Xia

Abstract Whole blood, as one of the most significant biological fluids, maintains the normal functioning of various organs and tissues in the human body. The physicochemical changes of different components in whole blood provide critical information for health management, disease detection, drug analysis, and public safety supervision. With the mutual penetration and innovation of multiple disciplines, various advanced technologies for whole-blood testing have been developed vigorously. Among them, electrochemical biosensors, known to be rapid, sensitive, capable of miniaturization, intelligence, reagentless, and washing-free, become a class of emerging technology to achieve the direct detection of multiple targets in complex biological media or even in vivo. In this chapter, we will briefly discuss the broad application prospects of electrochemical biosensors in detecting endogenous and exogenous targets in whole blood, and classify the target objects based on structural and functional properties. A detailed discussion of different analytes can be found in subsequent chapters.

Keywords Biosensors · Electrochemical detection · Whole blood · Disease diagnosis · Virus detection · Drug testing

1.1 Blood Analysis as a Powerful Database

Whole blood, as one of the most significant biological fluids, is mainly composed of plasma (mainly water and plasma proteins) and three types of blood cells (i.e., red blood cells, leucocytes, and platelets) [1], and has the functions of transportation, buffering and maintaining homeostasis [2, 3]. Generally, physiological or pathological changes in the body can alter the biochemical properties of endogenous components in the blood system, so whole-blood analysis can be regarded as a powerful database to provide critical information for health management and disease detection

L. Jing · Q. Li · S. Li · H. Li (✉) · F. Xia
State Key Laboratory of Biogeology and Environmental Geology, Faculty of Materials Science and Chemistry, China University of Geosciences, Wuhan 430074, China
e-mail: lihui-chem@cug.edu.cn

F. Xia et al. (eds.), *Electrochemical Biosensors for Whole Blood Analysis*,
https://doi.org/10.1007/978-981-99-5644-9_1

(Fig. 1.1, bottom left) [4, 5]. Medical studies have shown that specific biomolecules (i.e., biomarkers) released from body fluids are often closely related to the occurrence and progression of a certain disease [6–8], with which one can distinguish between individual health and pathological states. For example, patients with sickle-cell anemia are diagnosed by abnormal sickle shape of red blood cells [9]. Breast cancer patients can use immunohistochemical surrogate panels of progesterone receptor (PR), estrogen receptor (ER), human epidermal growth factor receptor 2 (HER2), and Ki-67 proliferation index (Ki67) for biological subtyping and personalized treatment [10, 11]. Up to now, the initial diagnosis of most diseases in clinical practice relies on blood testing items to provide judgment basis, including blood sugar [12], lipids [13], routine [14], and organ functional biochemical tests [15] (Table 1.1).

Beyond the endogenous substances secreted by the body itself, there may be exogenous substances (e.g., drugs [16], narcotics [17], and viruses [18]) entering the blood system through oral, inhalation, and injection (Fig. 1.1, bottom right). The presence or dosage of such substances has a dual impact on the physiological function of the body. For instance, doxorubicin (DOX), as a class of broad-spectrum anti-tumor drugs, has a killing effect on tumor cells of various growth cycles [19]. However, DOX is also highly cytotoxic and may cause side effects of myocardial damage once the dosage of a patient exceeds the scope of clinical administration [26]. Owing to the fact that exogenous substances are eventually excreted from the body or released into body fluids through various metabolic pathways, whole blood analysis can be applied as an effective research tool to promote the development of drug analysis [20], virus detection [21], and public safety supervision [22].

Despite the availability of blood analysis providing important information basis for multiple application fields, the common detection technologies used by hospitals and third-party testing institutions for two source components in whole blood include chemiluminescence [23], enzyme-linked immunosorbent assay (ELISA) [24], real-time fluorescence quantitative polymerase chain reaction (RT-qPCR) [25], and gene sequencing [26]. Despite the undoubted merits, the major drawbacks of these analytical techniques are that they are time-consuming, expensive cost, requiring large-scale testing equipment and professional operators, making them unsuitable for point-of-care (POC) diagnosis. Moreover, due to the extremely low concentration of analytes in the whole blood and the matrix interference effect of non-target substances, the specificity and sensitivity of various detection techniques are greatly reduced [27]. When being placed and interrogated in whole blood for continuous, real-time monitoring, sensing devices are facing a time-dependence, high-frequency challenge of testing. In other words, the hematology analyzer must exhibit a consistent performance in their high sensitivity and specificity over the period of testing time.

Over the past half-century, with the mutual penetration and innovation of disciplines such as molecular biology, analytical chemistry, nanomaterials, and information computing, various advanced technologies for whole blood testing have been developed vigorously [4, 28]. Among them, electrochemical biosensors, known to be rapid, sensitive, capable of miniaturization, intelligence, reagentless and washing free, become a class of emerging technology to achieve the direct detection of multiple

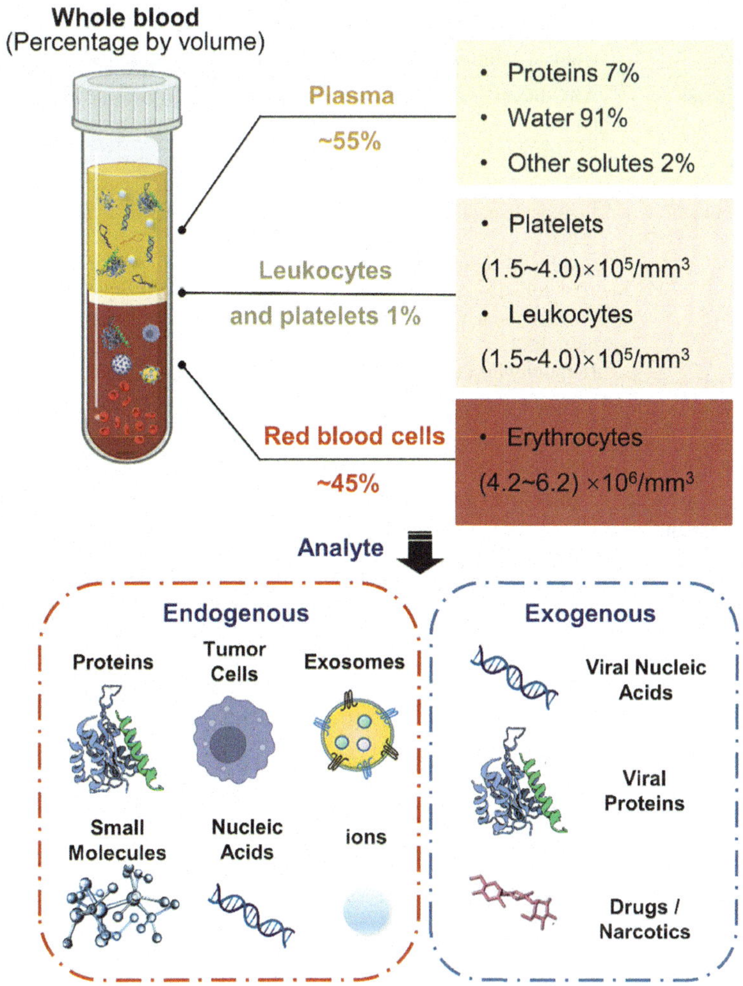

Fig. 1.1 Numerous analytes in whole blood can be used as biomarkers in disease diagnosis, virus detection, drug analysis, and other fields. In this chapter, we classify common blood analytes into exogenous and endogenous based on their source

targets in complex biological media or even in vivo [29–31], e.g., the commercial glucose meter has been successfully upgraded to realize the integrated diagnosis and treatment mode of continuous monitoring-intelligent drug delivery [12]. Given the application prospects of electrochemical biosensors, the boosted research works in recent years have focused on improving the detection performance of sensing platforms in complex whole-blood systems (e.g., specificity, sensitivity, and anti-fouling), including the use of hierarchically structural electrode [32], nanomaterial modification [33], framework nucleic acid design [34], enzyme amplification [35] and mathematical algorithm for data analysis [36], etc. (Fig. 1.2).

Table 1.1 Several common clinical blood testing items

Test items	Clinical cutoff values	Clinical detection methods
Blood groups	ABO blood group: A, B, AB, and O types RH blood group: RH (+) and RH (–) types	Gel microcolumn method
Routine blood	Leukocyte count: (3.5 ~ 9.5) × 10^9 cells/L Red blood cell count: (3.5 ~ 5.0) × 10^{12} cells/L Platelet count: (100 ~ 350) × 10^9 particles/L Hemoglobin: 110 ~ 150 g/L	Photoelectric colorimetry
Blood sugar	3.9 ~ 6.1 mmol/L	Glucose oxidase colorimetry/ Electrochemical method
Blood lipids	Total cholesterol: <5.7 mmol/L Triglyceride: <1.7 mmol/L	Enzymatic method (COD PAP assay)
Renal function	Urea: 2.78 ~ 7.14 mmol/L Creatinine: 59 ~ 104 μmol/L Uric acid: 210 ~ 416 μmol/L	Colorimetry/Enzymatic method
Liver function	Transaminase (Glutamic-pyruvic, Glutamic oxaloacetic): 5 ~ 40 U/L	Chemiluminescence method
	γ-glutamyl transpeptidase: 8 ~ 50 U/L	Enzymatic kinetic method
	Total Bilirubin: 5.1 ~ 22.2 μmol/L	Diazonium method
	Albumin: 35 ~ 55 g/L	Biuret reactive/Bromocresol green method
Cardiac function	Creatine kinase: 18 ~ 198 U/L Creatine kinase isoenzyme: <5 ng/mL Lactate dehydrogenase: 109 ~ 245 U/L Troponin-I: <0.2 μg/L Myoglobin: 10 ~ 80 μg/L	Chemiluminescence method
	B-type natriuretic peptide: <100 ng/L	ELISA method

* This table only lists the main important indicators of each test item. All clinical cutoff values are referenced from Peking Union Medical College Hospital

In order to facilitate readers to more intuitively understand the application distribution of various blood components for electrochemical detection over the past decade, based on the molecular structure characteristics, we classified the specific analytes for whole blood analysis into four subtypes, including small molecules, nucleic acids, proteins, and cells, with the exogenous analytes being excluded from the cells category. As shown in Fig. 1.3, for each target type, we have summarized the well-known specific analytes with their respective detection ranges and publication

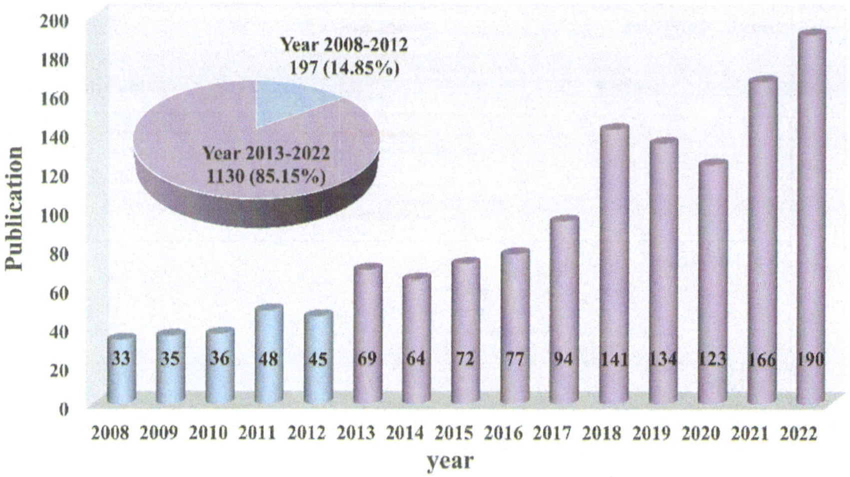

Fig. 1.2 Recent advances in electrochemical biosensing technologies demonstrate constantly increased research interests, as indicated by the annual publication number since 2008. The results are derived from Web of Science using the searching terms "blood" and "electrochemical biosensors." The boosted papers are seen for this field from the year 2013–2022, accounting for 85.15% of the total number of publications in the past 15 years

numbers achieved by electrochemical sensors, providing an overview and guidelines for the following chapters.

1.2 Electrochemical Biosensors for Endogenous Substances Detection in Whole Blood

Endogenous substances in the blood are a class of natural products formed in the synthetic and metabolic pathways of human body [37, 38], e.g., nucleic acids, proteins, cells, polypeptides, metabolites, etc. Since circulating endogenous molecules in whole blood can indicate the occurrence, development, and metastasis of a specific type of disease or cancer in the human body, biomarker-driven early diagnosis and personalized treatment can achieve the goal of "early detection, early prevention, and early treatment" [39]. Beyond these traditional and routine clinical blood tests, "liquid biopsy" technology has attracted great attention in precision medicine research on major diseases of global concern, such as cancer and Alzheimer's disease, with a non-invasive (minimally invasive), rapid, convenient, and repeatable sampling method [40, 41].

Cancer is one of the most widely concerned public health issues worldwide. According to the latest global cancer statistics of the International Agency for Research on Cancer (IARC), in the year 2020, more than 19 million new cancer cases have been diagnosed, as well as nearly 10 million cancer deaths occurred, of which

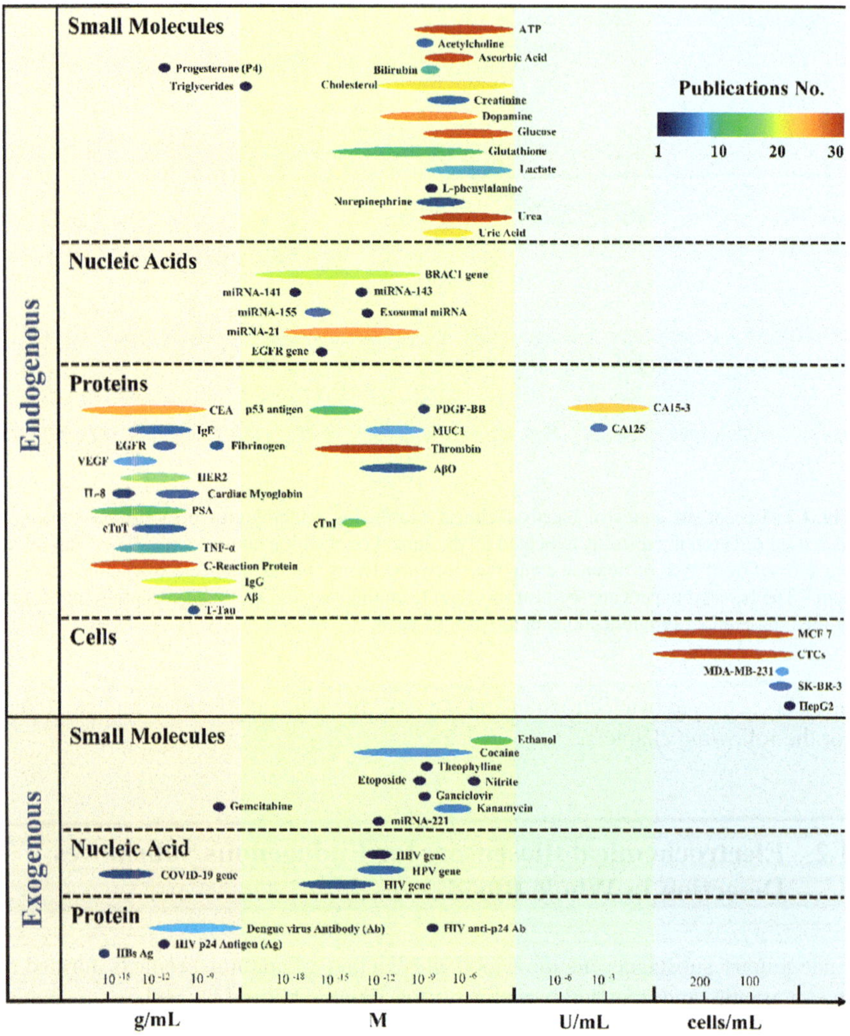

Fig. 1.3 Summary of reference detection ranges and publication numbers for proteins, small molecules, nucleic acids, and cells, which were detected by electrochemical biosensors in whole blood. Subsequent chapters will detail the recent advances of electrochemical biosensors for detecting various targets in whole blood (not limited to the examples in this figure). For instance, Chaps. 3, 4, 5, and 7 are mainly discussing the analysis of small molecules, Chap. 8 focuses on nucleic acid detection, Chaps. 9–11 represent examples of protein analysis, and Chap. 13 discusses the analysis of cells and exosomes

breast cancer and lung cancer account for the highest morbidity and mortality respectively (Fig. 1.4). In the clinical environment, the diagnosis and staging of various cancers mainly depend on imaging analysis and histopathology [28]. The former is difficult to perform accurate clinical staging and pathological staging of tumors, while the invasive sampling method used by the latter will cause unbearable pain to patients. Endogenous analytes are readily available from body fluids including blood, urine, saliva, and sweat, ensuring the availability of these biomarkers present therein as targets of interest to develop "liquid biopsy" techniques [40]. Compared with other body fluid samples, the types of candidate biomarkers circulating in the blood are more abundant, and their application is in line with the purpose of modern precision medicine. At present, the National Institute of Clinical Biochemistry (NACB) has announced the clinical guidelines for several tumor markers present in blood samples. For example, carcinoembryonic antigen (CEA) is considered as a biomarker of breast cancer and colon cancer [42, 43], CA125 is used for early diagnosis of ovarian cancer and lung cancer [44, 45], and CA15-3 and MCF-7 are recognized as independent biomarkers of breast cancer [28].

In clinical practice, the gold standards for cell and protein analysis are flow-cytometry and ELISA methods [46, 47], respectively, but their sensitivity is limited by the low capture efficiency of recognition elements. For nucleic acid analysis, apart from the classic PCR method, next-generation sequencing is the most cutting-edge technology [48], but the complexity of its operation is not applicable to clinical settings. Electrochemical biosensors have been explored as good candidates for point-of-care testing (POCT) devices due to their fast response, high sensitivity, and strong specificity, as they rely on specific recognition between biorecognition elements and targets [28, 49]. Depending on the target category, electrochemical biosensors provide quantitative or semi-quantitative analysis by selecting the appropriate bio-recognition element (e.g., DNA / RNA probe for nucleic acid detection [50, 51], antibody or aptamer for protein / cell detection [52, 53]). Given the multitude of biomarkers referred to as "general" tumor indicators (i.e., low specificity for cancer),

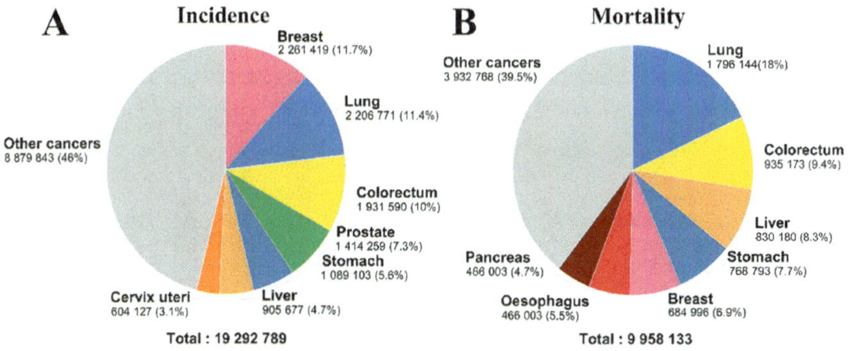

Fig. 1.4 Pie charts reveal the most common cancers among both sexes worldwide in 2020 for incidence (**A**) and mortality (**B**). Statistics data for plotting figures is available from the CANCER TODAY website (http://gco.iarc.fr/today)

simultaneous detection of multiple targets has become a hot spot in the field of early cancer detection in the past five years. For example, Gu et al. developed a new electrochemical aptamer-based sensor for the simultaneous detection of protein marker HER2 and cell marker MCF-7 in serum samples of breast cancer patients [53], the proposed sensor exhibited an extremely low detection limits of 19 fg/mL for HER2 and 23 cell/mL for MCF-7 (Fig. 1.5A).

Similarly, for other major diseases with limited clinical detection methods (e.g., Alzheimer's disease, AD), "liquid biopsy technology as an alternative can provide crucial diagnostic guidance information [54]. AD is a neurodegenerative disease with insidious onset and progressive development, which can lead to cognitive loss and even death from encephalatrophy [55, 56]. One of the key issues in diagnosing AD at the current stage is the lack of an accurate and reliable way to measure the early onset of mild symptoms, which allows up to 57.26% of AD patients to develop advanced stages before receiving medication [57]. Currently, AD is still an incurable disease, and a definitive diagnosis of AD can only be obtained by pathophysiological imaging of the patient's postmortem brain [58]. In 2014, the international working group identified amyloid-β (Aβ) and tau protein (p-tau) as independent diagnostic markers for Alzheimer's disease [59, 60]. Numerous medical studies have shown that the expression level of Aβ in cerebrospinal fluid or plasma of AD patients is decreased, while the expression level of tau protein is increased (Fig. 1.6) [61, 62]. Compared with the invasive sampling method of cerebrospinal fluid (CSF), blood testing is less invasive, easy to operate, and allows for more frequent sampling, making it the preferred test for hospitals to screen for potentially high-risk AD populations.

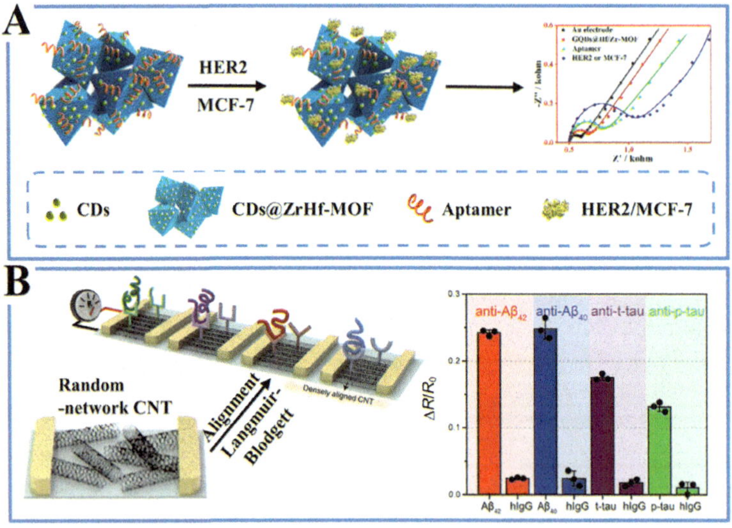

Fig. 1.5 Electrochemical sensors for detecting multiple disease biomarkers. **A** Schematic diagram of the CDs@ZrHf-MOF-based aptasensor for detecting HER2 and MCF-7 [53]. **B** Schematic illustration of a densely aligned CNT sensor array for identifying multiple AD biomarkers [52]

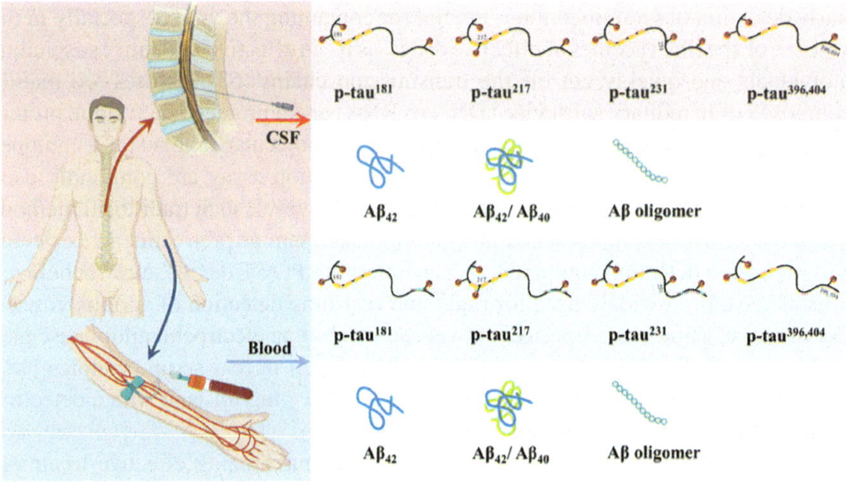

Fig. 1.6 AD core biomarkers present in human cerebrospinal fluid and plasma

So far, a variety of techniques have been used to detect AD core biomarkers in CSF and plasma, such as ELISA, fluorescence, mass spectrometry, electrochemical method, surface-enhanced Raman spectroscopy, etc. [63]. Taking an example for electrochemical technology, Kayoung et al. constructed an electrochemical biosensor using densely aligned carbon nanotubes as a signal conversion device for the clinical accurate and ultrasensitive diagnosis of Alzheimer's disease via detecting multiple core biomarkers (i.e., t-tau, p-tau181, Aβ42 and Aβ40) in patient plasma [52], with detection limits as low as the femtomolar level for each biomarker (Fig. 1.5B).

1.3 Electrochemical Biosensors for Exogenous Substances Detection in Whole Blood

Exogenous substances (e.g., viruses, drugs, toxins, etc.) enter the blood circulation system mainly through the mouth, nose, skin, and veins, thereby producing a good therapeutic effect on the human body or causing malignant consequences of organ functional damage. Accurate and real-time detection of exogenous substances in blood sample is of great scientific significance for the development of virus detection, drug analysis, and public safety supervision.

Viruses have outstanding resistance, rapid variability, and high transmissibility, which increases the likelihood and risk of a pandemic, especially in a world of accelerated globalization [64]. From the smallpox epidemic in 1520 to the current COVID-19, numerous viruses, such as influenza viruses, human immunodeficiency virus (HIV), human papillomavirus (HPV), Ebola virus (EBOV), or dengue virus, have had a significant impact, and challenge on public health and healthcare system.

Early detection of viral infection is crucial for containing the virus, especially in the absence of specific vaccines or effective drugs, as it can effectively identify suspicious individuals and quickly cut off the transmission chains [65]. Viruses are mainly composed of hereditary substance (DNA or RNA) and a protective envelope protein coat [66]. Based on this exclusive characteristic, molecular diagnostic techniques such as ELISA, PCR, and hemagglutination/inhibition assay are commonly used for virus identification or quantification [67–69]. However, such traditional methods exhibit several disadvantages, including complicated sample pretreatment operations and expensive detection equipment. As an emerging POCT device, electrochemical sensors have been widely used for rapid and real-time detection of various viruses. For instance, Zhou et al. reported a novel sandwich-type electrochemiluminescence immunosensor for the analysis of HIV-1 p24 antigen in real serum samples [18], achieving a linear range from 1.0×10^{-9} to 1.0×10^{-5} mg/mL and with a detection limit as low as 1.0×10^{-9} mg/mL (Fig. 1.7A).

Reasonable medication has a direct impact on maximizing effective treatment of diseases. Generally, small doses of drugs cannot achieve therapeutic effects, in retrospect, medication beyond the scope of clinical administration may have side effects on the human body. Therefore, accurate and real-time monitoring of drugs in body fluids is beneficial for doctors to develop personalized and accurate drug treatment plans for patients [16]. Antibiotics, as the most common antibacterial agents in clinical practice, such as vancomycin, kanamycin, and chloramphenicol, are capable of killing and inhibiting invasive pathogenic bacteria [70]. However, ordinary and excessive use of antibiotics will not only damage liver and kidney functions, but also easily increase bacterial resistance, causing an outbreak of "super bacteria" [71]. In 2017, Plaxco et al. first demonstrated that aptamer-based electrochemical sensors can be emplaced directly in the blood circulation of living animals to support continuous, real-time, multi-hour measurement of small molecule drugs [16], and successfully achieved high resolution pharmacokinetic profiles of DOX, kanamycin, gentamicin, and tobramycin (Fig. 1.7B). Furthermore, electrochemical biosensors can provide a powerful basis for the screening of new disease drugs by monitoring the concentration changes of specific biomarkers in the human body after drug administration. For example, Wang et al. analyzed the lethality of anticancer drugs DOX and tamoxifen on tumor cells by using the designed integrated electrochemical nanodevice [72].

Beyond the above-mentioned therapeutic drugs, the abuse of addictive drugs (e.g., cocaine, morphine, and marijuana) can seriously harm human health, even destroy family happiness and endanger social public safety [73]. Currently, the main materials used for drug testing include hair, urine, and blood [74], among which blood testing has the advantages of fast, convenient, and high accuracy. The establishment of rapid, sensitive, and on-site detection methods is essential for a comprehensive crackdown on drug crimes worldwide. Zuo et al. constructed a novel sandwich-type electrochemical approach based on a single aptamer sequence for the direct detection of cocaine in serum samples and other complex matrices [17], achieving the significant detection of cocaine at concentrations as low as 1 μM (Fig. 1.7C).

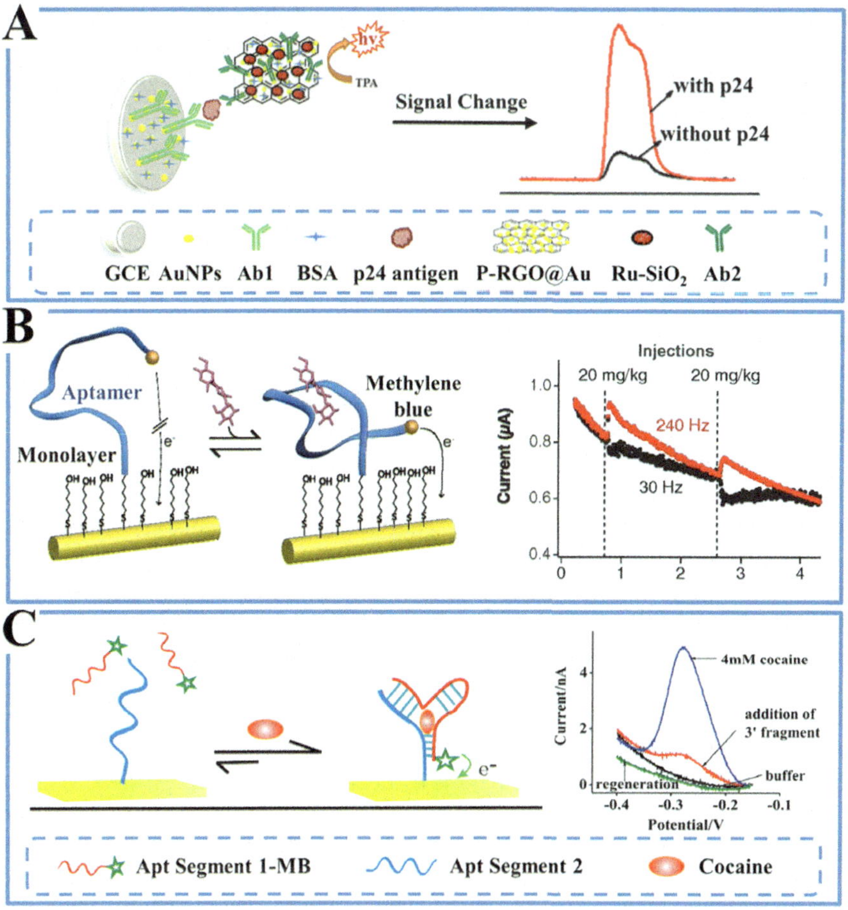

Fig. 1.7 Electrochemical biosensors for exogenous substances detection in whole blood. **A** Schematic diagram of an electrochemical luminescence immunosensor based on P-RGO@Au@Ru-SiO_2 composite for HIV-1 p24 antigen detection [18]. **B** Schematic illustration of an aptamer-based electrochemical sensing platform for continuous measurement of specific drugs directly in the living animal [16]. **c** Schematic illustration of a sandwich-type electrochemical approach based on a single aptamer sequence for detecting cocaine [17]

1.4 Conclusion and Outlook

In this chapter, we emphasized that whole blood analysis as a powerful database can provide guidance for disease diagnosis, virus detection, and drugs / narcotics testing. Among the techniques employed for whole blood analysis, electrochemical biosensors have attracted widespread attention due to their advantages of high specificity and sensitivity, convenient operations, low instrument cost, capable of miniaturization and intelligence. Here, we specifically summarized the respective detection ranges and corresponding publication numbers of electrochemical biosensors for well-known specific analytes under different target categories (i.e., small molecules, nucleic acids, proteins, and cells) in whole blood, thus providing a guideline for the following chapters.

Electrochemical biosensors have shown broad application prospects in the quantitative or semi-quantitative analysis of endogenous and exogenous components in the circulating blood system. Especially for the construction of personalized diagnosis and treatment systems, they have defaulted as the most promising candidates for POCT devices, which facilitate patients to monitor their health status at any time, so as to achieve the goal of "early detection, early diagnosis, and early treatment." Nevertheless, such electrochemical sensing devices still face several critical challenges before transitioning from experimental models to practical commercial products, including their sensitivity, specificity, multiplexing and implantation possibilities, etc. [28].

Although the physicochemical changes in biomarkers provide important information about disease status, their extremely low concentrations in blood samples make high sensitivity crucial for sensors to track the levels of target biomarkers. Meanwhile, for complex biological media, such as human serum or blood samples, the matrix interference effect caused by the existing non-target components would seriously affect the accuracy and sensitivity of the analytical results of electrochemical sensor (e.g., biofouling). In addition, the specificity of disease biomarkers should also be taken into account, because the vast majority of biomarkers are broad-spectrum and cannot be served as independent parameters to predict the occurrence and development of specific diseases. Therefore, using a single biomarker as diagnostic evidence for a disease often leads to misdiagnosis by doctors.

There is an urgent need to develop multiplexing electrochemical biosensors capable of simultaneously detecting multiple disease biomarkers. At present, multiplexing sensors can be achieved by two assembly strategies: one is a single electrode system with several biorecognition elements and redox tags, and the other is an array electrode system with multiple biorecognition elements and a single redox tag. The former has limited availability of multi-redox tags with different electrochemical interrogating windows, while the latter is not satisfactory in terms of circuit arrangement and signal resolution.

The construction of implantable biosensors for long-term monitoring of specific targets (e.g., disease biomarkers, small molecule drugs) in blood vessels would greatly advance the progress of precision medicine. When being placed and interrogated in the complex human circulatory system, implantable electrochemical biosensors require high-performance antifouling, anti-degradation, anti-corrosion, and biocompatibility [75]. Of course, many other factors such as battery life and the miniaturization and intelligence of biosensors are also major challenges to be overcome urgently.

References

1. Sperryn PN (1989) Blood. Brit J Sports Med 23:129–131
2. Hand H (2001) Blood and the classification of anaemia. Nurs Stand 15:45–53
3. Rickards CA (2015) Cerebral blood-flow regulation during hemorrhage. Compr Physiol 5:1585–1621
4. Peri AM, Harris PNA, Paterson DL (2022) Culture-independent detection systems for bloodstream infection. Clin Microbiol Infect 28:195–201
5. Rashid ZZ, Othman SN, Samat MNA, Ali UK, Wong KK (2020) Diagnostic performance of COVID-19 serology assays. Malays J Pathol 42:13–21
6. Cracowski JL (2012) Towards prognostic biomarkers in pulmonary arterial hypertension. Eur Respir J 39:799–801
7. Meyer A (2016) Developing psychiatric biomarkers: a review focusing on the error-related negativity as a biomarker for anxiety. Curr Opin Psychiatr 3:356–364
8. Anonymous (2001) Biomarkers and surrogate endpoints: preferred definitions and conceptual framework. Clin Pharmacol Ther 69:89–95
9. Kneut C (1980) Sickle-cell anemia. Issues Compr Pediatr Nurs 4:19–27
10. Nagini S (2017) Breast cancer: current molecular therapeutic targets and new players. Anti-Cancer Agent Me 17:152–163
11. Xu LF, Han F, Zhu L, Ding WL, Zhang KX, Kan CX, Hou NN, Li QY, Sun XD (2023) Advances in understanding the role and mechanisms of tumor stem cells in HER2-positive breast cancer treatment resistance. Int J Oncol 62:48–58
12. Li XL, Huang XS, Mo JS, Wang H, Huang QQ, Yang C, Zhang T, Chen HJ, Hang T, Liu FM, Jiang LL, Wu QN, Li HB, Hu N, Xi X (2021) A fully integrated closed-loop system based on mesoporous microneedles-iontophoresis for diabetes treatment. Adv Sci (Weinh) 8:2100827–2100841
13. Okpala I, Ren H, Ghebremeskel K, Ibegbulam O, Ugochukwu CC, Crawford M (2005) Blood lipid abnormalities in sickle cell disease are pancellular and clinically important. Blood 106:2339–2339
14. Silverstein WK, Weinerman AS, Born K, Dumba C, Moriates CP (2022) Reducing routine inpatient blood testing. BMJ 379:e070698
15. Du XW, Su XJ, Zhang WX, Yi SY, Zhang G, Jiang S, Li H, Li SG, Xia F (2022) Progress, opportunities, and challenges of troponin analysis in the early diagnosis of cardiovascular diseases. Anal Chem 94:442–463
16. Arroyo-Currás N, Somerson J, Vieira PA, Ploense KL, Kippin TE, Plaxco KW (2017) Real-time measurement of small molecules directly in awake, ambulatory animals. Proc Natl Acad Sci USA 114:645–650
17. Zuo XL, Xiao Y, Plaxco KW (2009) High specificity, electrochemical sandwich assays based on single aptamer sequences and suitable for the direct detection of small-molecule targets in blood and other complex matrices. J Am Chem Soc 131:6944–6945
18. Zhou LM, Huang JS, Yu B, Liu Y, You TY (2015) A novel electrochemiluminescence immunosensor for the analysis of HIV-1 p24 antigen based on P-RGO@Au@Ru-SiO_2 composite. ACS Appl Mater Interfaces 7:24438–24445
19. Khan A, Amidi O, Hess KR, Ewer MS, Yusuf W, Banchs J, Palaskas N (2019) Defining cardiotoxicity of doxorubicin and trastuzumab. J Clin Oncol 37:e23097–e23097
20. Kang JS, Lee MH (2009) Overview of therapeutic drug monitoring. Korean J Intern Med 24:1–10
21. Ruhanya V, Jacobs GB, Glashoff RH, Engelbrecht S (2017) Clinical relevance of total HIV DNA in peripheral blood mononuclear cell compartments as a biomarker of HIV-associated neurocognitive disorders (HAND). Viruses 9:324–340
22. Alhefeiti MA, Barker J, Shah I (2021) Roadside drug testing approaches. Molecules 26:3291–3318

23. Yang HR, Weng BR, Liu SJ, Kang NN, Ran JB, Deng ZS, Wang HM, Yang CY, Wang F (2022) Acid-improved DNAzyme-based chemiluminescence miRNA assay coupled with enzyme-free concatenated DNA circuit. Biosens Bioelectron 204:114060–114066
24. Tran HV, Piro B, Reisberg S, Huy Nguyen L, Dung Nguyen T, Duc HT, Pham MC (2014) An electrochemical ELISA-like immunosensor for miRNAs detection based on screen-printed gold electrodes modified with reduced graphene oxide and carbon nanotubes. Biosens Bioelectron 62:25–30
25. Schiller M, Wydra S, Kerl HU, Kick W (2020) Diagnosis of COVID-19 pneumonia despite missing detection of viral nucleic acid and initially inconspicuous radiologic findings. J Med Virol 92:2863–2865
26. Chang YS, Huang HD, Yeh KT, Chang JG (2017) Evaluation of whole exome sequencing by targeted gene sequencing and Sanger sequencing. Clin Chim Acta 471:222–232
27. Na J, Hong MH, Choi JS, Kwak H, Song S, Kim H, Chae Y, Cheong E, Lee JH, Lim YB, Choi HJ (2019) Real-time detection of markers in blood. Nano Lett 19:2291–2298
28. Jing L, Xie CY, Li QQ, Yang MQ, Li SG, Li H, Xia F (2022) Electrochemical biosensors for the analysis of breast cancer biomarkers: from design to application. Anal Chem 94:269–296
29. Wang SD, Liu YD, Zhu AW, Tian Y (2023) In Vivo electrochemical biosensors: recent advances in molecular design, electrode materials, and electrochemical devices. Anal Chem 95:388–406
30. Xu J, Fang YS, Chen J (2021) Wearable biosensors for non-invasive sweat diagnostics. Biosensors (Basel) 11:245–265
31. Liu JY, Tang YF, Cheng YH, Huang W, Xiang L (2022) Electrochemical biosensors based on saliva electrolytes for rapid detection and diagnosis. J Mater Chem B 11:33–54
32. Meng LY, Turner APF, Mak WC (2018) Positively-charged hierarchical PEDOT interface with enhanced electrode kinetics for NADH-based biosensors. Biosens Bioelectron 120:115–121
33. Zhou Y, Marar A, Kner P, Ramasamy RP (2017) Charge-directed immobilization of bacteriophage on nanostructured electrode for whole-cell electrochemical biosensors. Anal Chem 89:5734–5741
34. Su J, Liu WH, Chen SX, Deng WP, Dou YZ, Zhao ZH, Li JY, Li ZH, Yin H, Ding XT, Song SP (2020) A Carbon-based DNA framework nano-bio interface for biosensing with high sensitivity and a high signal-to-noise ratio. ACS Sens 5:3979–3987
35. Wang HF, Ma RN, Sun F, Jia LP, Zhang W, Shang L, Xue QW, Jia WL, Wang HS (2018) A versatile label-free electrochemical biosensor for circulating tumor DNA based on dual enzyme assisted multiple amplification strategy. Biosens Bioelectron 122:224–230
36. Andrianova M, Kuznetsov A (2020) Logic gates based on DNA aptamers. Pharmaceuticals (Basel) 13:417–465
37. Ludwig JA, Weinstein JN (2005) Biomarkers in cancer staging, prognosis and treatment selection. Nat Rev Cancer 5:845–856
38. Su J, Chen SX, Dou YZ, Zhao ZH, Jia XL, Ding XT, Song SP (2022) Smartphone-based electrochemical biosensors for directly detecting serum-derived exosomes and monitoring their secretion. Anal Chem 94:3235–3244
39. Wu XL, Li JH, Gassa A, Buchner D, Alakus H, Dong QZ, Ren N, Liu M, Odenthal M, Stippel D, Bruns C, Zhao Y, Wahba R (2020) Circulating tumor DNA as an emerging liquid biopsy biomarker for early diagnosis and therapeutic monitoring in hepatocellular carcinoma. Int J Biol Sci 16:1551–1562
40. Crowley E, Di Nicolantonio F, Loupakis F, Bardelli A (2013) Liquid biopsy: monitoring cancer-genetics in the blood. Nat Rev Clin Oncol 10:472–484
41. Khan S, Barve KH, Kumar MS (2020) Recent advancements in pathogenesis, diagnostics and treatment of alzheimer's disease. Curr Neuropharmacol 18:1106–1125
42. Borthwick NM, Wilson DW, Bell PA (1977) Carcinoembryonic antigen (CEA) in patients with breast cancer. Eur J of Cancer 13:171–176
43. Dhar P, Moore T, Zamcheck N, Kupchik HZ (1972) Carcinoembryonic antigen (CEA) in colonic cancer: use in preoperative and postoperative diagnosis and prognosis. JAMA 221:31–35
44. Zhang MH, Cheng SS, Jin Y, Zhao YQ, Wang Y (2021) Roles of CA125 in diagnosis, prediction, and oncogenesis of ovarian cancer. BBA-Rev Cancer 1875:188503–188513

45. Wang X, Wang MH, Feng L, Song J, Dong X, Xiao T, Cheng SJ (2022) Four-protein model for predicting prognostic risk of lung cancer. Front Med 16:618–626
46. Fleisher TA, Oliveira JB (2019) Flow Cytometry. In: Rich RR, Fleisher TA, Shearer WT, Schroeder HW, Frew AJ, Weyand CM (eds) Clin Immunol 1239–1251.e1231
47. Kinnamon DS, Heggestad JT, Liu J, Chilkoti A (2022) Technologies for frugal and sensitive point-of-care immunoassays. Annu Rev Anal Chem (Palo Alto Calif) 15:123–149
48. Dua Puri R (2016) Next generation sequencing in the clinic. Indian J Pediatr 83:281–282
49. Zhang TT, Du X, Zhang ZG (2022) Advances in electrochemical sensors based on nanomaterials for the detection of lipid hormone. Front Bioeng Biotechnol 10:993015–993022
50. De Castro ACH, Kochi LT, Flauzino JMR, Soares M, Alves VA, Da Silva LA, Madurro JM, Brito-Madurro AG (2022) Electrochemical biosensor for sensitive detection of hepatitis B in human plasma. Appl Biochem Biotechnol 194:2604–2619
51. Chen JY, Liu ZJ, Wang XW, Ye CL, Zheng YJ, Peng HP, Zhong GX, Liu AL, Chen W, Lin XH (2019) Ultrasensitive electrochemical biosensor developed by probe lengthening for detection of genomic dna in human serum. Anal Chem 91:4552–4558
52. Kim K, Kim MJ, Kim DW, Kim SY, Park S, Park CB (2020) Clinically accurate diagnosis of Alzheimer's disease via multiplexed sensing of core biomarkers in human plasma. Nat Commun 11:119–127
53. Gu CX, Guo CP, Li ZZ, Wang MH, Zhou N, He LH, Zhang ZH, Du M (2019) Bimetallic ZrHf-based metal-organic framework embedded with carbon dots: ultra-sensitive platform for early diagnosis of HER2 and HER2-overexpressed living cancer cells. Biosens Bioelectron 134:8–15
54. Zetterberg H, Schott JM (2022) Blood biomarkers for Alzheimer's disease and related disorders. Acta Neurol Scand 146:51–55
55. Hu S, Yang CW, Luo HM (2022) Current trends in blood biomarker detection and imaging for Alzheimer's disease. Biosens Bioelectron 210:114278–114288
56. Bondi MW, Edmonds EC, Salmon DP (2017) Alzheimer's disease: past, present, and future. J Int Neuropsychol Soc 23:818–831
57. Ittner LM, Götz J (2011) Amyloid-β and tau–a toxic pas de deux in Alzheimer's disease. Nat Rev Neurosci 12:65–72
58. Struyfs H, Molinuevo JL, Martin JJ, De Deyn PP, Engelborghs S (2014) Validation of the AD-CSF-index in autopsy-confirmed Alzheimer's disease patients and healthy controls. J Alzheimers Dis 41:903–909
59. Dubois B, Feldman HH, Jacova C, Hampel H, Molinuevo JL, Blennow K, Dekosky ST, Gauthier S, Selkoe D, Bateman R, Cappa S, Crutch S, Engelborghs S, Frisoni GB, Fox NC, Galasko D, Habert MO, Jicha GA, Nordberg A, Pasquier F, Rabinovici G, Robert P, Rowe C, Salloway S, Sarazin M, Epelbaum S, De Souza LC, Vellas B, Visser PJ, Schneider L, Stern Y, Scheltens P, Cummings JL (2014) Advancing research diagnostic criteria for Alzheimer's disease: the IWG-2 criteria. Lancet Neurol 13:614–629
60. Leuzy A, Ashton NJ, Mattsson-Carlgren N, Dodich A, Boccardi M, Corre J, Drzezga A, Nordberg A, Ossenkoppele R, Zetterberg H, Blennow K, Frisoni GB, Garibotto V, Hansson O (2021) 2020 update on the clinical validity of cerebrospinal fluid amyloid, tau, and phospho-tau as biomarkers for Alzheimer's disease in the context of a structured 5-phase development framework. Eur J Nucl Med Mol Imaging 48:2121–2139
61. Wang J, Qiao F, Shang SH, Li P, Chen C, Dang LJ, Jiang Y, Huo K, Deng MY, Wang JY, Qu QM (2018) Elevation of plasma amyloid-β level is more significant in early stage of cognitive impairment: a population-based cross-sectional study. J Alzheimers Dis 64:61–69
62. Brickman AM, Manly JJ, Honig LS, Sanchez D, Reyes-Dumeyer D, Lantigua RA, Lao PJ, Stern Y, Vonsattel JP, Teich AF, Airey DC, Proctor NK, Dage JL, Mayeux R (2021) Plasma p-tau181, p-tau217, and other blood-based Alzheimer's disease biomarkers in a multi-ethnic, community study. Alzheimers Dement 17:1353–1364
63. Hu S, Yang CW, Luo HM (2022) Current trends in blood biomarker detection and imaging for Alzheimer's disease. Biosens Bioelectron 210:114278–114288

64. Brazaca LC, Dos Santos PL, De Oliveira PR, Rocha DP, Stefano JS, Kalinke C, Abarza Muñoz RA, Bonacin JA, Janegitz BC, Carrilho E (2021) Biosensing strategies for the electrochemical detection of viruses and viral diseases-a review. Anal Chim Acta 1159:338384–338413
65. Moreno S, Berenguer J, Fuster-Ruizdeapodaca MJ, García Ontiveros M (2018) Early diagnosis. Enferm Infecc Microbiol Clin (Engl Ed) 36 Suppl 1:35–39
66. Ribeiro BV, TaR C, Oliveira EFGR, Ferreira LF, Franco DL (2020) Biosensors for the detection of respiratory viruses: a review. Talanta Open 2:100007–100022
67. Chen LL, Ruan FE, Liu MB, Zhou JF, Song WJ, Qin K (2019) A sandwich ELISA for detecting the hemagglutinin of avian influenza A (H10N8) virus. J Med Virol 91:877–880
68. Kim H, Kang NN, An KH, Kim D, Koo JH, Kim MS (2017) MRPrimerV: a database of PCR primers for RNA virus detection. Nucleic Acids Res 45:D475–D481
69. Eckels KH, Summers PL, Dubois DR (1989) Hepatitis A virus hemagglutination and a test for hemagglutination inhibition antibodies. J Clin Microbiol 27:1375–1376
70. Rowe AA, Miller EA, Plaxco KW (2010) Reagentless measurement of aminoglycoside antibiotics in blood serum via an electrochemical, ribonucleic acid aptamer-based biosensor. Anal Chem 82:7090–7095
71. Christaki E, Marcou M, Tofarides A (2020) Antimicrobial resistance in bacteria: mechanisms, evolution, and persistence. J Mol Evol 88:26–40
72. Wang HY, Ruan YF, Zhu LB, Shi XM, Zhao WW, Chen HY, Xu JJ (2021) An integrated electrochemical nanodevice for intracellular RNA collection and detection in single living cell. Angew Chem Int Ed Engl 60:13244–13250
73. Tsatsakis A, Docea AO, Calina D, Tsarouhas K, Zamfira LM, Mitrut R, Sharifi-Rad J, Kovatsi L, Siokas V, Dardiotis E, Drakoulis N, Lazopoulos G, Tsitsimpikou C, Mitsias P, Neagu M (2019) A mechanistic and pathophysiological approach for stroke associated with drugs of abuse. J Clin Med 8:1295–1330
74. Tamama K (2021) Advances in drugs of abuse testing. Clin Chim Acta 514:40–47
75. Russo MJ, Han M, Desroches PE, Manasa CS, Dennaoui J, Quigley AF, Kapsa RMI, Moulton SE, Guijt RM, Greene GW, Silva SM (2021) Antifouling strategies for electrochemical biosensing: mechanisms and performance toward point of care based diagnostic applications. ACS Sens 6:1482–1507

Chapter 2
Electrochemical Biosensors and the Signaling

Xuewei Du, Wanxue Zhang, Suyan Yi, Hui Li, Shaoguang Li, and Fan Xia

Abstract An electrochemical biosensor is a type of sensor that uses biological materials as the sensitive component, electrodes as the conversion element, and current, potential or resistance as detectable characteristics of the signal. In recent decades, electrochemical biosensors have been developed for the detection of many biological elements. Biosensors designed to detect a range of targets have been extensively studied in terms of their transduction, biometric, and electrochemical components. This chapter presents a discussion of bio-receptors and electrochemical techniques that can be used to detect biological elements. Biological elements include tissue, living cells, an enzyme, antibody or antigen, and nuclei. Electrochemical methods include electric current, electric potential, impedance. This critical review will discuss the two components of electrochemical biosensors, biological elements and electrochemical techniques, in order to provide an easy to understand introduction to electrochemical biosensors.

Keywords Biosensors · Biorecognition elements · Electrochemical signaling · Electrochemical methods · Voltammetry · Electrochemical Impedance Spectroscopy

2.1 Introduction

Electrochemical biosensors combine the low detection limits of electrochemical transducers with the high specificity of biometrics [1]. Therefore, the electrochemical biosensor has the characteristics of high sensitivity and specificity. At the same time, the electrochemical biosensor also has the advantages of low cost, convenient use, portability, simple structure, and so on. Currently, the electrochemical biosensor also boasts the advantages of being cost-effective, user-friendly, portable, and having a simplified structure. Therefore, it provides precise quantitative or semi-quantitative

X. Du · W. Zhang · S. Yi · H. Li · S. Li (✉) · F. Xia
State Key Laboratory of Biogeology and Environmental Geology, Faculty of Materials Science and Chemistry, China University of Geosciences, Wuhan 430074, China
e-mail: lisg@cug.edu.cn

F. Xia et al. (eds.), *Electrochemical Biosensors for Whole Blood Analysis*,
https://doi.org/10.1007/978-981-99-5644-9_2

analysis for various types of molecules, including small molecules, nucleic acids, proteins, cells, and more [2, 3].

Electrochemical biosensors employ recognition elements that have the ability to selectively react to targeted analytes, thereby reducing interference from other components present in the sample [3]. Biometric components may be antibodies or antigens, nucleic acids, enzymes, molecular imprinted polymers and peptide nucleic acids, etc. [3–5]. Sensor/transducer signals include current, potential, intensity, and phase of electromagnetic radiation, mass, conductivity, impedance, temperature, light, and viscosity [1, 3, 6, 7] (Fig. 2.1). The overall performance of electrochemical biosensors is determined by biometric components, sensors, and transduction methods [4].

Electrochemistry does not depend on the volume of the reaction, so rapid analysis of small samples can be achieved. (less than 1 μL) [8–10]. The electrochemical detection process requires no preparation or fewer experimental steps, specifically, electrochemical analysis of homogeneous samples, without the need to separate the antibody-antigen complex from the sample, is a step toward achieving electrochemical rapid detection. Because electrochemical biosensors are unaffected by chromophores, fluorophores, and particles in the sample composition, electrochemical sensors can detect colored or turbid samples, such as serum, plasma, or whole blood, without interference from lipoprotein, red blood cells, hemoglobin, and bilirubin [11, 12].

Electrochemical biosensors have made great progress in recent years in laboratory research, and translation from laboratory to clinic remains a challenge for many laboratories, including sensitivity, specificity, and reproducibility. The solution to these several key problems must also start from the basic origin of the electrochemical biosensors themselves, and promote the translation of electrochemical biosensors

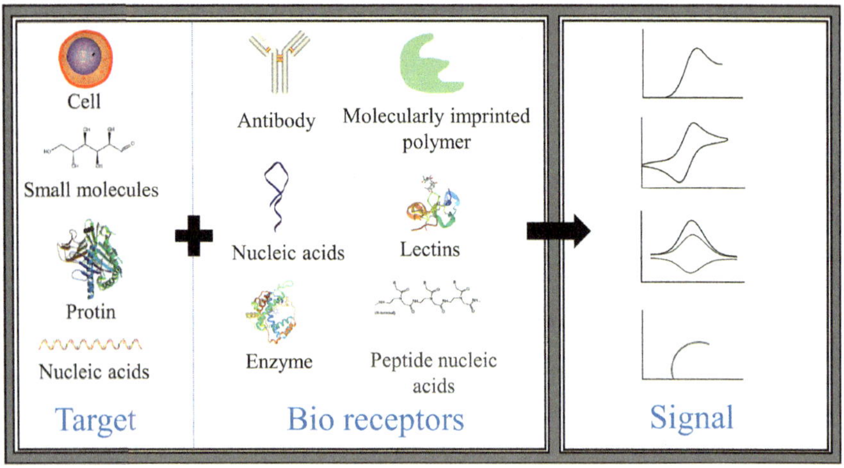

Fig. 2.1 Components of electrochemical biosensors

from laboratory to clinic by optimizing the recognition components, selecting the appropriate sensors and conduction methods, increasing the use of new materials and integrating interface technologies. In this section, we aim to provide a brief discussion of signaling and biorecognition elements.

2.2 Electrochemical Methods

Depending on the respective input and output signals, electrochemical biosensors can be classified as oltammetric/amperometric (typically generating a measurable current), impedimetric (monitoring resistance and reactance), potentiometric (the ratio of the potential energy of a unit charge in an electric field to the amount of charge it carries), conductometry (the velocity of charge flow in a substance), photo-electrochemical and electrochemiluminescent biosensors (measurable iuminescence signal), field effect transistor (FET, measurable ionic charge).

Voltammetric/amperometric is the most commonly used of these techniques, including cyclic voltammetry (CV), chronoamperometry (CA), differential pulse voltammetry (DPV), square wave voltammetry (SWV), linear sweep voltammetry (LSV) [13]. The classification of these techniques depends on how the question potential method is applied, which we will discuss in detail in the next section (Fig. 2.2). Among these techniques used for electrochemical biosensors, voltam-metry/amperometry accounts for 80%, followed by EIS and potentiometry, etc. We describe the principles of all these techniques and their corresponding sensor applications.

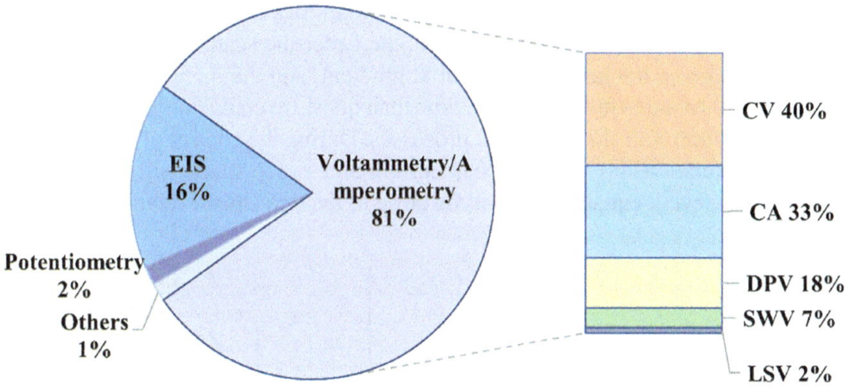

Fig. 2.2 Electrochemical techniques employed for biosensors. The different classifications of electrochemical techniques and their proportions, and the subclassifications of voltammetry/amperometry techniques and their proportions are shown in the figure. Both statistical analyses come from Web of Science, using the search terms "electrochemical biosensors" and "blood" from the year 2013 to the year 2023

2.2.1 *Voltammetry/Amperometry*

Voltammetry/amperage technology works by applying a potential to the working electrode and measuring the current resulting from the oxidation or reduction reaction of electrochemically active molecules at the working electrode. The redox-active molecules are either attached to the electrode surface or free in solution. The electron transport of attached redox molecules is limited by the mass transfer rate at the reaction molecule-electrode interface in terms of solution pH, ionic strength, temperature, solution viscosity, nano-limited domain space, etc. [13–18].

Voltammetric and amperometric biosensors can be classified according to the technique used to apply interrogation potentials, which is briefly discussed below:

(1) **Linear sweep voltammetry (LSV)**

In LSV, the potential is constant input, the current is output, and the potential is linear with respect to time. The peak current measured by the current–potential curve is linearly related to the concentration of the measured substance, which can be used for quantitative analysis and is more suitable for the determination of adsorbed substances (Fig. 2.3).

The number of electron transfers, the chemical reactivity of the redox molecule, and the scanning rate are all key factors affecting LSV. According to the Randles–Sevcik equation, the higher the number of electrons gained and lost in a reversible reaction, the higher the peak current. The larger the electrode area of the reaction, the higher the molarity of the reactant redox molecules and the corresponding increase in current.

In LSV, the current increases with the scan rate, while the potential varies with the scan rate. At a fixed scan rate, the potential shift depends on the type of redox reaction (reversible, quasi-reversible or irreversible). In a reaction process controlled by free diffusion, for a reversible reaction, the faster the scanning rate, the thinner the diffusion layer, the larger the potential gradient, and the increased current. The potential shift requires more response time than quasi-reversible or irreversible reactions or reactions with slower kinetic processes. During the control of the adsorption process, the relationship between peak current and sweep speed will not be that of the Randles–Sevcik equation but will be close to the one-square relationship.

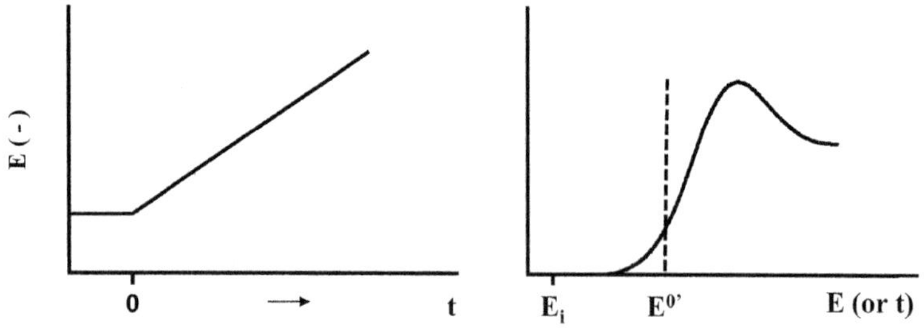

Fig. 2.3 The mechanism of LSV employed for electrochemical biosensors

LSV allows the study of materials in the general potential energy and time scale domains to gain insight into the thermodynamic and kinetic behavior of insertion reactions. In addition, LSV is highly sensitive to structural changes in the inserted compound. LSV is therefore a widely used method for the study of insertion compounds [19].

(2) **Cyclic voltammetry (CV)**

CV is a common method used in electrochemical research. CV controls the change in electrode potential, with the voltage fluctuating between the two values at a fixed rate, but returning to the minimum/maximum when the voltage reaches a maximum/minimum voltage reversal. The potential range includes both reduction and oxidation reactions (depending on the REDOX molecule used), and current–potential curves are recorded (Fig. 2.4). Therefore, CV is often used to study the properties, mechanism, and kinetic parameters of electrode reactions. It can also be used for the quantitative determination of reactant concentration, adsorption coverage on electrode surface, electrode active area, electrode reaction rate constant, exchange current density, reaction transfer coefficient, and other kinetic parameters.

For CV technology, scan rate and potential are key factors. The scan time must provide sufficient time for the complete redox reaction, and the potential must contain the complete redox reaction [20]. Therefore, the scanning time must be sufficient to allow meaningful chemical reactions to occur. And the scan range for CV measurements is highly dependent on the specific redox-active molecules.

(3) **Differential pulse voltammetry (DPV)**

DPV uses successive rising potential pulses in step wave mode, with a series of forward and reverse pulses as excitation signals. The electrolytic current $\triangle i$ within the cycle is obtained by subtracting the positive and negative pulse current within the cycle. With the increase of potential, the electrolytic current $\triangle i$ for several cycles is measured continuously and $\triangle i \sim E$ is plotted as pulse difference curve (Fig. 2.5). DPV can increase the sensitivity of electrochemical detection, and the application of step pulse potential can improve the effect of charge current generated by CV when the linear potential changes and the influence of non-Faraday current caused by the diffusion layer.

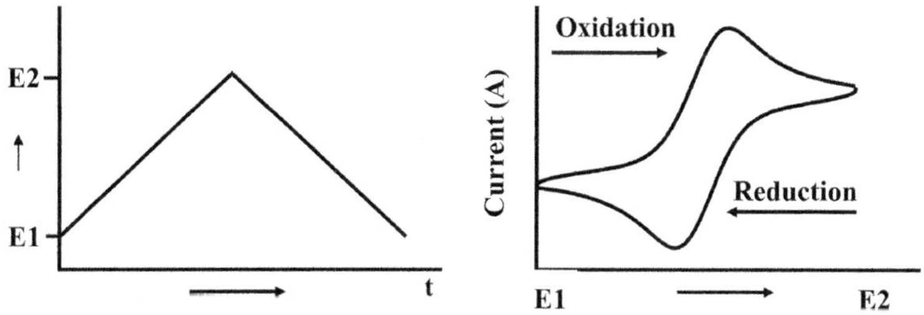

Fig. 2.4 The mechanism of CV employed for electrochemical biosensors

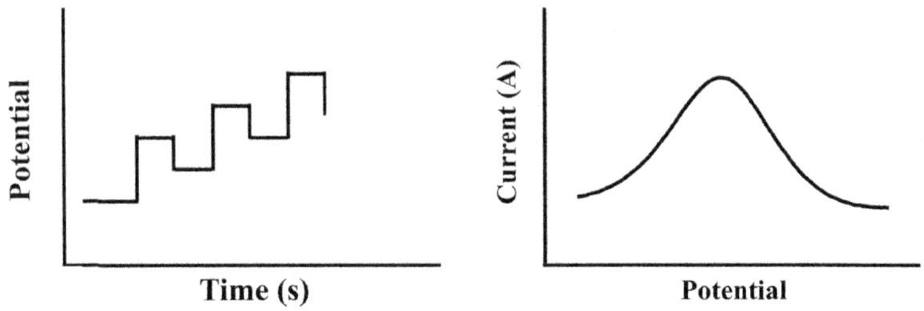

Fig. 2.5 The mechanism of DPV employed for electrochemical biosensors

(4) **Square wave voltammetry (SWV)**

SWV is a technique in which potentiostat superimposes potential in the form of symmetric square wave (forward and reverse) on the basis of step linear scanning. Forward and reverse potentials have the same duration and are applied at the same frequency. The forward and reverse pulses of current subtracted from each other will result in a differential current and potential curve (Fig. 2.6). The forward and reverse components of SWV are curves measured experimentally and are directly related to the oxidation and reduction reactions of redox molecules [21].

SWV can be used not only for quantitative analysis, but also for the study of chemical reaction mechanisms, kinetics and thermodynamics. SWV is characterized by its ability to distinguish between background (capacitive) current and Faraday current, as well as its rapid analysis rate. Compared to DPV, the consumption of electroactive components is lower, and in SWV scanning, the low concentration of dissolved oxygen in the solution has no time to diffuse to the electrode surface for reaction, so no nitrogen deoxygenation is required. And the SWV current is 3–4 times higher than the corresponding DPV response.

(5) **Chronoamperometry (CA)**

Chronoamperometry, applies a stabilizing potential to a working electrode and measures steady-state current as a function of time [22]. The CA regulation mandates an initial, as well as a high and low potential. After the initial potential, from the initial value to either a higher or lower potential. After a duration of τ (pulse width),

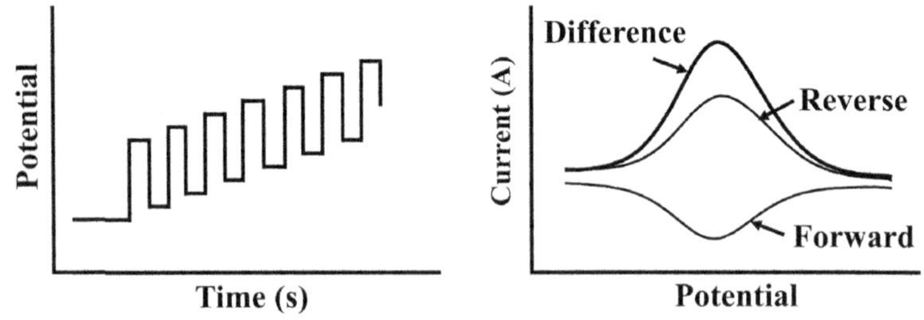

Fig. 2.6 The mechanism of SWV employed for electrochemical biosensors

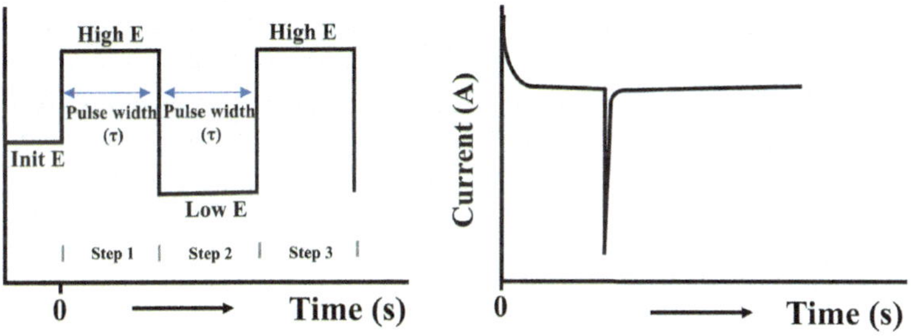

Fig. 2.7 The mechanism of CA employed for electrochemical biosensors

the potential will shift in the opposite direction (either from high to low potential or from low to high potential) and persist for τ time. The current response is contingent on initial and final potential values (Fig. 2.7).

The change in current is caused by an increase or decrease in the thickness of the diffusion layer on the electrode [23–26]. Notably, CA is a sensitive technique that does not require redox molecules. Compared to other detection methods, timing amperometry is usually used to quantify known analytes due to its better signal-to-noise ratio [27], and CA can be applied alone or in combination with other electrochemical techniques.

(6) **Chronocoulometry (CC)**

The CC method is a determination of the change of charge over time, it may be performed during a potential scan or a fixed potential step [28]. The CC method requires an initial potential (Initial E) and a final potential (Final E) and is a change of potential from an initial potential to a final potential with charge as a function of time, which will change to a third value (usually the initial potential value) after holding the second potential for a period of time τ. Thus, the potential step experiment of the CC method can be a single or double step (Fig. 2.8).

If the Faraday current resulting from electrolysis of molecular substances in solution does not occur at any of the potentials, the response is the current due to electrode charging, and the response decreases exponentially with the peak current. Usually,

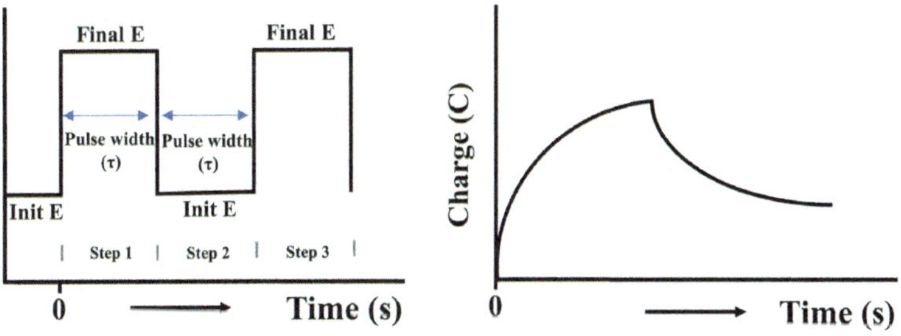

Fig. 2.8 The mechanism of CC employed for electrochemical biosensors

the Faraday response does not occur at the initial potential, but arises at the final potential. During the CC test, redox active molecules are electrolyzed as soon as they reach the surface of the working electrode; therefore, the magnitude of the CC current depends on the rate of transfer of electrochemically active material from the solution to the surface of the working electrode, i.e., the rate of diffusion of redox molecules.

CC measurements are not used for absolute concentration measurements, and are usually used to measure electrode area and diffusion coefficient.

The advantage of CC is that the signal increases with time, the response occurs in the second half of the potential, and the signal has a good signal-to-noise ratio. In addition, CC retains information about the initial response by accumulating charge during the test and can be used to detect adsorbed substances on the surface of the working electrode [29].

2.2.2 Electrochemical Impedance Spectroscopy (EIS)

EIS uses a small sine wave potential or current as a disturbance signal to generate an approximately linear response of the electrode system, thus measuring the impedance spectrum of the electrode system over a wide frequency range [30]. EIS can investigate intrinsic material characteristics or specific mechanisms that affect the conductivity, resistivity, or capacitance of an electrochemical system. EIS can measure the conductivity of a medium and indicate the interaction between the electrode surface and the target.

Electrochemical impedance sensing technology can be used to monitor changes in the electrical properties of the electrode surface caused by material modification or molecular identification. For example, changes in electrode conductance can be measured using EIS due to antigen/antibody binding and antibody-antigen reactions on the electrode surface (Fig. 2.9) [31].

2.2.3 Photoelectrochemical and Electrochemluminescent (ECL) Biosensors

The principle of photoelectrochemical (PEC) sensors relies on the process by which light irradiation is absorbed by the electrode material, generating a measurable current, also known as photocurrent. The photocurrent information is proportional to the concentration of the analyte, thus facilitating its application in analytical chemistry. Moreover, the correlation between photocurrent, wavelength, electrode potential, and solution composition can reveal valuable details about the properties, energetics, and kinetics of the optical processes involved.

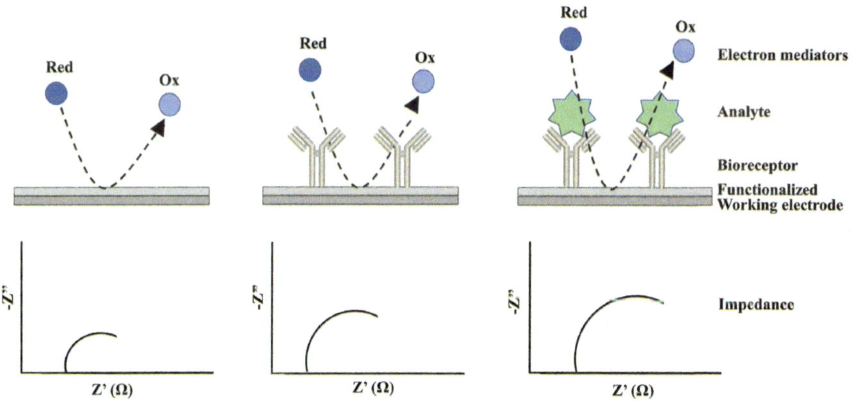

Fig. 2.9 The mechanism of EIS employed for electrochemical biosensors [32]. Copyright 2020 Springer Nature

The design principle of PEC is to enhance light absorption and optimize load separation efficiency. In the photoelectric conversion process, chemical information or analyte concentration is initially converted into a specific physical and chemical parameter. Subsequently, the heterogenous electron transfer resulting from the change in the dynamics of the interfacial charge transfer is detected as an electrical signal, to achieve the detection target.

PEC analysis can be classified into two types based on the electronic transmission method: photoanode and photogalvanic analysis. Based on the N-type semiconductor, the optical anode mode is used as a light-sensitive material and interacts with electron donors in electrolytes. In contrast, the optoelectronic cathode mode depends on the P-shaped semiconductor material that acts as an electron receptor in PEC.

ECL biosensors use the conversion of electrochemical energy into light at the electrode interface [33]. In an electrochemical reaction, luminescence is observed when the molecule is in an excited state. To initiate this phenomenon, a potential is applied to the electrode. The intensity of the emitted light is measured using a photodetector, typically a photomultiplier tube [34].

High responses are often achieved by the conversion of tags by repeater molecules, such as ruthenium complexes such as $Ru(bpy)_3^{2+}$ [35, 36]. The application of ECL to a variety of NPs is a recent development. This includes popular forms such as binary, core–shell, and doped QDs, as well as single-element NPs such as Si, Ge, Ag, Au. Metal oxide semiconductors, up-converting NPs, molecular nanoaggregates, and more complex hierarchical assemblies are also explored for their diverse ECL mechanisms and reaction pathways [37]. Compared with other luminescence methods, electrochemical luminescence instruments have been widely used due to their simplicity and low background signal.

Among all the methods described above, due to its flexibility in signal interpretation and ease of operation, voltammetry including CV, EIS, SWV, DPV, etc., is the most applicable technique that has been widely used in biosensors.

2.3 Biorecognition Elements for Electrochemical Biosensors

As the "core" component of electrochemical biosensors, Biorecognition elements have the essential function to promise the sensitivity and specificity of electrochemical biosensors. The unique structure of biorecognition elements influences the performance characteristics of electrochemical biosensors. Therefore, before conducting an in-depth analysis of electrochemical biosensors, we should first have a basic understanding of each biometric element. Loads of biorecognition elements exist, from natural existence to artificial design and synthesis. Here, we focus on several of the most commonly-used biometric elements.

2.3.1 Aptamers

Aptamers are short (usually 10–80 bases in length) single-stranded DNA or RNA oligonucleotides, with their 3D folded structures, which can identify target analytes through hydrogen bonding, electrostatic interaction, or base-stacking and other interactions, and combine with target analytes to form a stable complex [38]. The aptamer sequence of specific analytes can be manually isolated from oligonucleotide libraries through in vitro screening procedures, e.g., systematic evolution of ligands by exponential enrichment (SELEX) [39]. The main steps of the SELEX method include incubation, separation, amplification, and purification. Aptamers produced by this method can combine with a wide range of target analytes, including small molecules, ions, proteins, cells, viruses, etc. The recognition property of aptamers is similar to that of antibodies, but aptamers have some obvious advantages:

(1) Changes in temperature or the addition of exogenous reagents cause aptamers to undergo reversible conformational changes without losing their binding capacity [40, 41].
(2) Aptamers are produced by chemical synthesis, are easy to modify, and cost-effective.
(3) Since the production process of aptamers does not involve any biological system, batch-to-batch variations are passively minimized in comparison to antibody production.

With these advantages, biosensors based on electrochemical capacitors have been extensively studied. According to different design strategies, electrochemical aptamer sensors can be divided into three categories: (1) "aptamer-target" direct detection mode; (2) "target induced strand-displacement" competitive detection mode; (3) sandwich detection mode.

In direct detection mode, the aptamer can bind directly to the target and induce a conformational change in the probe, thus changing the electron transfer dynamics between the redox reporter molecule and the electrode, resulting in measurable changes in electrical signals. Some aptamers do not undergo conformational changes after binding directly to the target, but binding to the target will also change the electron transfer dynamics between the redox reporter molecule and the electrode in the solution, resulting in measurable changes in electrical signals [42].

The competitive detection mode depends on the competition between the complementary aptamer sequence and the target [43]. Specifically, the complementary strand of the aptamer with redox reporter-labeled is fixed to the electrode and forms a dsDNA conformation with the aptamer strand. When the target analyte exists, the aptamer binds to the target is present, the complex falls off, and the electrochemical signal changes.

Sandwich electrochemical aptamer sensors typically have higher selectivity than single aptamer engagement, but require at least two aptamer recognition sites for the target analyte, and the two aptamers need to have different recognition sites. In sandwich electrochemical sensors, one aptamer modifies the signal molecule as the electrical signal element and the other aptamer with the target as the capture element [44].

2.3.2 DNAzymes

DNAzymes are oligonucleotide molecules that can bind target analytes with high specificity and enzymatic properties, which can be obtained through in vitro selection. Target-responsive DNAzymes can serve as excellent biological recognition elements because they can be rationally designed to utilize DNA complementarity for controlled capture and release. In biological systems, DNA enzymes may carry the risk of nuclease degradation. The current solution to this problem is to directly select DNAzymes from biological samples containing nucleases or to make chemical modifications to prevent exonuclease digestion. To date, DNAzyme biosensors have been developed using various amplification strategies such as hybridization chain reaction (HCR) and catalytic hairpin assembly (CHA) as promising recognition elements [45].

2.3.3 Peptides

Peptides are short chains of amino acids linked by peptide bonds and consist of 20 amino acids based on different sequence information [46]. Their building blocks are the same as proteins, so the peptide with a specific sequence can display the target recognition ability. According to different sources, peptides can be divided into natural peptides and artificial peptides. The standard synthetic protocol for artificial peptides is called Fmoc (Fluorenylmethyloxycarbonyl) and t-Boc (tert-butoxycarbonyl) solid-phase peptide synthesis (SPPS) [47, 48]. Because of its better chemistry, conformational stability, and high specificity, easy modification, versatility, and chemical versatility. Peptide-based biosensors have been developed for electrochemical detection of various targets such as cells, proteases, nucleic acids, metal ions, and more. Generally, the combination of peptides and the target analyte will not produce measurable signals. Therefore, Peptide-based electrochemical biosensors modify redox reporter molecules on peptides to convert target information into measurable electrochemical signals.

2.3.4 Peptide Nucleic Acids (PNAs)

Peptide nucleic acids are a unique DNA analogue [49]. The backbone used is N-(2-aminoethyl) glycine, with pyrimidines and purines bound to it via methylene carbonyl groups. Similar to DNA, PNA also binds to specific RNA and DNA sequences based on Watson Crick's complementary pairing principle. Compared to hybridization between nucleic acids, PNA has the following advantages:

(1) Because PNA molecules have no electrical charge, there is no charge repulsion between RNA or DNA strands and the electrically neutral PNA strand. And the PNA/DNA(RNA) duplex has extraordinary thermal stability and is less affected by medium ion strength [50, 51].
(2) DNA/PNA hybrid complexes exhibit high sensitivity to the presence of a single mismatched base pair, allowing accurate discrimination of single base mismatches. Therefore, peptide nucleic acid biosensors are mainly used to analyze single nucleotide polymorphisms (SNP) [49].
(3) PNA is a synthetic molecule with an amide bond backbone, which distinguishes it from phosphodiester bonds present in DNA. This chemical composition makes it highly resistant to hydrolysis by protease or nuclease enzymes, providing superior stability [52].

Due to its distinctive characteristics, PNA has recently become a popular choice for building electrochemical biosensors. According to the different signal generation mechanisms, PNA-based electrochemical biosensors are mainly divided into two types: direct signal detection mode and indirect signal detection mode. The electrical signal of the direct detection model sensor is derived from changes in guanine oxidation signals before and after PNA and DNA hybridization [53]. Indirect detection uses PNA as a capture probe, introducing an electroactive label during sensor construction, and sensor signal changes are provided by the label's direct redox signal.

2.3.5 *Proteins*

Proteins have also been integrated into electrochemical sensors as biological recognition elements, with the gradual development of phage, bacterial, and ribosome-related technologies. For example, metal ion detection uses the interaction between metals and proteins to cause changes in their function and/or conformation. Specifically, the recognition proteins are attached to an electrode surface. Upon binding of metal ions, it changes to a more compact configuration, which opens the electrode surface for metal ions to enter and thus increases the capacitance of the electrode surface. However, this strategy requires high protein specificity [54–56]. Recently, a protein-based electrochemical biosensor has been developed for the detection of proteins and peptides. This type of sensor is based on the basic principle of conformational changes caused by proteins binding to target analytes and applied redox reporter to modify proteins to indicate the occurrence of binding events [57]. The signal transduction of this sensor is closely related to the placement of redox reporters on proteins.

2.3.6 *Antibodies*

Antibodies, also known as immunoglobulins, are highly soluble glycoproteins involved in the immune system's defense mechanisms. Antibodies have a Y-shaped structure and are composed of two parts, each consisting of a light and a heavy chain. The larger molecular weights are called heavy chains, the smaller molecular weights are called light chains, and the light and heavy chains are joined together by disulfide bonds [58, 59]. The light (V_L) and heavy (V_H) chain region of the antibody with a large sequence variation of about 110 amino acids near the N-terminus is called the variable region, which accounts for 1/4 and 1/2 of the heavy and light chains, respectively. The variable region determines the specificity of the antibody and is the site where the antibody recognizes and binds the antigen [60].

Various types of antibody-based biological recognition components, including monoclonal antibodies and derivatives thereof, such as antigen-binding fragments, single-chain variable fragments, or single-chain antibodies, may serve as capture or signal probes in the development of immunosensor applications [61]. Antibody-based sandwich electrochemical sensors typically employ a surface-bound antibody as the capture probe. Once attached to the target antigen, a reporter molecule, labeled as a secondary antibody, is introduced into the detection system. It binds to the same antigen, creating an electrochemical signal that depends primarily on the identification of distinct epitopes by the two antigens [62].

2.3.7 Enzymes

Enzymes are a type of protein or RNA created by living cells that are highly specialized and highly effective catalysts. The catalytic efficiency of enzymes depends on the integrity of both the primary and spatial structure of enzyme molecules. Enzyme activity may be lost due to the molecular denaturation of enzymes or depolymerization of subunits. Enzymes are classified as biological macromolecules with molecular weights ranging from 10,000 to millions of macromolecules. Enzymatic biosensors, as they are called, employ enzymes as recognition elements, which are highly specific to their substrates. Enzyme electrochemical biosensor uses enzymes as biometric molecules to capture signals generated by the reaction between the target and the enzyme through various chemical signal transducers. Enzyme electrochemical biosensors can be divided into catalytic systems or enzyme affinity systems according to the principle of biological selectivity. Usually, the signals are proportional to the concentration of the target, so as to achieve quantitative analysis of the target.

Enzymatic sensors have wide applications in detecting different substances in biological specimens, including glucose, cholesterol, or lactic acid, and in analyzing toxicity levels in environmental monitoring, food inspection, quality assurance, and biomedical and pharmaceutical sensing [63–68].

2.3.8 Lectins

Lectin is a glycoprotein or glycobinding protein that can be isolated from a variety of sources, including plants, invertebrates, and higher animals. Because of its ability to agglutinate red blood cells, it is referred to as lectin, commonly used as plant lectin [69, 70]. Lectins' ability to recognize complex carbohydrate structures found in glycoproteins and glycolipids, mainly located in cell membranes, specifically glycogroups present on the surface. Lectins have the unique ability to bind exclusively to a specific type of glycogen group [71–74]. This forms the basis for the development of lectin sensors that can directly examine intact cell surfaces, as well as glycolipids, glycoproteins, and sugars present in membranes [75–77]. There are some disadvantages, such as the need to label lectins, which can reduce the performance of the biosensor.

2.3.9 Epitopes

Epitopes are short continuous peptide chains [78–81] that specifically recognize the corresponding antibodies and have the advantage of being easy to obtain, synthesize and modify, which provides a convenient option for antigen–antibody interactions.

Epitopes can specifically recognize the corresponding antibody, and because epitopes are easy to obtain, synthesize and modify, it provides a scheme for the detection of antibodies in the blood. It is worth noting that the rapid response time of epitopes to targeted antibodies is up to minute, so epitope sensors are a powerful means of antibody analysis [82–85].

2.3.10 Molecularly Imprinted Polymers (MIPs)

MIPs are widely used as recognition elements for biosensors due to their ease of synthesis, cost-effectiveness, flexibility in the design of recognition elements for target analytes, and the ability to mimic natural enzyme analogs [86]. MIPs are synthetic matrices used as detection materials in biosensor fabrication due to their exceptional recyclability, thermal stability, and specificity in comparison to biorecognition components. Based on the function, charge, and morphology of the target, imprinted polymer receptors exhibit selective binding, causing changes in their structural characteristics and signals, enabling effective detection [87–89]. However, several challenges remain for MIP technologies, including low affinity, incomplete removal of template molecules, low conductivity and electrocatalytic activity, and inefficient mass transfer.

2.4 Conclusion and Outlook

Electrochemical biosensors with high specificity and sensitivity for target detection are of great importance for clinical prevention, diagnosis, and prognosis of related diseases. The application of electrochemical biosensors focuses on the selection of appropriate test methods and the identification of the original components, which can effectively improve the detection performance of the sensor. Common electrochemical biosensors can be classified into voltammetric/ampere, impedance, potential, conductivity, photoelectrochemical and electrochemiluminescent, field effect transistors, etc. Biorecognition elements may be tissue, living cells, an enzyme, antibody or antigen, and nuclei. Due to the different detection methods and characteristics of biometric originals, the corresponding sensors have different research priorities. Although electrochemical sensor detection has made great progress in recent years, there are still some challenges in translation from laboratory to clinic. For example, most sensors can only detect a single target, instability, and poor immunity to interference.

The simultaneous detection of multiple markers helps In the accurate diagnosis of diseases, while ability of electrochemical biosensors to identify multiple analytes makes them very popular in practical applications. The senso's the ability to achieve peak separation and improved sensitivity to specific analytes is critical to the success of this application. Peak separation is closely related to the detection method, nanomaterials, and the signal molecule used. The use of nanomaterials and signal molecules facilitated the precise segregation of various analytes, allowing accurate quantitative measurements. CV, SWV, and DPV can be altered by nanomaterials or by the use of multiple signal molecules, allowing complete differentiation of multiple peak currents, and enabling simultaneous detection of multiple targets [90, 91].

The simultaneous use of multiple different potential signal molecules is a prerequisite for the simultaneous use of multiple targets, and the signal molecules commonly used in electrochemical sequencing sensors currently including methylene blue, ferrocene, anthraquinone, thionine, nile blue, woster's blue, multicolor electrochemiluminescence, luminol, isoluminol, and their derivatives, acridinium ester derivatives, ALP, and HRP has been also explored for electrochemical biosensors. Methylene blue is the most widely used signaling molecule in electrochemical biosensors. Most current electrochemical studies use a single redox molecule, i.e., MB or Fc [92, 93]. In addition, the position of the redox marker affects DNA interfacial electrochemistry. Specifically, limitations in ion accessibility affect the current of the redox marker located at the bottom of the DNA monolayer, while the redox marker at the top is unaffected [94]. It is important to note that there are generally three requirements for the simultaneous use of multiple reporter molecules. First, their potential does not overlap, allowing both to be monitored simultaneously. Second, both are fairly stable. Finally, their physical properties are very similar, so they respond consistently to environmental changes that cause drift. Further improvements and development of nanomaterials and signal molecules will provide sensitive and selective substrates by which multianalyte detection can be performed.

The design of novel nucleic acid probes and the development of related interfacial sensing principles are important factors driving the development of sensing technologies. In the field of electrochemical biosensing, a lot of related research work has been done by many excellent groups at home and abroad. Self-assembled single-molecule layers on the sensor surface, which act as molecular closure layers, can eliminate or reduce the non-specific adsorption of DNA probes on the electrode surface to improve the sensitivity, specificity, and lifetime of electrochemical sensors. Researchers have greatly enhanced antifouling properties by using alkanethiol molecules with different hydrophilic portions [95], alkanethiol molecules with different hydrophobic portions [96], and mimetic phosphatidylcholine [93] to improve detection performance (e.g., stability, sensitivity, and specificity).

The combination of multifunctional materials, recognition elements, and electrochemical methods enhances selectivity and stability. To achieve this, researchers are exploring innovative materials and diverse signal molecule potentials, redox peak potential/current, amplitude, scan rate, bionic film, etc. These techniques have been used to improve one or both of these deficiencies. Although electrochemical biosensors face some challenges, they have a significant clinical impact in diagnosing and treating diseases. In addition, these sensors play a crucial role in collecting data for biomedical research.

References

1. Lowe CR (1989) Biosensors. Philos Trans R Soc Lond B Biol Sci 324:487–496
2. Li F, Li Q, Zuo X, Fan C (2020) DNA framework-engineered electrochemical biosensors. Sci China Life Sci 63:1130–1141
3. Ronkainen NJ, Halsall HB, Heineman WR (2010) Electrochemical biosensors. Chem Soc Rev 39:1747–1763
4. Majdinasab M, Mitsubayashi K, Marty JL (2019) Optical and electrochemical sensors and biosensors for the detection of quinolones. Trends Biotechnol 37:898–915
5. Ahmad OS, Bedwell TS, Esen C, Garcia-Cruz A, Piletsky SA (2019) Molecularly imprinted polymers in electrochemical and optical sensors. Trends Biotechnol 37:294–309
6. Smutok O, Katz E (2022) Biosensors: electrochemical devices-general concepts and performance. Biosensors (Basel) 13:44
7. Banakar M, Hamidi M, Khurshid Z, Zafar MS, Sapkota J, Azizian R, Rokaya D (2022) Electrochemical biosensors for pathogen detection: an updated review. Biosensors (Basel) 12:927
8. Wang M, Yang Y, Min J, Song Y, Tu J, Mukasa D, Ye C, Xu C, Heflin N, Mccune JS, Hsiai TK, Li Z, Gao W (2022) A wearable electrochemical biosensor for the monitoring of metabolites and nutrients. Nat Biomed Eng 6:1225–1235
9. Surya SG, Majhi SM, Agarwal DK, Lahcen AA, Yuvaraja S, Chappanda KN, Salama KN (2020) A label-free aptasensor FET based on Au nanoparticle decorated Co(3)O(4) nanorods and a SWCNT layer for detection of cardiac troponin T protein. J Mater Chem B 8:18–26
10. Hemmig E, Temiz Y, Gökçe O, Lovchik RD, Delamarche E (2020) Transposing lateral flow immunoassays to capillary-driven microfluidics using self-coalescence modules and capillary-assembled receptor carriers. Anal Chem 92:940–946
11. Zamani M, Furst AL, Klapperich CM (2021) Strategies for engineering affordable technologies for point-of-care diagnostics of infectious diseases. Acc Chem Res 54:3772–3779
12. Timilsina SS, Jolly P, Durr N, Yafia M, Ingber DE (2021) Enabling multiplexed electrochemical detection of biomarkers with high sensitivity in complex biological samples. Acc Chem Res 54:3529–3539
13. Nano A, Furst AL, Hill MG, Barton JK (2021) DNA electrochemistry: charge-transport pathways through DNA films on gold. J Am Chem Soc 143:11631–11640
14. Compagnone D, Francia GD, Natale CD, Neri G, Seeber R, Tajani A (2017) Chemical sensors and biosensors in Italy: a review of the 2015 literature. Sensors (Basel) 17:868
15. Saputra HA, Chung JH, Yoon SH, Seo KD, Park DS, Shim YB (2022) Disposable amperometric immunosensor with a dual monomers-based bioconjugate for granzyme B detection in blood and cancer progress monitoring of patients. Biosens Bioelectron 198:113846
16. Abi A, Ferapontova EE (2012) Unmediated by DNA electron transfer in redox-labeled DNA duplexes end-tethered to gold electrodes. J Am Chem Soc 134:14499–14507
17. Schuster GB (2000) Long-range charge transfer in DNA: transient structural distortions control the distance dependence. Acc Chem Res 33:253–260
18. Ma T, Grzędowski AJ, Doneux T, Bizzotto D (2022) Redox-controlled energy transfer quenching of fluorophore-labeled DNA SAMs enables in situ study of these complex electrochemical interfaces. J Am Chem Soc 144:23428–23437
19. Guo Y (1992) The kinetics of the reduction processes of PbO film on Pb in H_2SO_4—II. Linear sweep voltammetry (lsv). Electrochim Acta 37:495–499
20. Brinker M, Huber P (2022) Wafer-scale electroactive Nanoporous silicon: large and fully reversible Electrochemo-mechanical actuation in aqueous electrolytes. Adv Mater 34:e2105923
21. Mirceski V, Skrzypek S, Stojanov L (2018) Square-wave voltammetry. ChemTexts 4
22. Bănică F-G (2012) Electrochemical affinity and nucleic acid sensors. In: Chemical sensors and biosensors, pp 347–366
23. Molina A, González J, Laborda E, Compton RG (2013) On the meaning of the diffusion layer thickness for slow electrode reactions. Phys Chem Chem Phys 15:2381–2388

24. Nikonenko VV, Vasil'eva VI, Akberova EM, Uzdenova AM, Urtenov MK, Kovalenko AV, Pismenskaya NP, Mareev SA, Pourcelly G (2016) Competition between diffusion and electroconvection at an ion-selective surface in intensive current regimes. Adv Colloid Interface Sci 235:233–246
25. Grank J (1979) 3. Qualitative properties of steady states of reaction-diffusion equations and systems. In: The mathematics of diffusion, pp 31–64
26. Chandrasekhar S (1943) Stochastic problems in physics and astronomy. Rev Mod Phys 15:1–89
27. Cao Z, Li C, Yang X, Wang S, Zhang X, Zhao C, Xue B, Gao C, Zhou H, Yang Y, Shen Z, Sun F, Wang J, Qiu Z (2022) Rapid quantitative detection of live Escherichia coli based on chronoamperometry. Biosensors (Basel) 12:845
28. Lingane PJ, Anson FC, Osteryoung RA (1966) Chronopotentiometric investigation of the reduction of chromate in alkaline solutions. J Electroanal Chem 12:250–253
29. Faulknerjohn LR (2019) Electrochemical methods fundamentals and applications, 2nd edn
30. Brug GJ, Van Den Eeden ALG, Sluyters-Rehbach M, Sluyters JH (1984) The analysis of electrode impedances complicated by the presence of a constant phase element. J Electroanal Chem Interfacial Electrochem 176:275–295
31. Strong ME, Richards JR, Torres M, Beck CM, La Belle JT (2021) Faradaic electrochemical impedance spectroscopy for enhanced analyte detection in diagnostics. Biosens Bioelectron 177:112949
32. Leva-Bueno J, Peyman SA, Millner PA (2020) A review on impedimetric immunosensors for pathogen and biomarker detection. Med Microbiol Immunol 209:343–362
33. Farka Z, Juřík T, Kovář D, Trnková L, Skládal P (2017) Nanoparticle-based immunochemical biosensors and assays: recent advances and challenges. Chem Rev 117:9973–10042
34. Richter MM (2004) Electrochemiluminescence (ECL). Chem Rev 104:3003–3036
35. Liu Y, Zhang H, Li B, Liu J, Jiang D, Liu B, Sojic N (2021) Single biomolecule imaging by electrochemiluminescence. J Am Chem Soc 143:17910–17914
36. Han D, Goudeau B, Manojlovic D, Jiang D, Fang D, Sojic N (2021) Electrochemiluminescence loss in photobleaching. Angew Chem Int Ed Engl 60:7686–7690
37. Deng S, Ju H (2013) Electrogenerated chemiluminescence of nanomaterials for bioanalysis. Analyst 138:43–61
38. Thaler M, Luppa PB (2019) Highly sensitive immunodiagnostics at the point of care employing alternative recognition elements and smartphones: hype, trend, or revolution? Anal Bioanal Chem 411:7623–7635
39. Blind M, Blank M (2015) Aptamer selection technology and recent advances. Mol Ther-Nucl Acids 4:e223
40. Mok W, Li YF (2008) Recent progress in nucleic acid aptamer-based biosensors and bioassays. Sensors 8:7050–7084
41. Chen AL, Yang SM (2015) Replacing antibodies with aptamers in lateral flow immunoassay. Biosens Bioelectron 71:230–242
42. Li S, Ferrer-Ruiz A, Dai J, Ramos-Soriano J, Du X, Zhu M, Zhang W, Wang Y, Herranz MA, Jing L, Zhang Z, Li H, Xia F, Martin N (2022) A pH-independent electrochemical aptamer-based biosensor supports quantitative, real-time measurement in vivo. Chem Sci 13:8813–8820
43. Lu Y, Li XC, Zhang LM, Yu P, Su L, Mao LQ (2008) Aptamer-based electrochemical sensors with aptamer-complementary DNA oligonucleotides as probe. Anal Chem 80:1883–1890
44. Zhang DW, Sun CJ, Zhang FT, Xu L, Zhou YL, Zhang XX (2012) An electrochemical aptasensor based on enzyme linked aptamer assay. Biosens Bioelectron 31:363–368
45. Yun W, Cai DZ, Jiang JL, Wang XF, Liao JS, Zhang PC, Sang G (2016) An ultrasensitive electrochemical biosensor for uranyl detection based on DNAzyme and target-catalyzed hairpin assembly. Microchim Acta 183:1425–1432
46. Liu QT, Wang JF, Boyd BJ (2015) Peptide-based biosensors. Talanta 136:114–127
47. Mitchell AR (2008) Bruce Merrifield and solid-phase peptide synthesis: a historical assessment. Biopolymers 90:175–184
48. Fischer PM (2003) The design, synthesis and application of stereochemical and directional peptide isomers: a critical review. Curr Protein Pept Sci 4:339–356

49. Egholm M, Buchardt O, Christensen L, Behrens C, Freier SM, Driver DA, Berg RH, Kim SK, Norden B, Nielsen PE (1993) NA hybridizes to complementary oligonucleotides obeying the Watson-Crick hydrogen-bonding rules. Nature 365:566–568
50. Schwarz FP, Robinson S, Butler JM (1999) Thermodynamic comparison of PNA/DNA and DNA/DNA hybridization reactions at ambient temperature. Nucleic Acids Res 27:4792–4800
51. Liu JY, Tiefenauer L, Tian SJ, Nielsen PE, Knoll W (2006) PNA-DNA hybridization study using labeled streptavidin by voltammetry and surface plasmon fluorescence spectroscopy. Anal Chem 78:470–476
52. Briones C, Moreno M (2012) Applications of peptide nucleic acids (PNAs) and locked nucleic acids (LNAs) in biosensor development. Anal Bioanal Chem 402:3071–3089
53. Ahmadi M, Ahour F (2020) An electrochemical biosensor based on a graphene oxide modified pencil graphite electrode for direct detection and discrimination of double-stranded DNA sequences. Anal Methods-UK 12:4541–4550
54. Corbisier P, Van Der Lelie D, Borremans B, Provoost A, De Lorenzo V, Brown NL, Lloyd JR, Hobman JL, Csoregi E, Johansson G, Mattiasson B (1999) Whole cell- and protein-based biosensors for the detection of bioavailable heavy metals in environmental samples. Anal Chim Acta 387:235–244
55. Bontidean I, Berggren C, Johansson G, Csoregi E, Mattiasson B, Lloyd JA, Jakeman KJ, Brown NL (1998) Detection of heavy metal ions at femtomolar levels using protein-based biosensors. Anal Chem 70:4162–4169
56. Bontidean I, Lloyd JR, Hobman JL, Wilson JR, Csoregi E, Mattiasson B, Brown NL (2000) Bacterial metal-resistance proteins and their use in biosensors for the detection of bioavailable heavy metals. J Inorg Biochem 79:225–229
57. Kang D, Sun S, Kurnik M, Morales D, Dahlquist FW, Plaxco KW (2017) New architecture for reagentless, protein-based electrochemical biosensors. J Am Chem Soc 139:12113–12116
58. Crivianu-Gaita V, Thompson M (2016) Aptamers, antibody scFv, and antibody Fab' fragments: an overview and comparison of three of the most versatile biosensor biorecognition elements. Biosens Bioelectron 85:32–45
59. Chiu ML, Goulet DR, Teplyakov A, Gilliland GL (2019) Antibody structure and function: the basis for engineering therapeutics. Antibodies (Basel) 8:55
60. Sharma S, Byrne H, O'kennedy RJ (2016) Antibodies and antibody-derived analytical biosensors. Essays Biochem 60:9–18
61. Arshavsky-Graham S, Heuer C, Jiang X, Segal E (2022) Aptasensors versus immunosensors-Which will prevail? Eng Life Sci 22:319–333
62. Conroy PJ, Hearty S, Leonard P, O'kennedy RJ (2009) Antibody production, design and use for biosensor-based applications. Semin Cell Dev Biol 20:10–26
63. Clark LC Jr, Lyons C (1962) Electrode systems for continuous monitoring in cardiovascular surgery. Ann N Y Acad Sci 102:29–45
64. Vargas E, Teymourian H, Tehrani F, Eksin E, Sanchez-Tirado E, Warren P, Erdem A, Dassau E, Wang J (2019) Enzymatic/immunoassay dual-biomarker sensing chip: towards decentralized insulin/glucose detection. Angew Chem Int Ed Engl 58:6376–6379
65. Bui TT, Park S-Y (2016) A carbon dot–hemoglobin complex-based biosensor for cholesterol detection. Green Chem 18:4245–4253
66. Bi Y, Ye L, Mao Y, Wang L, Qu H, Liu J, Zheng L (2019) Porous carbon supported nanoceria derived from one step in situ pyrolysis of Jerusalem artichoke stalk for functionalization of solution-gated graphene transistors for real-time detection of lactic acid from cancer cell metabolism. Biosens Bioelectron 140:111271
67. Malhotra BD, Chaubey A (2003) Biosensors for clinical diagnostics industry. Sens and Actuators B Chem 91:117–127
68. Chang J, Li H, Hou T, Li F (2016) Paper-based fluorescent sensor for rapid naked-eye detection of acetylcholinesterase activity and organophosphorus pesticides with high sensitivity and selectivity. Biosens Bioelectron 86:971–977
69. Sharon N, Lis H (2004) History of lectins: from hemagglutinins to biological recognition molecules. Glycobiology 14:53R-62R

70. Lis H, Sharon N (1998) Lectins: carbohydrate-specific proteins that mediate cellular recognition. Chem Rev 98:637–674
71. Wu H, Lee CJ, Wang H, Hu Y, Young M, Han Y, Xu FJ, Cong H, Cheng G (2018) Highly sensitive and stable zwitterionic poly(sulfobetaine-3,4-ethylenedioxythiophene) (PSBEDOT) glucose biosensor. Chem Sci 9:2540–2546
72. Long L, Hu Y, Xie L, Sun F, Xu Z, Hu J (2021) Constructing a bacterial cellulose-based bacterial sensor platform by enhancing cell affinity via a surface-exposed carbohydrate binding module. Green Chem 23:9600–9609
73. Takahashi T, Nakano Y, Onomoto K, Murakami F, Komori C, Suzuki Y, Yoneyama M, Ui-Tei K (2018) LGP2 virus sensor regulates gene expression network mediated by TRBP-bound microRNAs. Nucleic Acids Res 46:9134–9147
74. Li S, Liu Y, Ma Q (2019) Nanoparticle-based electrochemiluminescence cytosensors for single cell level detection. TRAC Trend Anal Chem 110:277–292
75. Hirabayashi J, Kuno A, Tateno H (2011) Lectin-based structural glycomics: a practical approach to complex glycans. Electrophoresis 32:1118–1128
76. Hirabayashi J, Yamada M, Kuno A, Tateno H (2013) Lectin microarrays: concept, principle and applications. Chem Soc Rev 42:4443–4458
77. Krishnamoorthy L, Mahal LK (2009) Glycomic analysis: an array of technologies. ACS Chem Biol 4:715–732
78. Barbato G, Bianchi E, Ingallinella P, Hurni WH, Miller MD, Ciliberto G, Cortese R, Bazzo R, Shiver JW, Pessi A (2003) Structural analysis of the epitope of the anti-HIV antibody 2F5 sheds light into its mechanism of neutralization and HIV fusion. J Mol Biol 330:1101–1115
79. Palacios-Rodriguez Y, Gazarian T, Rowley M, Majluf-Cruz A, Gazarian K (2007) Collection of phage-peptide probes for HIV-1 immunodominant loop-epitope. J Microbiol Methods 68:225–235
80. Brunel FM, Zwick MB, Cardoso RM, Nelson JD, Wilson IA, Burton DR, Dawson PE (2006) Structure-function analysis of the epitope for 4E10, a broadly neutralizing human immunodeficiency virus type 1 antibody. J Virol 80:1680–1687
81. Gorny MK, Moore JP, Conley AJ, Karwowska S, Sodroski J, Williams C, Burda S, Boots LJ, Zolla-Pazner S (1994) Human anti-V2 monoclonal antibody that neutralizes primary but not laboratory isolates of human immunodeficiency virus type 1. J Virol 68:8312–8320
82. Mclaurin J, Cecal R, Kierstead ME, Tian X, Phinney AL, Manea M, French JE, Lambermon MH, Darabie AA, Brown ME, Janus C, Chishti MA, Horne P, Westaway D, Fraser PE, Mount HT, Przybylski M, St George-Hyslop P (2002) Therapeutically effective antibodies against amyloid-beta peptide target amyloid-beta residues 4–10 and inhibit cytotoxicity and fibrillogenesis. Nat Med 8:1263–1269
83. Juszczyk P, Paraschiv G, Szymanska A, Kolodziejczyk AS, Rodziewicz-Motowidlo S, Grzonka Z, Przybylski M (2009) Binding epitopes and interaction structure of the neuroprotective protease inhibitor cystatin C with β-amyloid revealed by proteolytic excision mass spectrometry and molecular docking simulation. J Med Chem 52:2420–2428
84. Iuraşcu M-I, Marroquin Belaunzanar O, Cozma C, Petrausch U, Renner C, Przybylski M (2016) An HLA-B27 homodimer specific antibody recognizes a discontinuous mixed-disulfide epitope as identified by affinity-mass spectrometry. J Am Soc Mass Spectr 27:1105–1112
85. Stefanescu R, Born R, Moise A, Ernst B, Przybylski M (2011) Epitope structure of the carbohydrate recognition domain of asialoglycoprotein receptor to a monoclonal antibody revealed by high-resolution proteolytic excision mass spectrometry. J Am Soc Mass Spectr 22:148–157
86. Song Y, Liu H, Wan L, Wang Y, Hou H, Wang L (2013) Direct electrochemistry of cytochromecbased on Poly(Diallyldimethylammonium Chloride)- graphene nanosheets/gold nanoparticles hybrid nanocomposites and its biosensing. Electroanalysis 25:1400–1409
87. Zuo J, Zhao X, Ju X, Qiu S, Hu W, Fan T, Zhang J (2016) A new molecularly imprinted polymer (MIP)-based electrochemical sensor for monitoring cardiac troponin I (cTnI) in the serum. Electroanalysis 28:2044–2049
88. Babamiri B, Salimi A, Hallaj R (2018) A molecularly imprinted electrochemiluminescence sensor for ultrasensitive HIV-1 gene detection using EuS nanocrystals as luminophore. Biosens Bioelectron 117:332–339

89. Rahman MA, Kumar P, Park D-S, Shim Y-B (2008) Electrochemical sensors based on organic conjugated polymers. Sensors 8:118–141
90. Atta NF, El-Kady MF, Galal A (2010) Simultaneous determination of catecholamines, uric acid and ascorbic acid at physiological levels using poly(N-methylpyrrole)/Pd-nanoclusters sensor. Anal Biochem 400:78–88
91. Noroozifar M, Khorasani-Motlagh M, Taheri A (2010) Preparation of silver hexacyanoferrate nanoparticles and its application for the simultaneous determination of ascorbic acid, dopamine and uric acid. Talanta 80:1657–1664
92. Li H, Arroyo-Currás N, Kang D, Ricci F, Plaxco KW (2016) Dual-reporter drift correction to enhance the performance of electrochemical aptamer-based sensors in whole blood. J Am Chem Soc 138:15809–15812
93. Li H, Dauphin-Ducharme P, Arroyo-Currás N, Tran CH, Vieira PA, Li S, Shin C, Somerson J, Kippin TE, Plaxco KW (2017) A Biomimetic phosphatidylcholine-terminated monolayer greatly improves the in vivo performance of electrochemical aptamer-based sensors. Angew Chem Int Ed Engl 56:7492–7495
94. Silva SM, Tavallaie R, Gonçales VR, Utama RH, Kashi MB, Hibbert DB, Tilley RD, Gooding JJ (2018) Dual signaling DNA electrochemistry: an approach to understand DNA interfaces. Langmuir 34:1249–1255
95. Zhang Z, Wang Y, Mei Z, Wang Y, Li H, Li S, Xia F (2022) Incorporating hydrophobic moieties into self-assembled monolayers to enable electrochemical aptamer-based sensors deployed directly in a complex matrix. ACS Sens 7:2615–2624
96. Li S, Wang Y, Zhang Z, Wang Y, Li H, Xia F (2021) exploring end-group effect of alkanethiol self-assembled monolayers on electrochemical aptamer-based sensors in biological fluids. Anal Chem 93:5849–5855

Chapter 3
Electrochemical Biosensors for Ions Detection

Kai Zhang, Xiaojin Zhang, and Fan Xia

Abstract As one of the most important biological fluids, whole blood provides important information for health management and disease monitoring. Ions are an important component of blood and carry key information for early diagnosis of diseases. Therefore, a highly selective, sensitive, accurate, and reliable technology is needed to detect the ion content in blood to monitor the health status. Among them, electrochemical biosensors are famous for their advantages in detecting ions in blood, such as quickness, sensitivity, reagent-free, and cleaning-free. In order to achieve the sensitivity and selectivity of blood analysis, electrochemical biosensors have explored many strategies, including the use of hierarchical electrodes, nanomaterial modification, and enzyme amplification. Here, we aim to comprehensively review all these advances, challenges, and opportunities in electrochemical biosensors for the detection of ions in blood.

Keywords Biosensors · Electrochemical detection · Electrodes · Nanomaterials · Blood · Serum · Metal ions · Inorganic phosphate · Nitrite · Nitrate

3.1 Introduction

In the past two decades, monitoring the concentration of ions in the body has provided key information for disease monitoring and health management, and has been the core of many clinical studies, especially the analysis of metal ions in the body, which plays a key role in health monitoring [1–3]. For example, K^+, Na^+, and Ca^{2+} can reflect important life activities in the human body [4]. Fe^{2+} and Zn^{2+} in physiological fluids

---

K. Zhang
Faculty of Materials Science and Chemistry, China University of Geosciences, Wuhan 430074, China

X. Zhang (✉) · F. Xia
State Key Laboratory of Biogeology and Environmental Geology, Engineering Research Center of Nano-Geomaterials of Ministry of Education, Faculty of Materials Science and Chemistry, China University of Geosciences, Wuhan 430074, China
e-mail: zhangxj@cug.edu.cn

F. Xia et al. (eds.), *Electrochemical Biosensors for Whole Blood Analysis*,
https://doi.org/10.1007/978-981-99-5644-9_3

such as whole blood are important metal enzyme ligands that affect enzyme activity [5]. As^{3+}, Cr^{6+}, and Hg^{2+} can cause cancer and kidney, liver, and nervous system failure [6, 7]. Although these ions can provide important information about disease status, their concentration in blood is extremely low and the matrix environment is complex, so it is a great challenge for early clinical detection and diagnosis.

At present, the detection methods of ions in physiological fluids include inductively coupled plasma mass spectrometry (ICP-MS), atomic fluorescence, atomic absorption, and ultraviolet–visible (UV–Vis) spectroscopy [8–10]. However, the chemical steps of the above methods are tedious, and the requirements of expensive reagents and instruments limit their use. In this context, electrochemical biosensors are known for their fast, sensitive, reagent-free, and clean-free performance. They can provide various analytical capabilities to realize the detection of small molecules, ions, nucleic acids, proteins, and cells. In order to achieve sensitive and selective blood analysis, many strategies of electrochemical biosensors have been explored, including the use of hierarchical electrodes, nanomaterial modification, and enzyme amplification. The detection methods include amperometric method [11, 12], potentiometric method [13], and impedance method [14]. The electrochemical biosensors for ions detection are summarized in Table 3.1. Here, we aim to provide an overview of electrochemical biosensors for blood analysis in terms of ion detection, including the sensitivity, specificity, and selectivity of biosensors, and their application in blood detection. Finally, we also discuss the prospects and challenges of these biosensors being converted into clinical application or commercialization.

3.2 Cu^{2+} Detection

Copper ion (Cu^{2+}) is the most important cofactors of metalloproteins and one of the essential trace elements for organisms. It regulates the function of proteins and produces many enzymes that are vital to life [15]. However, the excess or deficiency of Cu^{2+} will cause a disturbance in cellular homeostasis, leading to severe damage to the central nervous system and diseases related to neurodegeneration (such as Alzheimer's disease and Wilson's disease) [16]. In recent years, it has also been found that elevated levels of Cu^{2+} can lead to various types of human cancer and infant liver damage. Therefore, it is urgent to develop convenient and specific ion biosensors to detect Cu^{2+} in environmental matrix and biological fluid sensitively. So far, various materials and technologies have been developed to detect trace Cu^{2+}, such as nanomaterials [17, 18], biomaterials [19], organic molecules [20], optical-based instrument technology [21], and ICP-MS [22]. They have greatly improved the ability to detect Cu^{2+} in physiological fluids. However, these traditional methods typically require a large number of samples, and accurate detection of Cu^{2+} becomes a huge challenge when the target quantity is limited and scarce.

Xia et al. described the application of the mixed structure of nanochannel and ion channel for the first time in the unmarked, ultra-sensitive detection of Cu^{2+} in blood (Fig. 3.1) [23]. Using polyglutamic acid (PGA) as a probe, modification was carried

Table 3.1 Analytical performance parameters for detecting ions in whole blood using electrochemical-based biosensors in the previously reported

Target	Sensing material	Detection mode	Target samples	Linear range	Detection limit	Assay time	References
Cu^{2+}	Polyglutamic acid-modified porous anodic alumina	Voltammetry	Human blood	5×10^{-7}–5×10^{5} μM	0.337 fM	800 s	[23]
Hg^{2+}	DNA probes modified anodic aluminium oxide	Voltammetry	Human serum	0.1–100,000 nM	1 nM	4 h	[30]
Pb^{2+}	AgNF@GCE	Cyclic voltammetry	Human serum	10–700 ppb	0.74 ppb	120 s	[33]
Simultaneous detection of Na^{+}, K^{+} and Cl^{-}	Solid-state ion-selective electrode and gold electrode	Square-wave voltammetry and cyclic voltammetry	Human serum	Cl^{-} 25–200 mM Na^{+} 50–200 mM K^{+} 2–10 mM	/	120 s	[37]
Zn^{2+}	Polyelectrolyte-based carbon fiber sensor	Square-wave anodic stripping voltammetry	Human blood	1×10^{-7}–1×10^{-5} M	4.6×10^{-8} M	60 s	[40]
Simultaneous detection of Cu^{2+}, Pb^{2+}, Hg^{2+} and Cd^{2+}	Al_2NiCoO_5 nanoflakes/ glassy carbon electrode sensor	Anodic stripping differential pulse voltammetry	Human serum	1.0×10^{-5}–1.0 ppm	Pb^{2+} 0.00154 ppb Hg^{2+} 0.00232 ppb Cu^{2+} 0.00261 ppb Cd^{2+} 0.00114 ppb	25 s	[43]
As^{3+}	PANI@$BiVO_4$/SPCE	Differential pulse anodic stripping voltammetry	Human serum	0.01–300 ppb	0.0072 ppb	50 s	[47]
Cr^{6+}	PANI/Hep/MTA/MGCE	Differential pulse voltammetry	Rabbit blood	10^{-7}–10^{-4} ppm	0.87 μM	/	[55]

(continued)

Table 3.1 (continued)

Target	Sensing material	Detection mode	Target samples	Linear range	Detection limit	Assay time	References
Ca^{2+}	Alizarin red S at a gold *nanoparticle-modified* glassy carbon electrode	Cyclic voltammetry	Human serum	2.0×10^{-7}–1.2×10^{-4} M	2.57×10^{-8} M	120 s	[59]
Inorganic phosphate	Molybdenum phosphide-modified electrode	Differential pulse voltammetry	Human blood	0.10–20.0 mM	0.030 mM	1 h	[65]
Nitrite and nitrate	NaR-SOD1-CNT-PPy-Pt	Cyclic voltammetry	Human plasma, whole blood, and saliva	NO_2^- 100 nM–1 mM NO_3^- 500 nM–10 mM	NO_2^- 50 nM NO_3^- 200 nM	/	[71]
Blood urea nitrogen	Thread-based micro-fluidic system	Cyclic voltammetry	Human blood	0.1–10.0 mM	/		[75]

out in a nanochannel array. In the presence of Cu^{2+}, the nitrogen atom of PGA amide group and the oxygen atom of free carboxyl group can combine with Cu^{2+} to form a Cu^{2+}-PGA chelate. The increase in the volume of chelated Cu^{2+}-PGA, resulting in different free mass transfer regions for ion transfer in the nanochannel. Therefore, different ionic currents can be expected in the presence of Cu^{2+}. The experimental results showed that Cu^{2+}-PGA chelation reached equilibrium within 800 s, and the ion current decreased significantly with the increase of Cu^{2+} concentration. The reduced ion current indicates that the spatial potential resistance of ion transport in the nanochannel increases. Surprisingly, this platform can successfully detect as low as 3.37×10^{-10} μM of ultra-low concentration Cu^{2+}. In order to evaluate the applicability of this method to detect Cu^{2+} in real samples, after adding different amounts of Cu^{2+} in blood, the recovery rate of the measured values was 86–104%, indicating that this method has strong anti-interference ability for the detection of Cu^{2+} in blood.

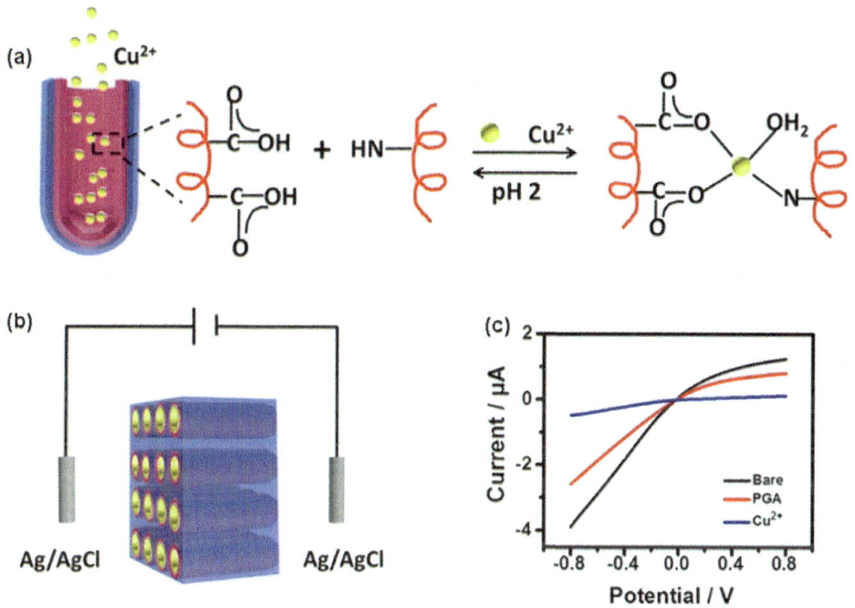

Fig. 3.1 **a** Schematics of Cu^{2+} detection principle. **b** I-V measuring device. **c** I-V curve. The scanning rate is 100 mV/s. The black line is the bare nanochannel array, the red line is PGA-modified nanochannel array and the blue line is the addition of Cu^{2+}. (Reprinted with permission from Ref. [23]. Copyright 2017 American Chemical Society)

3.3 Hg^{2+} Detection

Mercury ion (Hg^{2+}) is a common pollutant and poses a serious threat to human health. Because Hg^{2+} is easy to accumulate in organisms, even at very small doses, Hg^{2+} may cause nerve injury, physiological, and respiratory failure, and induce gene polymorphism [24, 25]. In view of this, the development of trace level Hg^{2+} detector is crucial for sensitively and selectively detecting Hg^{2+} pollution in seafood and water resources. Similarly, high-sensitivity detection of Hg^{2+} in blood is in urgent need of a simple and fast sensor. Among the current detection methods [26–29], electrochemical biosensors have many outstanding advantages, such as high sensitivity, low price, rapid response, easy field operation, and good anti-interference.

Xia et al. were inspired by the synergistic effect of biological nanopores, designed a biomimetic solid-state nanochannel with hydrophobic coating on the outer surface ($None@OS_{Hydrophobic}$) and hydrophilic DNA probe on the inner wall ($DNA@IW_{Hydrophilic}$) (Fig. 3.2) [30]. Hydrophilic DNA is the probe and Hg^{2+} is the target. The experimental results indicated that the limit of detection (LOD) of $DNA@IW_{Hydrophilic} + None@OS_{Hydrophobic}$ nanochannel was reaching 1 nM. The reason for ultra-high sensitivity is attributed to the synergy: $DNA@IW_{Hydrophilic}$ induces a specific sensing target and $None@OS_{Hydrophobic}$ leads to a decrease in the effective diameter of nanochannels. To demonstrate the nanochannel is applicable in complex samples, the concentration of Hg^{2+} in serum was tested by gold standard (ICP-MS) and nanochannel, respectively, and use MedCalc software for Deming regression to evaluate the correlation between ICP-MS and nanochannel methods. The slope of the curve is 1.093, indicating a good logical correlation between the two methods. The output Hg^{2+} concentrations measured by the two methods are similar, indicating that the nanochannel can maintain sensitivity in complex samples, such as serum.

3.4 Pb^{2+} Detection

Lead ion (Pb^{2+}) is considered to be a systemic toxicant even at a low level. It is related to a variety of health problems, including cardiovascular disease [31], and intellectually deficits [32]. Children are more susceptible to lead poisoning than adults. Considering the harm to the human body, the detection of Pb^{2+} becomes very important, and it is urgent to detect trace Pb^{2+} rapidly.

Feng et al. developed a multi-step in-situ preparation method of shape-controlled silver nanoflower on a glassy carbon electrode (AgNF@GCE), and extended AgNF@GCE to anodic stripping voltammetric analysis of Pb^{2+} (Fig. 3.3) [33]. The sensor does not need complex off-line sample pretreatment or labeling and has low LOD (0.74 ppb Pb^{2+}), good linearity. The relative standard deviation of 8 repeated measurements of 50 ppb Pb^{2+} was 2.8%. The sensor is tolerable from the interference of other metal ions such as Na^+, K^+ Fe^{2+}, Ni^{2+}, Mg^{2+}, and Cu^{2+}. When using

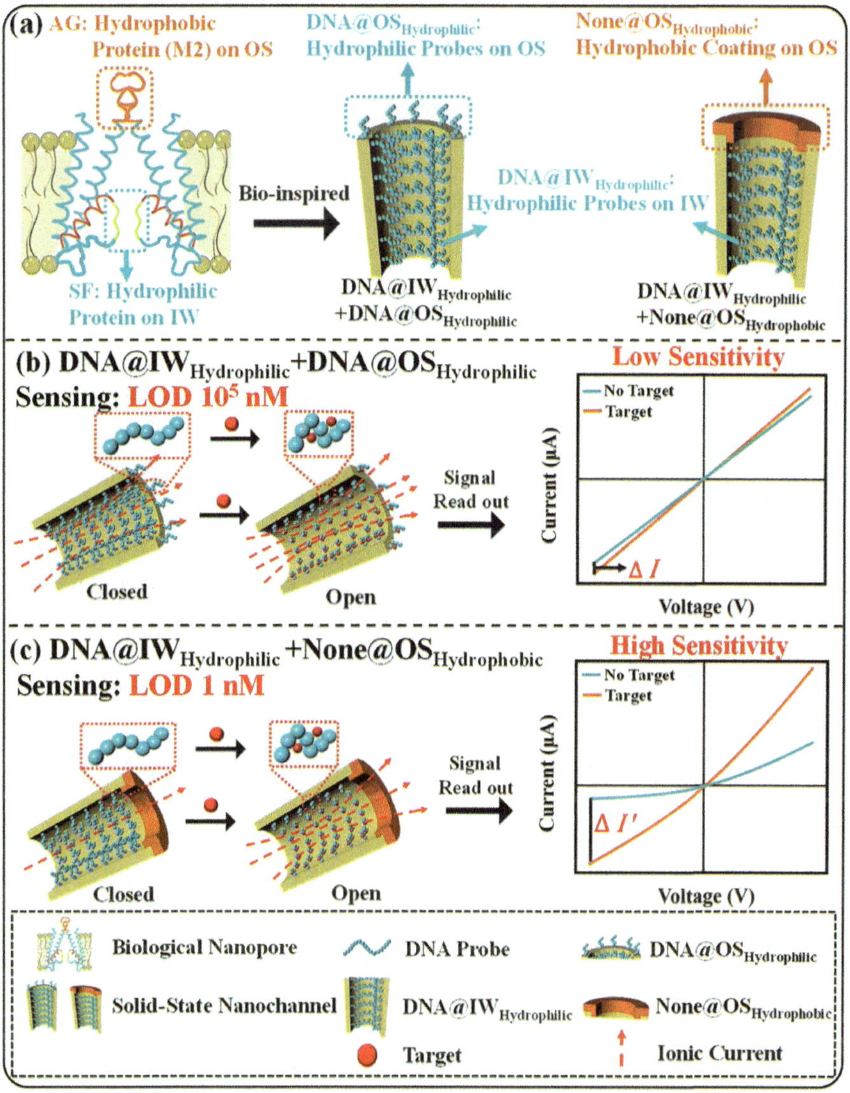

Fig. 3.2 Schematics of solid-state nanochannels and their sensing process. **a** Biological nanopore and solid-state nanochannels. Inspired by different functional proteins distributed at various partitions of biological nanopore, solid-state nanochannels with DNA@IW$_{Hydrophilic}$ + DNA@OS$_{Hydrophilic}$ and DNA@IW$_{Hydrophilic}$ + None@OS$_{Hydrophobic}$ were designed and fabricated. **b**, **c** Schematics of nanochannel sensing target. (Reprinted with the permission from Ref. [30]. Copyright 2022 Wiley–VCH GmbH)

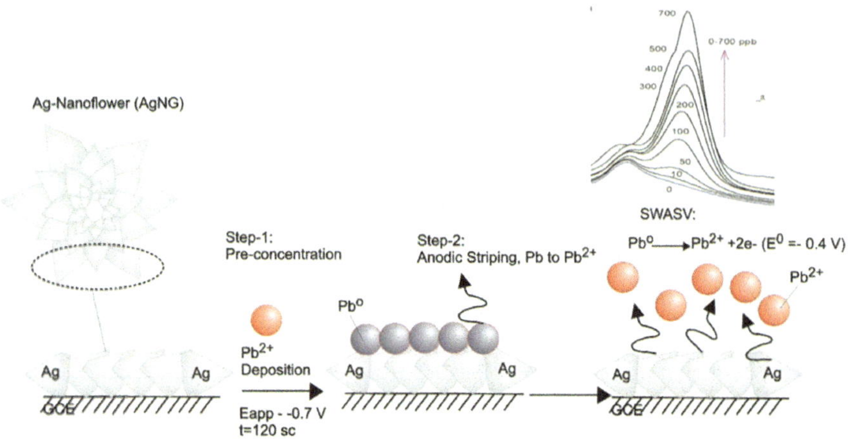

Fig. 3.3 AgNF@GCE for Pb2 + analysis through the pre-concentration, stripping and calibration. (Reprinted with the permission from Ref. [33]. Copyright 2020 Elsevier B.V.)

AgNF@GCE to detect Pb^{2+} in serum, the recoveries of 50, 100, and 150 ppb Pb^{2+} were calculated as 97.2%, 96%, and 96%, respectively. These results demonstrate the applicability and reliability of AgNF@GCE detection Pb^{2+} in serum.

3.5 Simultaneous Detection of Cl^-, Na^+ and K^+

Chloride ion (Cl^-), Sodium ion (Na^+) and potassium ion (K^+) can maintain the physiological function of acid–base balance, water metabolism, osmotic pressure, and neuromuscular system in the body. Therefore, the metabolic function of human body can be evaluated by measuring Cl^-, Na^+, and K^+ [34, 35]. The concentrations of Cl^-, Na^+, and K^+ in normal serum are 98–106 mM, 135–145 mM, and 3.5–5 mM respectively. High concentration will cause hyperkalemia, hypernatremia, and hyperchloremia, while low concentration will cause hypoxemia, hyponatremia, and hypochloremia [36]. Their sensitive, precise, and simultaneous monitoring, could allow early diagnosis of various physiological and pathological events, as well as tracking medical outcomes for correct treatment interventions.

Wang et al. developed simple electrodes for rapid detection of Cl^-, Na^+, and K^+ in serum [37]. A mixture of ion-selective membrane (ISM) and 7,7,8,8-tetracyanoquinoline methane (TCNQ) was covered on a planar gold electrode to form a solid-state ion selective electrode (ISE) for detecting Na^+ and K^+. A planar thin-film gold electrode was used to detect Cl^-. The linear detection range of Cl^-, Na^+, and K^+ standard samples is 25–200 mM, 50–200 mM, and 2–10 mM, and the recovery rate is 101%, 100%, and 96%, respectively. The interference experiments show that the electrode has good selectivity. In addition, the electrode has the characteristics of low relative standard deviation, high detection sensitivity, and simple process conditions. The linear detection range is applicable to the concentration of Na^+, K^+, and Cl^- in serum.

3.6 Zn^{2+} Detection

Zinc ion (Zn^{2+}) are essential trace element for the normal function of various enzymes, hormones and transcription-related factors [38]. Zinc homeostasis is necessary for the normal function of the antioxidant system, insulin secretion/action, and immune system. Studies on children and adults have shown that the Zn^{2+} level of severe patients is abnormally low, because inflammation and infection are related to the reduction of serum Zn^{2+} level. Zn^{2+} supplementation may be a beneficial treatment strategy for critically ill patients [39]. However, to make the strategy safely and effective, it is necessary to continuously monitor the serum Zn^{2+} level.

Compared with spectral methods, anodic stripping voltammetry (ASV) is a promising method for the determination of trace metal ions. The analysis includes a negative bias pre-concentration step, which reduces the target metal ions to metal form, accumulates them on the electrode surface, and then carries out a positive scanning dissolution step to oxidize the metal back to its ionic form. ASV is more time-saving and cost-effective than spectral methods. It provides the LOD in the nanomolar range, which is sufficient for clinical measurement. Skorb et al. prepared an electrochemical sensing platform based on self-assembled polyelectrolyte multilayer film on the surface of carbon fiber (CF) electrode for detecting Zn^{2+} (Fig. 3.4) [40]. The working electrode was modified by layer-by-layer assembly of poly(sodium 4-styrenesulfonate) (PSS) and polyethyleneimine (PEI) eliminating the toxicity of the solution and providing a stable stripping voltammetric measurement. The reusable, solid, stable, and sustainable Ag/AgCl reference electrode is composed of 32 PEI-KCl/PSS-KCl bimolecular layers, which are adsorbed on CF/silver paste and separated from the external solution through polyvinyl chloride (PVC) membrane. The sensor interface based on polyelectrolyte can prevent the biological liquid from adsorbing onto the protein molecules on the CF surface, thus improving the sensitivity of Zn^{2+} detection to 2.2 μA/M with the linear detection range of 1×10^{-7} to 1×10^{-5} M and the LOD of 4.6×10^{-8} M. This method can accurately detect the concentration of Zn^{2+} in blood without digestion steps.

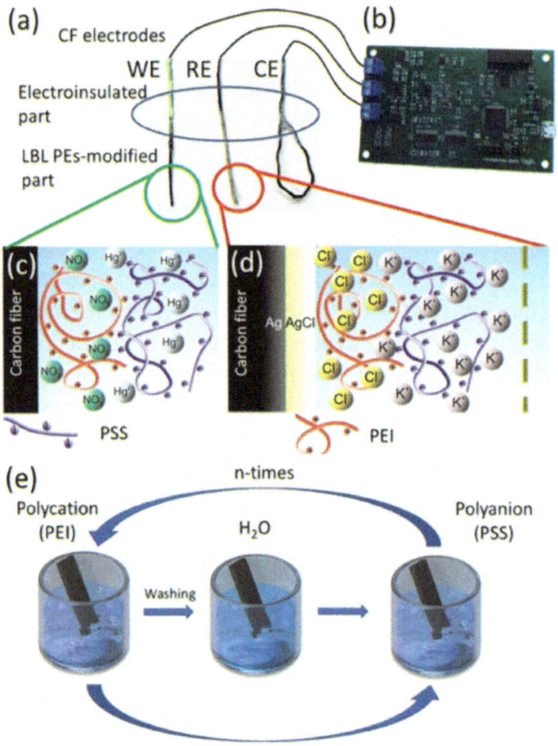

Fig. 3.4 **a** Electrochemical platform setup. **b** Connected to the minipotentiostat. **c** Modification of the working electrode (WE). **d** Modification of the reference electrode (RE). (Reprinted with permission from Ref. [40]. Copyright 2020 American Chemical Society)

3.7 Simultaneous Detection of Cu^{2+}, Pb^{2+}, Hg^{2+}, and Cd^{2+}

Heavy metal ions play a vital role in the metabolic function of organisms at trace concentrations and are considered fatal at higher concentrations. The physiological standard concentration of Pb^{2+} in human body is 50 ppb, and exceeding this concentration will cause nervous system diseases such as low birth weight, hyperactivity disorder, and inattention [41]. High concentrations of Cd^{2+} (normal physiological concentration: 5 ppb) and Hg^{2+} (normal physiological concentration: 0.2 ppb) in blood will cause various cancers and directly harm the nervous system. Low concentrations of Cu^{2+} (normal physiological range: 100 ppb to 150 ppb) may be related to nutritional deficiency and kidney disease, while high concentrations above 150 ppb may lead to liver diseases, such as rheumatoid arthritis, lymphoma, and reactive thyroid leukemia [42].

Electrochemical analysis provides a low-cost platform for detecting heavy metal ions in blood, and the necessity of simultaneous sensing has aroused great concern.

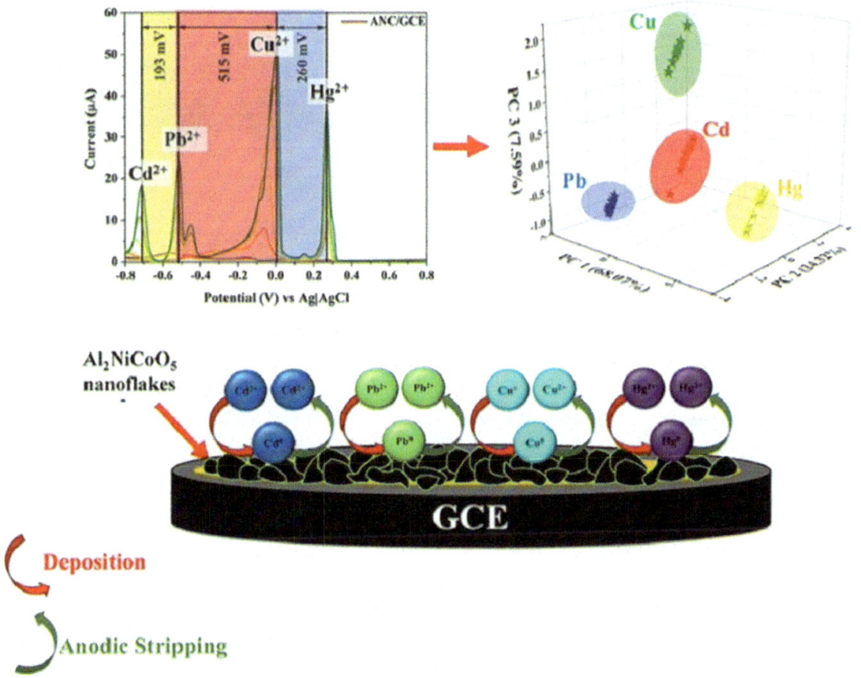

Fig. 3.5 Heavy metal ions sensing mechanism of the ANC/GCE sensor. (Reprinted with the permission from Ref. [43]. Copyright 2022 The Authors)

Badhulika et al. prepared Al_2NiCoO_5 (ANC) nanoflakes for the simultaneous detection of Cu^{2+}, Pb^{2+}, Hg^{2+}, and Cd^{2+} in serum (Fig. 3.5) [43]. Anodic stripping differential pulse voltammetry analysis showed that the ANC-modified glassy carbon electrode (ANC/GCE) had good sensitivity, selectivity, and stability, and could simultaneously detect heavy metal ions. The redox pairs of Ni^{2+}/Ni^{3+} and Co^{2+}/Co^{3+} redox couples and their surface oxygen functional groups enable ANC nanosheets to exhibit high electrocatalytic performance. The LOD of the ANC/GCE sensor for Cd^{2+}, Hg^{2+}, Cu^{2+}, and $Pb^{2+}$is 0.00114 ppb, 0.00232 ppb, 0.00261 ppb, and 0.00154 ppb, which is far below the hazard limit of the concentration of heavy metal ions in serum.

3.8 As^{3+} Detection

Arsenic ion (As^{3+}) has high toxicity, causing serious biological damage to human beings [44]. As^{3+} can lead to kidney, liver and nervous system failure [45]. Among the various oxidation states of arsenic, As^{3+} is the most toxic and stable [46]. Therefore, monitoring the level of As^{3+} in blood is very important for preventing As^{3+} poisoning. The World Health Organization (WHO) stipulates that the content of As^{3+} in blood

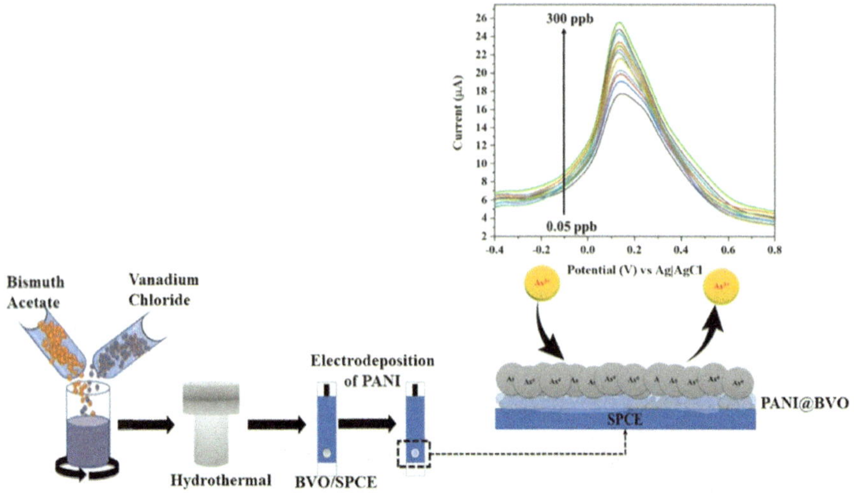

Fig. 3.6 As^{3+} sensing mechanism of PANI@BiVO4/SPCE. (Reprinted with the permission from Ref. [47]. Copyright 2020 Elsevier B.V.)

should be less than 50 ppb. Therefore, a highly sensitive, selective, accurate, and reliable technology used for detecting As^{3+} in blood is necessary.

Durai et al. reported a simple hydrothermal method to synthesize ultra-thin bismuth vanadate ($BiVO_4$) coated with polyaniline (PANI) by electrodeposition on screen-printed carbon electrode (SPCE) for detecting trace As^{3+} in biological samples (Fig. 3.6) [47]. The sensor can detect As^{3+} by differential pulse anodic stripping voltammetry (DPASV) technique with high sensitivity (6.06 $\mu A/ppb/cm^2$) and low LOD (0.0072 ppb) in the linear range of 0.01 to 300 ppb. The improvement of the sensing ability can be attributed to the synergistic effect of $BiVO_4$ and PANI, which provide high electrocatalytic active sites and high conductivity, respectively. The sensor has significant selectivity for detecting As^{3+} in serum.

3.9 Cr^{6+} Detection

Chromium (Cr) is a common heavy metal and is mainly present in two oxidized states: Cr^{3+} and Cr^{6+} [48]. Cr^{3+} and Cr^{6+} can accumulate in human blood and tissues, which is difficult to metabolize and clear [49]. Previous studies have shown that the toxicity of Cr^{6+} is 1000 times that of Cr^{3+}, because Cr^{6+} has high bioavailability and solubility, and has toxic effects on skin, digestive system, and respiratory system [50, 51]. Therefore, detecting the concentration of Cr^{6+} in the blood has profound significance for accurately determining the status of chromium poisoning [52].

Electrochemical method has the advantages of good stability, low cost, simple operation, and fast response, making it an effective method for detecting Cr^{6+} [53].

However, it is well known that when the traditional electrode come into contact with fibrin, platelets, whole blood, and blood cells, they will adhere to the surface of the electrode to form a biological dirt layer, which hinders the electron transfer between the probe and the electrode. Therefore, it is not easy to directly detect Cr^{6+} in blood using electrochemical method [54]. Mao et al. developed an efficient blood compatible sensor capable of detecting Cr^{6+} based on the magnetic glassy carbon electrode (MGCE) modified by tannic, heparin, and PANI-modified Fe_3O_4 nanoparticles (magnetic tannic modified adsorbent, MTA) (Fig. 3.7) [55]. PANI has the advantages of special proton transfer mechanism, corrosion resistance, and low cost [56]. The presence of imine functional groups and plentiful cationic amines enables PANI to adsorb Cr^{6+} oxygen anion through electrostatic interaction. Due to the high content of adjacent phenolic hydroxyl groups in tannin, the sensor has good chelating affinity for Cr^{6+} [57]. The experimental results show that PANI/Hep/MTA/MGCE has high sensitivity for Cr^{6+} detection in the range of 10^{-7} to 10^{-4} ppm, and the LOD is 0.87 μM. In particular, the sensor not only shows excellent electrochemical performance for the detection of Cr^{6+}, but also has good anti-biological pollution characteristics due to the presence of heparin, providing a guarantee for the direct detection of Cr^{6+} in blood.

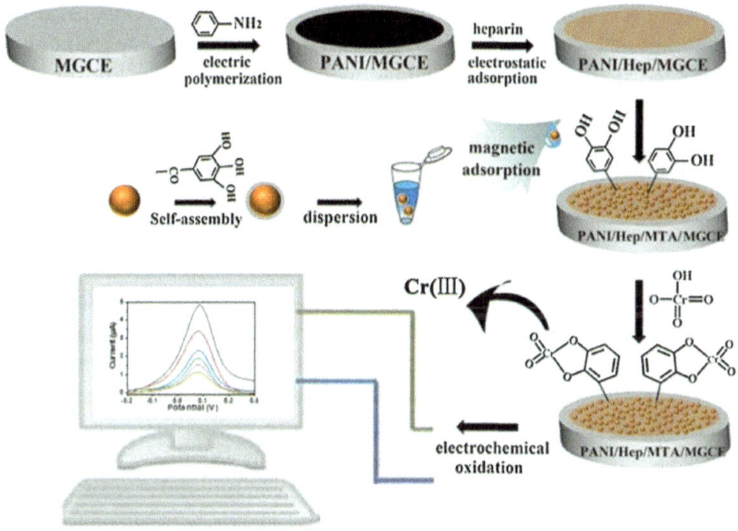

Fig. 3.7 Schematics of the antifouling electrode for Cr^{6+} detection. (Reprinted with the permission from Ref. [55]. Copyright 2020 Elsevier B.V.)

3.10 Ca^{2+} Detection

Calcium ion (Ca^{2+}) is essential in serum and living cells. Mitochondrial calcium homeostasis plays a key role in the regulation of cell survival and aerobic metabolism [58]. Coagulation, reduction of nerve and muscle excitability, and maintenance of cardiac rhythm are related to Ca^{2+} in serum. Therefore, it is important to know its concentration.

Yang et al. proposed a simple method for preparing highly dispersed and small gold nanoparticles (AuNPs) by pulse electrodeposition on glassy carbon electrodes [59]. The electrocatalytic activity of AuNPs can be controlled by changing the concentration of $HAuCl_4$ and adjusting the number of potential pulses. Trace Ca^{2+} was determined by studying the absorption wave of alizarin red S (ARS) and Ca-ARS complex on the electrode. In 0.1 M KOH solution, the LOD for the detection of Ca^{2+} in serum is 2.57×10^{-8} M, and the recovery rate is between 91.8% and 103.4%.

3.11 Inorganic Phosphate Detection

Phosphate is a fundamental component of almost all living organisms. It is one of the most commonly used electrolytes and participates in many important physiological processes, such as blood pH regulation and ATP synthesis. Phosphorus in the blood is usually present in the form of organic and inorganic phosphorus. Abnormal blood phosphate concentration may be related to vitamin D deficiency, bone diseases, Fanconi syndrome, hyperparathyroidism, and minerals [60]. Therefore, the determination of phosphate is very important for early screening of diseases in clinical medicine.

The detection methods of phosphate in blood mainly include ion chromatography, dye method, phosphomolybdic acid colorimetry, electrochemical method, and enzymatic method. Among them, electrochemical method has the advantages of low cost, simple operation, automation, online measurement, and miniaturization [61]. The reported electrochemical strategies for phosphate detection include enzyme electrode, dissolution voltammetry, and potentiometric method based on ion-selective electrode [62, 63]. Some strategies are to use the classical reaction of ammonium molybdate and phosphate to detect the product phosphomolybdate, to achieve indirect detection of phosphate, or to build an ion-selective electrode based on the interaction of phosphate and molybdenum to achieve indirect detection of phosphate [64].

Li et al. found that in the presence of phosphate, a phenomenon similar to "catalytic hydrogen evolution wave" appeared on the molybdenum phosphide (MoP) modified electrode, that is, a catalytic hydrogen evolution wave occurs before the normal hydrogen evolution reaction [65]. The catalytic hydrogen wave is generated from a structure similar to phosphomolybdic acid, which is formed by the interaction of molybdenum oxide and phosphate on the surface of MoP-modified electrode, resulting in the adjustment of interfacial catalytic activity and the change of surface

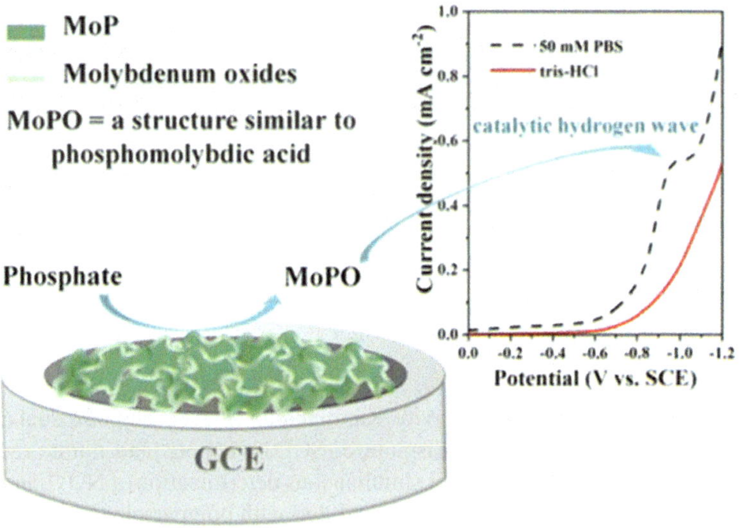

Fig. 3.8 Schematics of phosphate detection based on a catalytic hydrogen wave at a molybdenum phosphide modified electrode. (Reprinted with the permission from Ref. [65]. Copyright 2019 American Chemical Society)

structure (Fig. 3.8). They constructed a phosphate electrochemical sensor with a linear range of 0.10 ~ 20.0 mmol/L, the LOD of 0.030 mmol/L and the recovery rate of 94% ~ 107%. When this sensor is used to detect phosphate in blood, other substances will not form significant reduction peaks that interfere with phosphate detection.

3.12 Nitrite and Nitrate Detection

Nitric oxide (NO) is an important messenger molecule that regulates biological processes, such as vasorelaxation, intercellular communication, and immune function [66]. Due to its short half-life and fast metabolism, NO is easily oxidized to nitrite (NO_2^-) and nitrate (NO_3^-) [67]. NO_2^- and NO_3^- act as endocrine reservoirs to produce NO in tissues damaged by ischemia or hypoxia [68]. The concentration of NO is usually determined by measuring the plasma concentration of NO_2^- and NO_3^- [69]. The concentration of NO_2^- and NO_3^- in blood are possibly altered in vasodilation, physiological hypoxia signals, cellular response to ischemic stress, and cellular respiratory regulation [70]. The detection of NO_2^- and NO_3^- is very important in human physiology, because it provides valuable information about bioavailability, NO production, and prognosis of various diseases.

The electrochemical biosensors have been proven to be a practical advantage for measuring NO_2^- and NO_3^-, such as simple operation, low manufacturing cost, fast

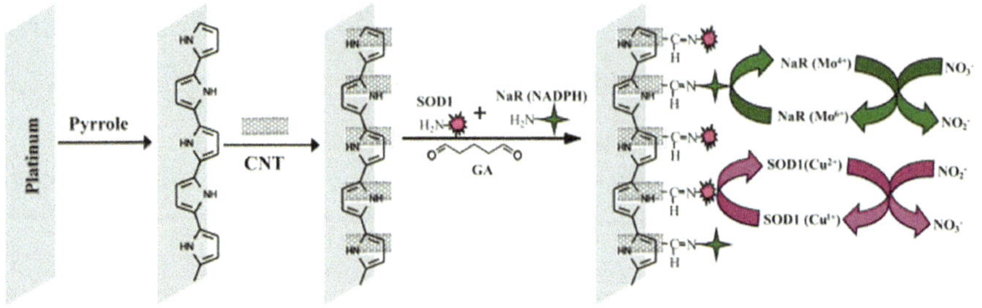

Fig. 3.9 Schematics of NaR-SOD1-CNT-PPy-Pt electrode for the simultaneous detection of NO_2^- and NO_3^-. (Reprinted with the permission from Ref. [71]. Copyright 2013 Elsevier B.V.)

response, real-time and high sensitivity. Karunakaran et al. developed a dual enzyme biosensor using zinc superoxide dismutase (SOD1), copper, and nitrate reductase (NaR) co-immobilized electrode for simultaneous determination of NO_2^- and NO_3^- (Fig. 3.9) [71]. The electrode surface is modified with polypyrrole (PPy) an ordered conductive polymer chains, and also provides a porous substrate for the fixation of SOD1 and NaR [72]. Carbon nanotubes (CNT) modify PPy matrix to form CNT-PPy nanocomposite, which can obtain additional surface area to fix more SOD1 and NaR and can also be used as molecular wires to accelerate the electron transfer between the active site and the bottom electrode, thus improving the sensitivity of the biosensor. Using cellulose acetate (CA) membrane to eliminate interference from other substances in biological samples. The experimental results show that the electrocatalytic activity of SOD1 for NO_2^- oxidation is linear between 100 nM to 1 mM with a sensitivity of 98.5 ± 1.7 nA/μM/cm^2 and the LOD of 50 nM. The co-immobilized NaR shows a linear NO_3^- detection from 500 nM to 10 mM with an LOD of 200 nM and a sensitivity of 84.5 ± 1.56 nA/μM/cm^2. When measuring NO_2^- and NO_3^- in blood, the two-enzyme biosensor shows good accuracy and reproducibility, making it a clinical diagnostic tool for monitoring NO_2^- and NO_3^-.

3.13 Blood Urea Nitrogen Detection

Blood urea nitrogen (BUN) in physiological fluids is an important biological indicator for monitoring renal function. The BUN level of patients with renal failure is high. Additionally, studies have shown that BUN levels are associated with heart failure [73, 74]. The standard method for determining the BUN concentration is the enzymatic ultraviolet absorption method. This method is clear and simple, but it is limited by the large amount of samples required and the limited detectable linear range.

Urea is difficult to detect using electrochemical methods due to the electrochemical inertness of uric acid. One strategy is to convert urea into ammonium ions in the presence of urease, and ammonium ion has electrochemical activity and is

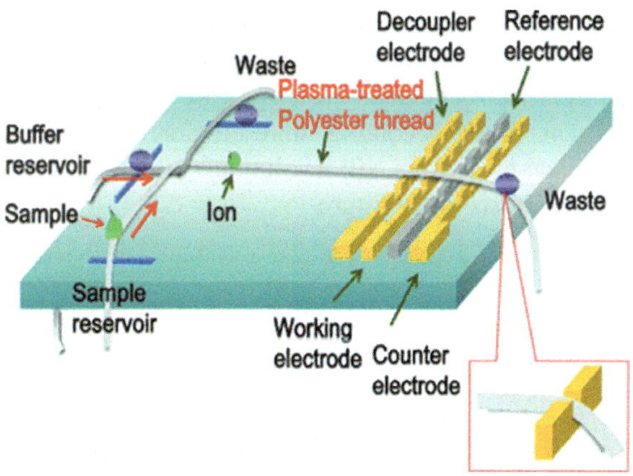

Fig. 3.10 Schematics of the thread-based microfluidic device with variable injection volume and 3D electrochemical electrode. (Reprinted with the permission from Ref. [75]. Copyright 2014 Springer-Verlag Berlin Heidelberg)

easy to detect. Yang et al. developed a thread-based microfluidic device with three-dimensional (3D) electrode and variable volume injection capability for the detection of BUN in blood by capillary electrophoresis electrochemistry (CE-EC) (Fig. 3.10) [75]. After mixing urea with urease, the marked redox peak confirms that urea can be electrochemically detected on the polyester line. The experimental results indicate that the sensor has an excellent linear response to detect urea in the range of 0.1–10.0 mM, which is sufficient to detect BUN in actual samples (1.78–7.12 mM). The sensor has the advantages of low cost, fast detection, one-time but high performance, and can be used to detect BUN in blood.

3.14 Conclusion

The detection of ions in blood by electrochemical biosensors has become an emerging technology in personalized medicine, because they can be directly used for the analysis of clinical samples, and have the advantages of rapid detection, good accuracy, requiring minimal sample volume, no sample pretreatment and good stability as compared with the traditional methods. Electrochemical methods such as square wave voltammetry, electrochemical impedance spectroscopy, cyclic voltammetry, differential pulse voltammetry, and current analysis have been used for ions detection. The application of nanomaterials in biosensors provides an opportunity to establish a new generation of biosensor technology and is developing toward single molecule biosensors with high sensitivity, high accuracy, and high throughput biosensor arrays.

However, electrochemical biosensors still have some challenges that require further efforts to achieve clinical applications, such as sample preparation, micro-/nanofluidics platform, sensitivity, specificity, dynamic range, rapid turnaround time, multiplexing, continuous monitoring, need to transport samples to the electrode surface, analyte loss, reproducibility, and calibration. Future studies may focus on developing polymer biosensors. Because the electrosynthesis of polymer nanowires or nanotubes significantly improves the sensing performance of conductive polymers, electrodeposition of polymer nanostructures is another promising method in the future. Electrochemical biosensors are helping to integrate electronics with biology. Biosensors are becoming more efficient, more economical, and smaller. Electrochemical biosensors will revolutionize the fields of food safety, healthcare, defense, and diagnosis in the future.

References

1. Crowley E, Di Nicolantonio F, Loupakis F, Bardelli A (2013) Liquid biopsy: monitoring cancer-genetics in the blood. Nat Rev Clin Oncol 10:472–484
2. Kim JJ, Kim YS, Kumar V (2019) Heavy metal toxicity: an update of chelating therapeutic strategies. J Trace Elem Med Biol 54:226–231
3. Ring G, O'Mullane J, O'Riordan A, Furey A (2016) Trace metal determination as it relates to metallosis of orthopaedic implants: evolution and current status. Clin Biochem 49:617–635
4. Jozanovic M, Sakac N, Karnas M, Medvidovic-Kosanovic M (2021) Potentiometric sensors for the determination of anionic surfactants-a review. Crit Rev Anal Chem 51:115–137
5. Shi HH, Jiang SF, Liu B, Liu ZC, Reis NM (2021) Modern microfluidic approaches for determination of ions. MicrochemJ 171:106845
6. Kozlowski H, Kolkowska P, Watly J, Krzywoszynska K, Potocki S (2014) General aspects of metal toxicity. Curr Med Chem 21:3721–3740
7. Daniya W, Saleviter S, Fen YW (2018) Development of surface plasmon resonance spectroscopy for metal ion detection. Sensor Mater 30:2023–2038
8. Johnson DK, Combs SM, Parsen JD, Jolley ME (2002) Lead analysis by anti-chelate fluorescence polarization immunoassay. Environ Sci Technol 36:1042–1047
9. Kandemir S, Dogru MI, Orun I, Dogru A, Altas L, Erdogan K, Orun G, Polat N (2010) Determination of heavy metal levels, oxidative status, biochemical and hematological parameters in cyprinus carpio L., 1758 from Bafra (Samsun) fish lakes. J Anim Vet Adv 9:617–622
10. Beeram S, Merchant ME, Sneddon J (2012) Studies of metals in crocodilians by spectrochemical methods. Appl Spectrosc Rev 47:144–162
11. Shokrekhodaei M, Quinones S, Fazel Z, Nazeran H (2019) Signal processing of amperometric biosensors for emerging blood glucose measurement/monitoring. Adv Electr Electron Eng 17:64–73
12. Zhang P, Sun TT, Rong SZ, Zeng DD, Yu HW, Zhang Z, Chang D, Pan HZ (2019) A sensitive amperometric AChE-biosensor for organophosphate pesticides detection based on conjugated polymer and Ag-rGO-NH_2 nanocomposite. Bioelectrochemistry 127:163–170
13. Lai CY, Foot PJS, Brown JW, Spearman P (2017) A urea potentiometric biosensor based on a thiophene copolymer. Biosensors-Basel 7:13
14. Qin JF, Sun H, Hao HY, Jia LJ, Yao CZ, Wang QQ, Yang HY (2019) Impedimetric biosensor for the detection of bacterial lipopolysaccharide based on lectin-functionalized gold nanoparticle-graphene composites. Sensor Mater 31:2917–2929
15. Solomon EI, Heppner DE, Johnston EM, Ginsbach JW, Cirera J, Qayyum M, Kieber-Emmons MT, Kjaergaard CH, Hadt RG, Tian L (2014) Copper active sites in biology. Chem Rev 114:3659–3853
16. Song YJ, Qu KG, Xu C, Ren JS, Qu XG (2010) Visual and quantitative detection of copper ions using magnetic silica nanoparticles clicked on multiwalled carbon nanotubes. Chem Commun 46:6572–6574
17. Li L, Feng J, Fan YY, Tang B (2015) Simultaneous imaging of Zn^{2+} and Cu^{2+} in living cells based on DNAzyme modified gold nanoparticle. Anal Chem 87:4829–4835

18. Jin JC, Wu J, Yang GP, Wu YL, Wang YY (2016) A microporous anionic metal-organic framework for a highly selective and sensitive electrochemical sensor of Cu^{2+} ions. Chem Commun 52:8475–8478
19. Zhao Z, Chen HD, Zhang H, Ma LN, Wang ZX (2017) Polyacrylamide-phytic acid-polydopamine conducting porous hydrogel for rapid detection and removal of copper(II) ions. Biosens Bioelectron 91:306–312
20. Jo J, Lee HY, Liu WJ, Olasz A, Chen CH, Lee D (2012) Reactivity-based detection of copper(II) ion in water: oxidative cyclization of azoaromatics as fluorescence turn-on signaling mechanism. J Am Chem Soc 134:16000–16007
21. Jin LH, Han CS (2014) Ultrasensitive and selective fluorimetric detection of copper ions using thiosulfate-involved quantum dots. Anal Chem 86:7209–7213
22. Su SW, Chen BB, He M, Hu B (2014) Graphene oxide-silica composite coating hollow fiber solid phase microextraction online coupled with inductively coupled plasma mass spectrometry for the determination of trace heavy metals in environmental water samples. Talanta 123:1–9
23. Zhao XP, Wang SS, Younis MR, Xia XH, Wang C (2018) Asymmetric nanochannel-ionchannel hybrid for ultrasensitive and label-free detection of copper ions in blood. Anal Chem 90:896–902
24. Safavi A, Farjami E (2011) Construction of a carbon nanocomposite electrode based on amino acids functionalized gold nanoparticles for trace electrochemical detection of mercury. Anal Chim Acta 688:43–48
25. Xie H, Wang Q, Chai YQ, Yuan YL, Yuan R (2016) Enzyme-assisted cycling amplification and DNA-templated in-situ deposition of silver nanoparticles for the sensitive electrochemical detection of Hg^{2+}. Biosens Bioelectron 86:630–635
26. Kristian KE, Friedbauer S, Kabashi D, Ferencz KM, Barajas JC, O'Brien K (2015) A simplified digestion protocol for the analysis of Hg in fish by cold vapor atomic absorption spectroscopy. J Chem Educ 92:698–702
27. Lin Y, Yang Y, Li YX, Yang L, Hou XD, Feng XB, Zheng CB (2016) Ultrasensitive speciation analysis of mercury in rice by headspace solid phase microextraction using porous carbons and gas chromatography-dielectric barrier discharge optical emission spectrometry. Environ Sci Technol 50:2468–2476
28. Wu L, Long Z, Liu LW, Zhou Q, Lee YI, Zheng CB (2012) Microwave-enhanced cold vapor generation for speciation analysis of mercury by atomic fluorescence spectrometry. Talanta 94:146–151
29. Sahu D, Sahoo G, Mohapatra P, Swain SK (2019) Dual activities of nano silver embedded reduced graphene oxide using clove leaf extracts: Hg^{2+} sensing and catalytic degradation. ChemistrySelect 4:2593–2602
30. Liu LX, Luo CH, Zhang JH, He X, Shen Y, Yan B, Huang Y, Xia F, Jiang L (2022) Synergistic effect of bio-inspired nanochannels: hydrophilic DNA probes at inner wall and hydrophobic coating at outer surface for highly sensitive detection. Small 18:2201925
31. Lanphear BP, Rauch S, Auinger P, Allen RW, Hornung RW (2018) Low-level lead exposure and mortality in US adults: a population-based cohort study. Lancet Public Health 3:E177–E184
32. Lanphear BP, Hornung R, Khoury J, Yolton K, Baghurstl P, Bellinger DC, Canfield RL, Dietrich KN, Bornschein R, Greene T, Rothenberg SJ, Needleman HL, Schnaas L, Wasserman G, Graziano J, Roberts R (2005) Low-level environmental lead exposure and children's intellectual function: an international pooled analysis. Environ Health Persp 113:894–899
33. Swetha P, Chen JN, Kumar AS, Feng SP (2020) High index facets-Ag nano flower enabled efficient electrochemical detection of lead in blood serum and cosmetics. J Electroanal Chem 878:114657
34. Berend K, van Hulsteijn LH, Gans ROB (2012) Chloride: the queen of electrolytes? Eur J Intern Med 23:203–211
35. Cinti S, Fiore L, Massoud R, Cortese C, Moscone D, Palleschi G, Arduini F (2018) Low-cost and reagent free paper-based device to detect chloride ions in serum and sweat. Talanta 179:186–192

36. Pirovano P, Dorrian M, Shinde A, Donohoe A, Brady AJ, Moyna NM, Wallace G, Diamond D, McCaul M (2020) A wearable sensor for the detection of sodium and potassium in human sweat during exercise. Talanta 219:121145
37. Chen LD, Wang WJ, Wang GJ (2021) Electrochemical detection of electrolytes using a solid-state ion-selective electrode of single-piece type membrane. Biosensors-Basel 11:109
38. Rahman S, Waheed S (2009) Validation of INAA and AAS analysis protocols for whole blood: a study on cardiovascular and malignant hypertensive patients. J Radioanal Nucl Chem 279:915–921
39. Lu TH, Huang JF, Sun IW (2002) Perfluorinated anion-exchange polymer mercury film electrode for anodic stripping voltammetric determination of zinc(II): effect of model organic compounds. Anal Chim Acta 454:93–100
40. Nikolaev KG, Kalmykov EV, Shavronskaya DO, Nikitina AA, Stekolshchikova AA, Kosareva EA, Zenkin AA, Pantiukhin IS, Orlova OY, Skalny AV, Skorb EV (2020) ElectroSens platform with a polyelectrolyte-based carbon fiber sensor for point-of-care analysis of Zn in blood and urine. ACS Omega 5:18987–18994
41. Gidlow DA (2015) Lead toxicity. Occup Med-Oxford 65:348–356
42. Carpenter WE, Lam D, Toney GM, Weintraub NL, Qin ZY (2013) Zinc, copper, and blood pressure: human population studies. Med Sci Monit 19:1–8
43. Durai L, Badhulika S (2022) Stripping voltammetry and chemometrics assisted ultra-selective, simultaneous detection of trace amounts of heavy metal ions in aqua and blood serum samples. Sensor Actuat Rep 4:100097
44. Kovalchuk O, Titov V, Hohn B, Kovalchuk I (2001) A sensitive transgenic plant system to detect toxic inorganic compounds in the environment. Nat Biotechnol 19:568–572
45. Gumpu MB, Sethuraman S, Krishnan UM, Rayappan JBB (2015) A review on detection of heavy metal ions in water-an electrochemical approach. Sensor Actuat B-Chem 213:515–533
46. Cassone G, Chille D, Foti C, Giuffre O, Ponterio RC, Sponer J, Saija F (2018) Stability of hydrolytic arsenic species in aqueous solutions: As^{3+} vs. As^{5+}. Phys Chem Chem Phys 20:23272–23280
47. Durai L, Badhulika S (2020) Ultra-selective, trace level detection of As^{3+} ions in blood samples using PANI coated $BiVO_4$ modified SPCE via differential pulse anode stripping voltammetry. Mater Sci Eng C-Mater Biol Appl 111:110806
48. Qi Y, Jiang M, Cui YL, Zhao L, Liu SJ (2015) Novel reduction of Cr(VI) from wastewater using a naturally derived microcapsule loaded with rutin-Cr(III) complex. J Hazard Mater 285:336–345
49. Collins BJ, Stout MD, Levine KE, Kissling GE, Melnick RL, Fennell TR, Walden R, Abdo K, Pritchard JB, Fernando RA, Burka LT, Hooth MJ (2010) Exposure to hexavalent chromium resulted in significantly higher tissue chromium burden compared with trivalent chromium following similar oral doses to male F344/N rats and female B6C3F1 mice. Toxicol Sci 118:368–379
50. Ou YQ, Bloom MS, Nie ZQ, Han FZ, Mai JZ, Chen JM, Lin S, Liu XQ, Zhuang J (2017) Associations between toxic and essential trace elements in maternal blood and fetal congenital heart defects. Environ Int 106:127–134
51. Zhao R, Li X, Sun BL, Li YZ, Li YM, Yang R, Wang C (2017) Branched polyethylenimine grafted electrospun polyacrylonitrile fiber membrane: a novel and effective adsorbent for Cr(VI) remediation in wastewater. J Mater Chem A 5:1133–1144
52. Pan Y, Guan DX, Zhao D, Luo J, Zhang H, Davison W, Ma LQ (2015) Novel speciation method based on diffusive gradients in thin-films for in situ measurement of Cr-VI in aquatic systems. Environ Sci Technol 49:14267–14273
53. Qi SP, Zhao B, Tang HQ, Jiang XQ (2015) Determination of ascorbic acid, dopamine, and uric acid by a novel electrochemical sensor based on pristine graphene. Electrochim Acta 161:395–402
54. Chen H, Shao SB, Yu YQ, Huang YY, Zhu XT, Zhang SY, Fan J, Yin GY, Chi B, Wan MM, Mao C (2020) A dual-responsive biosensor for blood lead detection. Anal Chim Acta 1093:131–141

55. Fang D, Xu TT, Fang LY, Chen H, Huang YY, Zhang HY, Miao ZY, Mao C, Chi B, Xu H (2021) A blood compatible, high-efficient sensor for detection of Cr(VI) in whole blood. Sensor Actuat B-Chem 329:129219
56. Zhai DY, Liu BR, Shi Y, Pan LJ, Wang YQ, Li WB, Zhang R, Yu GH (2013) Highly sensitive glucose sensor based on Pt nanoparticle/polyaniline hydrogel heterostructures. ACS Nano 7:3540–3546
57. Xu QH, Wang YL, Jin LQ, Wang Y, Qin MH (2017) Adsorption of Cu(II), Pb(II) and Cr(VI) from aqueous solutions using black wattle tannin-immobilized nanocellulose. J Hazard Mater 339:91–99
58. Szabadkai G, Duchen MR (2008) Mitochondria: the hub of cellular Ca^{2+} signaling. Physiology 23:84–94
59. Yang JX, He YB, Song XL, Lai LN, Li JB, Guo BP (2012) A novel sensitive biosensor for Ca^{2+} ion based on gold nanoparticles modified electrode by pulsed electrodeposition. Anal Lett 45:2436–2444
60. Khoshniat S, Bourgine A, Julien M, Weiss P, Guicheux J, Beck L (2011) The emergence of phosphate as a specific signaling molecule in bone and other cell types in mammals. Cell Mol Life Sci 68:205–218
61. Berchmans S, Issa TB, Singh P (2012) Determination of inorganic phosphate by electroanalytical methods: a review. Anal Chim Acta 729:7–20
62. Talarico D, Arduini F, Amine A, Moscone D, Palleschi G (2015) Screen-printed electrode modified with carbon black nanoparticles for phosphate detection by measuring the electroactive phosphomolybdate complex. Talanta 141:267–272
63. Xu K, Kitazumi Y, Kano K, Shirai O (2018) Phosphate ion sensor using a cobalt phosphate coated cobalt electrode. Electrochim Acta 282:242–246
64. Talarico D, Cinti S, Arduini F, Amine A, Moscone D, Palleschi G (2015) Phosphate detection through a cost-effective carbon black nanoparticle-modified screen-printed electrode embedded in a continuous flow system. Environ Sci Technol 49:7934–7939
65. Zhang JX, Bian YX, Liu D, Zhu ZW, Shao YH, Li MX (2019) Detection of phosphate in human blood based on a catalytic hydrogen wave at a molybdenum phosphide modified electrode. Anal Chem 91:14666–14671
66. Lowenstein CJ, Dinerman JL, Snyder SH (1994) Nitric oxide-a physiological messenger. Ann Intern Med 120:227–237
67. Szabo C (1996) Physiological and pathophysiological roles of nitric oxide in the central nervous system. Brain Res Bull 41:131–141
68. Lundberg JO, Gladwin MT, Ahluwalia A, Benjamin N, Bryan NS, Butler A, Cabrales P, Fago A, Feelisch M, Ford PC, Freeman BA, Frenneaux M, Friedman J, Kelm M, Kevil CG, Kim-Shapiro DB, Kozlov AV, Lancaster JR, Lefer DJ, McColl K, McCurry K, Patel RP, Petersson J, Rassaf T, Reutov VP, Richter-Addo GB, Schechter A, Shiva S, Tsuchiya K, van Faassen EE, Webb AJ, Zuckerbraun BS, Zweier JL, Weitzberg E (2009) Nitrate and nitrite in biology, nutrition and therapeutics. Nat Chem Biol 5:865–869
69. Vitturi DA, Patel RP (2011) Current perspectives and challenges in understanding the role of nitrite as an integral player in nitric oxide biology and therapy. Free Radic Biol Med 51:805–812
70. Pattillo CB, Bir S, Rajaram V, Kevil CG (2011) Inorganic nitrite and chronic tissue ischaemia: a novel therapeutic modality for peripheral vascular diseases. Cardiovasc Res 89:533–541
71. Madasamy T, Pandiaraj M, Balamurugan M, Bhargava K, Sethy NK, Karunakaran C (2014) Copper, zinc superoxide dismutase and nitrate reductase coimmobilized bienzymatic biosensor for the simultaneous determination of nitrite and nitrate. Biosens Bioelectron 52:209–215
72. Reiter S, Habermuller K, Schuhmann W (2001) A reagentless glucose biosensor based on glucose oxidase entrapped into osmium-complex modified polypyrrole films. Sensor Actuat B-Chem 79:150–156
73. Cauthen CA, Lipinski MJ, Abbate A, Appleton D, Nusca A, Varma A, Goudreau E, Cowley MJ, Vetrovec GW (2008) Relation of blood urea nitrogen to long-term mortality in patients with heart failure. Am J Cardiol 101:1643–1647

74. Manerba A, Lombardi C, Vizzardi E, Maiandi C, Milesi G, Bugatti S, Bettari L, Romeo A, Metra M, Cas LD (2010) Role of blood urea nitrogen variations in patients with chronic heart failure. Eur Heart J 31:1053–1054
75. Yang YA, Lin CH, Wei YC (2014) Thread-based microfluidic system for detection of rapid blood urea nitrogen in whole blood. Microfluid Nanofluid 16:887–894

Chapter 4
Electrochemical Biosensors for Neurotransmitters Detection

Qitao Zhou, Jing Pan, Hui Li, Shaoguang Li, and Fan Xia

Abstract Neurotransmitters (NTs) are related to various brain functions. Furthermore, abnormal NT levels are implicated in many psychotic, physical, and neurodegenerative diseases, including schizophrenia, depression, dementia, Parkinson's disease, Alzheimer's disease, Huntington's disease, and so on. Thus, their highly sensitive and selective detection possesses great clinical significance. Among all the existing monitoring means for NTs, the electrochemical method is one of the most promising, considering its simple, low cost, and fast analysis process. Herein, the recent advances in electrochemical NT biosensors are introduced. Five typical NTs are successively discussed, including epinephrine, dopamine, norepinephrine, γ-aminobutyric acid, and serotonin. The challenges and solutions for the electrochemical detection of single NT and simultaneous detection of multi-NTs are also investigated. Finally, the outlook of future electrochemical NT sensors is illustrated.

Keywords Electrochemical biosensors · neurotransmitters · epinephrine · dopamine · norepinephrine · γ-aminobutyric acid · serotonin

4.1 Introduction

Neurotransmitters (NTs), which are endogenous chemical messengers between neurons, are of great importance for brain function and related diseases diagnosis. The first known NT (acetylcholine) was discovered in the 1921s by Nobel laureate, Otto Loewi [1]. Until now, more than one hundred NTs have been reported and they can be classified based on the molecular species (Fig. 4.1) [2]. The amino acids' NTs include glutamic acid, glycine, gamma-aminobutyric acid, and tyrosine [3]. The NTs belonging to amines contain dopamine, serotonin, epinephrine, nor-epinephrine, histamine, and acetylcholine. Besides, there are many NTs that are soluble gases,

Q. Zhou · J. Pan (✉) · H. Li · S. Li · F. Xia
State Key Laboratory of Biogeology and Environmental Geology, Engineering Research Center of Nano-Geomaterials of the Ministry of Education, Faculty of Materials Science and Chemistry, China University of Geosciences, Wuhan 430074, China
e-mail: panjing@cug.edu.cn

F. Xia et al. (eds.), *Electrochemical Biosensors for Whole Blood Analysis*,
https://doi.org/10.1007/978-981-99-5644-9_4

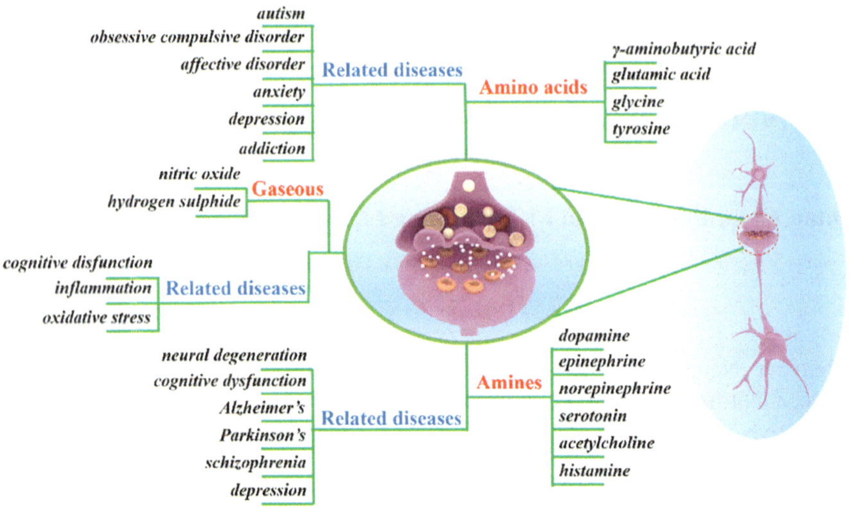

Fig. 4.1 The classification and illustration of NTs as well as the related diseases

such as nitric oxide (NO) and hydrogen sulfide (H_2S). It needs to be mentioned that, there still many NTs are beyond these three kinds, for example, acetylcholine and choline.

NTs establish the human brain–body integration. They affect and control heart rate and muscle tone, as well as adjustment of learning, sleeping, memory, consciousness, mood, and appetite. Changes in the concentration of NTs in the central nervous system have been correlated with numerous psychotic, neurodegenerative diseases, and physical illnesses. Different NTs would affect different brain function and their deregulation could result in different disorders. For example, biogenic amine NTs could regulate many physiological processes, including sleep and blood pressure regulation. Thus, their abnormalities could lead to various neurological neuropsychiatric disorders, such as neural degeneration, cognitive dysfunction, Alzheimer's, Parkinson's and schizophrenia (Fig. 4.1) [3]. Amino acids are the basic component of some peptides as well as some proteins and play important roles in the metabolization of other components [2]. The deregulation of amino acids NTs has a close relationship with various neuropsychiatric disorders, such as autism, affective disorder, anxiety, depression, obsessive–compulsive disorder, and addiction. While, the gaseous NTs NO and H_2S are strong relaxants for the gastrointestinal smooth muscles and physiologic vasorelaxants, respectively [2].

In consequence, the accurate detection of the NTs in human blood is significant to the early diagnosis, state monitoring, and prognosis for the related diseases. Among all the methods for NTs determination, the electrochemical biosensor is one of the most promising on account of its simple, low cost, and fast analysis process. In addition, the high selectivity can be ensured by employing the proper recognition part of the electrochemical sensor and could be beneficial to the detection of biogenic amines, which are very similar and hard to be distinguished. Taking these into consideration, this chapter discusses the recent advances of electrochemical NT biosensors associated with not only single NT (five typical NTs) detection but also simultaneous detection for multi-NTs.

4.2 Electrochemical Biosensors for Mono-NTS

4.2.1 Epinephrine (EP)

Epinephrine (EP), which is a hormone also called adrenaline, can enhance cardiac output and increase blood glucose concentration. Normally, the blood EP is at the nanomolar level. The abnormal EP concentration could impact the regulation of heart rate, glycogen metabolism, lipolysis, blood pressure and the immune system [4]. Its persistent abnormality could bring a series of diseases, including cerebral malaises, schizophrenia, and Parkinsonism [5]. Besides, the blood EP concentration can be artificially controlled for immediate control of bleeding or the treatment of some diseases, such as allergies, emphysema, bronchitis, and bronchial asthma. Considering the misuse of EP in sports to improve athletes' performance, the detection of blood EP concentration is also significant in sports fields. Thus, building an electrochemical EP sensor with high selectivity and sensitivity has a wide application value [6].

However, there are challenges for EP determination by electrochemical biosensors. The sluggish electrode kinetics of the redox reaction of EP on traditional electrodes lead to high overpotential as well as poor electrocatalytic activity [7]. Additionally, the electrochemical signals of some coexistent interferences would overlap with the EP signal. The most commonly used method to resolve the challenges is to design nanostructured electrodes. Compared with the corresponding bulk electrodes, nanostructured electrodes usually display better electrocatalytic performance toward EP electrooxidation and a lower limit of detection (LOD) for EP detection [3]. For instance, Sulka prepared a highly ordered nanoporous Au electrode (Fig. 4.2A) and realized the EP detection with a low LOD of 2.42×10^{-6} M as well as a linear detection range of 20–100 μM [8]. Fei et al. synthesized ordered mesoporous carbon/nickel oxide nanocomposite and built EP electrochemical sensor with a wide linear detection range of $8.0 \times 10^{-7} - 5.0 \times 10^{-5}$ M and a low LOD of 8.5×10^{-8} M [9]. Furthermore, this sensor exhibits excellent selectivity at the interference of uric acid and achieve successful detection of EP in spiked human blood.

In addition, using the proper recognition part could further enhance the selectivity and sensitivity of the electrochemical EP sensor. For example, molecularly imprinted polymers (MIPs) could realize high selective and stable recognition toward the target. Considering this, Zhou et al. utilized 3D MIP arrays, which were built on the surface of ZnO nanorod arrays, to synthesize an electrochemical EP sensor [10]. On account of the high exposure and large specific surface area, the 3D MIP arrays exhibit

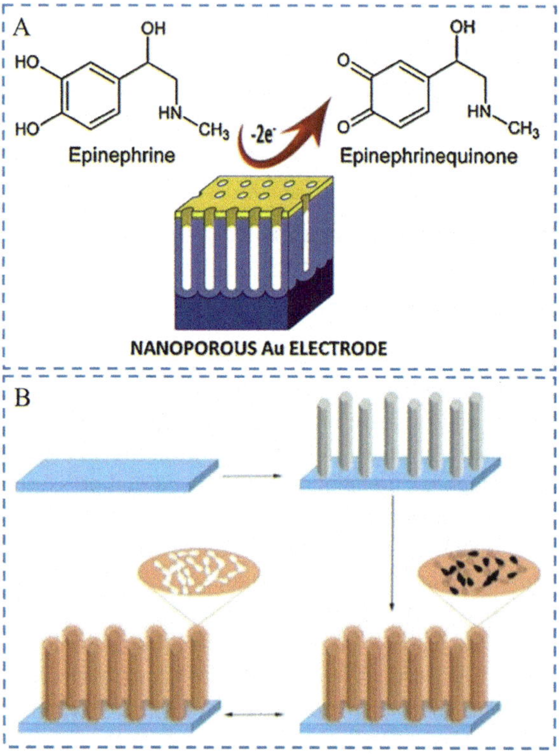

Fig. 4.2 **A** Illustration of the ordered nanoporous Au electrode, which can be employed for the EP detection [8]. **B** The preparation process of the 3D MIP arrays, which can help to build an electrochemical EP sensor [10]

excellent accessibility to the analyte and result in efficient signal as well as high sensitivity. Based on the differential pulse voltammetry method, two linear dynamic ranges (1–10 M and 10–800 M) can be achieved by this electrochemical EP sensor.

4.2.2 *Dopamine (DA)*

DA is a kind of neuromodulator, which can modulate the other NTs' function by means of volume transmission [11]. It plays an important role in metabolism, central nerve function, endocrine function, plasma glucose levels, neuronal elasticity, neuro-immune, and emotions regulation. The abnormal DA concentration could generate neuro-immune system dysfunctions [12].

However, the DA detection by electrochemical method is challenging for actual samples, considering the high concentrations of electroactive interfering substances (uric acid, ascorbic acid, and so on) [13]. Just like electrochemical EP sensor, the electrochemical EP sensors often use modified working electrodes to improve the detection performance [14]. Furthermore, the interaction between the functional materials and the supporting electrode could impact the DA detection result. For instance, employing screen-printed graphitic electrode (SPE) as supporting material, the 2D hexagonal boron nitride (2D-hBN) nanosheets (Fig. 4.3A-B) modification could obtain reduced DA electrochemical oxidation potential by ~90 mV [15]. While, when the supporting material is replaced by a glassy carbon electrode, the same modification would result in a completely different result and the DA oxidation potential can be increased by ~80 mV. Using 2D-hBN modified SPE, the LOD of 0.65 μM for DA can be realized in the presence of uric acid.

Considering the advantage of low blood consumption for clinic application, the sample volume is also important for the detection of markers in the blood. Recently, microfluidic device, which can reduce the sample consumption compared with traditional electrochemical devices, has attracted lots of attention to DA detection. For

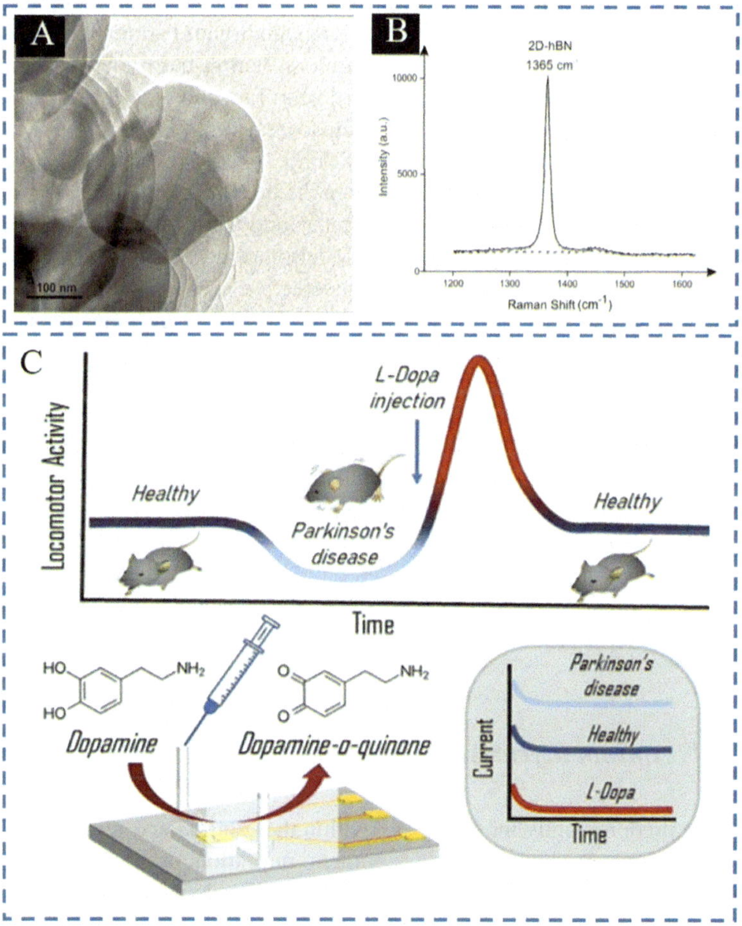

Fig. 4.3 **A** TEM image and **B** Raman spectra of the 2D-hBN nanosheets, which can be used for DA electrochemical detection [15]. **C** An electrochemical DA sensor based on a microfluidic device [16]

instance, Senel et al. fabricated an electrochemical DA sensor based on a microfluidic device (Fig. 4.3C) [16]. This kind of electrochemical sensor can operate within the DA concentration range of 0.1–1000 nM with only ~2.4 μL sample volume. Besides, the changes of DA concentrations in plasma and cerebrospinal fluid of a Parkinson's disease mouse model as well as the following treatment based on drug L-3, 4-dihydroxyphenylalanine can be monitored.

4.2.3 Norepinephrine (NEP)

NEP, which is also called noradrenaline, is a metabotropic NT secreted by the adrenal medulla. Abnormal NEP may result in many diseases, including neuroblastoma, ganglion neuronal, coronary heart disease, paraganglioma, Parkinson' disease, and multiple sclerosis. The addition of NEP in blood within the proper concentration range can increase blood pressure, and raises heart rate and blood sugar levels, just like the function of EP [17]. Besides, it can help to treat organic heart disease, myocardial infarction hypertension, and bronchial asthma. Therefore, the development of a quantitative determination method for NEP in the blood is necessary [18].

$FeMoO_4$ possesses a bimetallic character and can form Fe (II)-dioxygen complexes, which can help the catalytic oxidation of NEP, thus can be used as a functional material for NEP electrochemical detection [19]. Specifically, employing $FeMoO_4$ nanorods to NEP electrochemical sensor, the LOD and detection range can be achieved as 3.7×10^{-9} M and 5.0×10^{-8} M – 2.0×10^{-4} M, respectively [19].

Besides the specific compound, the 1D structure of the nanorod can benefit the electron transport of the electrode and obtain better NEP detection performance. A similar phenomenon is also found in MoO_3 nanowires and a low LOD of 0.11 μM is displayed for NEP detection (Fig. 4.4A) [20]. In addition, MIPs can be used to enhance the binding capacity of the electrochemical sensors for NEP, thus, the selectivity, current response, and response time can be simultaneously improved. For instance, the electrochemical sensor based on MIP-coated palladium nanoparticles (Fig. 4.4B) exhibits a wide linear range of 0.5–80.0 μM and a low LOD of 0.1 μM for NEP detection [21].

4.2.4 Γ-Aminobutyric Acid (GABA)

GABA is a predominantly inhibitory NT within the central nervous system [22]. Its normal secretion can regulate the development, migration, proliferation, as well as differentiation of neurons. While, abnormal GABA levels would bring a variety of neurological disorders, including epilepsy, Huntington's, and Parkinson's, diseases [18]. Thus, the detection of GABA concentration has clinical significance. However, it is difficult to quantify GABA by electrochemical methods on account of its nonredox-active nature [23]. In consequence, proper recognition part of the sensor

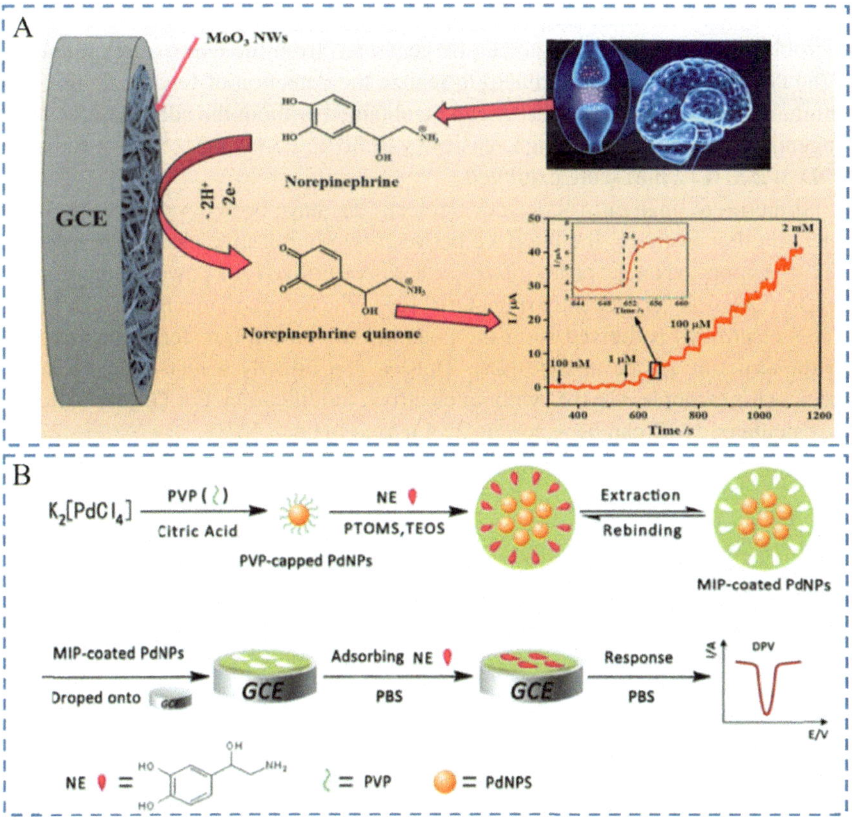

Fig. 4.4 **A** Illustration and detection performance of electrochemical NEP sensor based on MoO_3 nanowires [20]. **B** The preparation and detection process of electrochemical NEP sensor based on MIP-coated PdNPs [21]

or recognition method should be designed for electrochemical GABA sensors. For example, electrochemical active compounds can be formed by some materials, which can interact with GABA. Said et al. utilized a mixture of an alkylthiol reagent and orthophthalaldehyde to interact with GABA and formed an electrochemical GABA sensor. As a result, a wide detection range of 250 nmol L^{-1}–100 μmol L^{-1} and low LOD of 98 nmol L^{-1} can be achieved for GABA detection. In addition, the sensor exhibits the capability to detect GABA or a mixture of GABA and Glutamate in complex solutions, such as human serum and urine [24].

Another method is using enzymes to form a secondary electroactive product (for instance, hydrogen peroxide, H_2O_2) by selective oxidization of GABA [25]. Nevertheless, this method requires a complex pretreatment process and externally applied reagents, which is not suitable for continuous and real-time GABA monitoring. To resolve this problem, Arumugam et al. built a microbiosensor microarray to achieve continuous ex-vivo GABA measurement in brain slices (Fig. 4.5A) [26].

This microarray is fabricated by a glutamate (Glu) microbiosensor and a GABA microbiosensor. A-ketoglutarate can be generated from the Glu oxidation reaction at the Glu microbiosensor and help to realize the detection of GABA. In this way, simultaneous measuring of GABA can be obtained without the addition of external reagents. As a result, a very high sensitivity of 36 ± 2.5 pA mM^{-1} cm^{-2} and a low LOD of 2 ± 0.12 mM were exhibited.

Furthermore, many novel devices or electrodes have been developed for GABA detection. For instance, the interface between two immiscible electrolyte solutions (ITIES) supported on the orifice of a pipet has been utilized as a powerful platform for GABA detection very recently (Fig. 4.5B) [27]. The detection mechanism with pipet ITIES electrodes is derived from the potential-driven ion transfer, which depends on the structure of the analyte [28]. Thus, it can directly detect nonredox active species (for example, GABA) without electrode modification. For GABA detection, the calculated LOD can be as low as 22.4 μM at pH ≈ 7 [27].

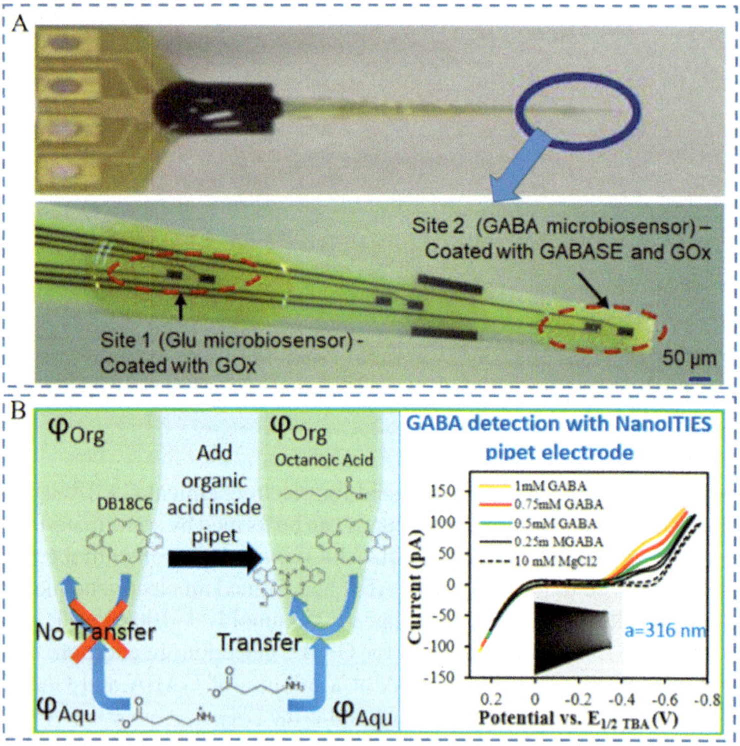

Fig. 4.5 **A** The optical image of the probe of microbiosensor microarray for continuous ex-vivo GABA measurement in brain slices [26]. **B** The detection mechanism and detection performance of ITIES pipet electrode-based electrochemical GABA sensor [27]

4.2.5 Serotonin (ST)

Serotonin (ST), which is also called 5-hydroxytryptamine, plays important roles in a series of human body functions, including emotions, appetite, and autonomic functions [29]. The detection of ST in body fluids (whole blood, serum, urine, and cerebrospinal fluid) is of great importance because it is the biomarker of several diseases, such as anxiety, depressive disorders, irritable bowel, carcinoid tumors, diabetes, and sleep trouble [30]. In consideration of low concentration and the presence of interfering biological substances in the actual sample, accurate electrochemical determination of ST is almost impossible for bare working electrodes, and surface modification is necessary [31]. For instance, an electrochemical sensor based on graphene oxide (GO) and chitosan has been established for ST detection (Fig. 4.6A) [32]. The use of GO can help the immobilization of antibodies by covalent bonding. While the addition of chitosan is crucial to the nanomaterial film's stability. The fabricated sensor exhibits a low LOD of 3.2 nM and a detection range of 10 nM–100 μM. In addition, it can selectively detect ST in human serum, artificial tears, urine, and saliva samples as well as in the presence of different interferences. Similarly, Au nanorattles can work with reduced GO and achieve ST detection with LOD of 3.87 ($\pm$0.02) $\times$ 10^{-7} (RSD < 4.2%) M [33]. This kind of sensor also shows its clinical applicability and can be applied to monitor ST in a variety of real samples, including blood serum, and urine.

Considering that the real-time monitoring of ST could provide more detailed information, the in-vivo electrochemical ST sensors become one of the research hotspots recently. For example, Bao et al. fabricated a NeuroString sensor, which is constructed by a metal-complexed polyimide in a graphene/nanoparticle skeleton embedded in an elastomer (Fig. 4.6B) [34]. The prepared stretchable neurochemical biological interface endows the NeuroString sensor with the capacity to in-vivo and real-time monitor in the brain of a behaving mouse and dynamically detect ST in the gut without perturbing peristaltic movements and undesired stimulations.

It needs to be mentioned that besides traditional electrochemical ST sensors, other kinds of electrochemical devices have also been developed for ST detection and obtained outstanding performance. For example, field-effect transistors [35].

The performance of the existing electrochemical NTs biosensors can be referred to Table 4.1.

4.3 Simultaneous Detection of Multi-NTs

With respect to mono-biomarker detection, the simultaneous detection of multi-biomarkers could realize more accurate diagnosis with lower cost and shorter detection time [36]. While some NTs could interfere with each other during electrochemical sensing. For instance, DA, EP, and NEP are all catecholamines NTs and have very similar structures [37, 38]. As a result, the simultaneous determination of them

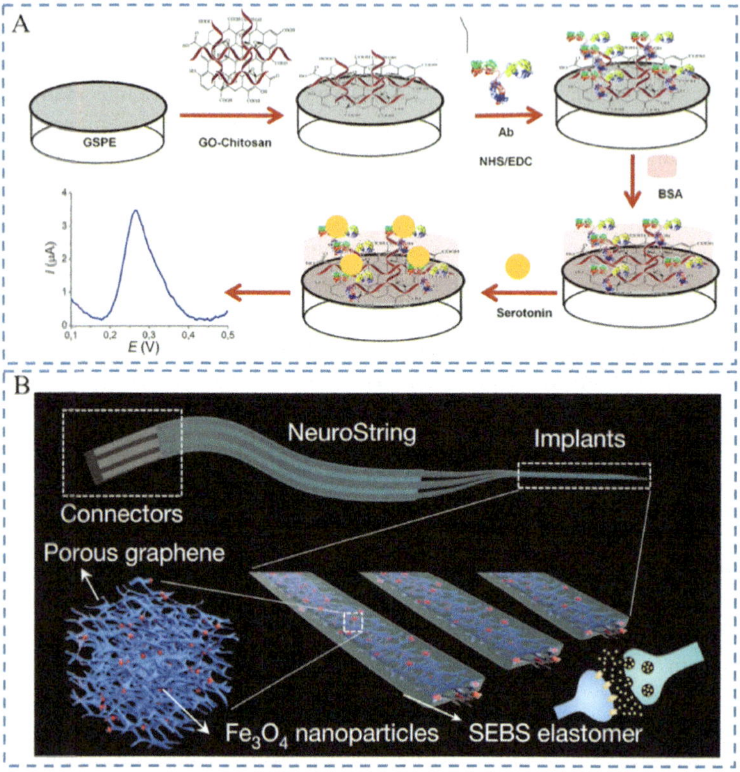

Fig. 4.6 **A** The preparation and detection process of the electrochemical ST sensor based on GO and chitosan [32]. **B** The schematic of the NeuroString sensor for in-vivo ST detection [34]

in real samples with high selectivity is of great difficulty. On the other hand, catecholamines NTs play important roles in many biological functions and have a close relationship with various diseases, including Alzheimer's and Parkinson's. In addition, the concentration of catecholamines NTs is an indicator of some reward-related behaviors related to drink, food, and drug addiction [39]. Therefore, electrochemical biosensors, which can realize simultaneous and highly selective detection of catecholamines NTs, have clinical significance. Recently, Si et al. employed MIPs as a recognition part of the biosensor to enhance the selectivity for catecholamines NTs (Fig. 4.7A). Furthermore, the MIPs-based array has been built to achieve multi-catecholamines NTs detection [38]. Besides. nanomaterials could be been employed to realize the effective detection of catecholamines NTs [40]. For instance, the electrochemical biosensor based on nanoporous gold microneedle could realize the highly sensitive and stable detection of catecholamines NTs in interstitial fluid [41]. The LOD for NEP detection could be as low as 100 nM and the response time can be lower than 4 s.

Table 4.1 The performance of the existing electrochemical biosensors for mono-NTs detection

NTs	LOD	Detection range	References
EP	2.42×10^{-6} M	20–100 μM	[8]
	8.5×10^{-8} M	$8.0 \times 10^{-7} - 5.0 \times 10^{-5}$ M	[9]
	–	1–10 M; 10–800 M	[10]
DA	0.65 μM	–	[15]
	–	0.1–1000 nM	[16]
NEP	3.7 × 10–9 M	$5.0 \times 10^{-8} - 2.0 \times 10^{-4}$ M	[19]
	0.11 μM	–	[20]
	0.1 μM	0.5–80.0 μM	[21]
GABA	98 nmol L^{-1}	250 nmol L^{-1} – 100 μmol L^{-1}	[24]
	2 ± 0.12 mM	36 ± 2.5 pA mM^{-1} cm^{-2}	[26]
	22.4 μM	–	[27]
ST	3.2 nM	10 nM–100 μM	[32]
	3.87 (±0.02) × 10–7 (RSD < 4.2%) M	–	[33]

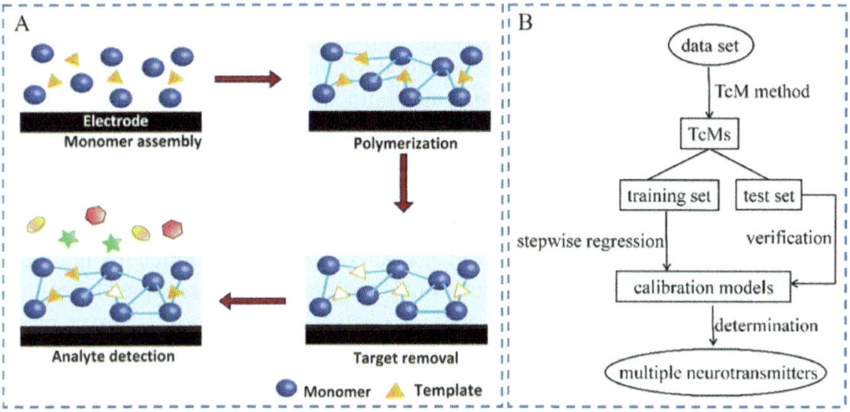

Fig. 4.7 **A** The illustration of the working principle of the MIPs-based sensor for multi-catecholamines NTs detection [38]. **B** The overall analysis procedures of the TcM-based method for the quantitative determination of NTs [42]

Besides catecholamines NTs, the simultaneous detection of other NTs in blood-related samples has also been realized. For example, DA, EP, NE, and ST in human blood serum can be detected by an electrochemical sensor based on GO modified electrode [42]. Although the four NTs exhibit overlapped voltammetric signals when employing DPV measurement, the Tchebichef curve moments (TcM) calculated from the voltammogram results can contribute to the quantitative determination of the NTs (Fig. 4.7B). The results suggest LOD for DA, EP, NEP, and ST as 74 nM,

104 nM, 84 nM, and 97 nM, respectively. In actual samples, NTs may also coexist with other biomarkers. A flexible electrochemical film sensor built by PEDOT-titania-poly(dimethylsiloxane) realized the simultaneous detection of EP, DA, ascorbic acid, and uric acid with a sensitivity of 63 $\mu A\ mM^{-1}\ cm^{-2}$ [43]. Being lightweight, flexible, and portable, this kind of sensor exhibits the potential to be applied in wearable and implantable devices.

4.4 Conclusion and Outlook

The NT levels in human blood can be important indexes for many neurological diseases. Electrochemical NTs biosensor, which has rapid response time, low cost, and high sensitivity, is promising for clinical applications. Until now, there are great signs of progress made for electrochemical NTs biosensors. Five typical NTs (EP, DA, NEP, GABA, and ST) exhibit different properties and the development of corresponding electrochemical sensors should resolve different challenges. For example, EP, which can enhance the cardiac output and increase blood glucose concentration, has sluggish electrode kinetics of the corresponding redox reaction on traditional electrodes. The most commonly used method to resolve the challenges is to design nanostructured electrodes. Similar to EP, NEP is a metabotropic NT secreted by the adrenal medulla. The addition of NEP in blood within the proper concentration range can increase blood pressure. Some specific compounds and 1D nanomaterials could benefit to the NEP detection. DA is a kind of neuromodulator, whose actual samples usually exhibit high concentrations of electroactive interfering substances (uric acid, ascorbic acid, and so on). Utilizing optimized interaction between the functional materials and the supporting electrode would be an effective way to improve the electrochemical detection performance of DA. GABA is a predominantly inhibitory NT within the central nervous system, which is difficult to be quantified by electrochemical methods on account of its nonredox-active nature. In consequence, proper recognition part of the sensor or recognition method should be designed for electrochemical GABA sensors. ST plays important roles in a series of human body functions, including emotions, appetite, and autonomic functions. In consideration of low concentration and presence of interfering biological substances in the actual sample, accurate electrochemical determination of ST is almost impossible for bare working electrodes, and surface modification is necessary. Compared with electrochemical detection of single NT, simultaneous detection of multi-NTs could realize more accurate diagnosis but it is much more difficult to achieve. For instance, DA, EP, and NEP are all catecholamines NTs and have very similar structures. Their characteristic signals could interfere with each other during electrochemical sensing. This challenge can be also resolved by the use of the proper recognition part of the sensor.

Although there are abundant electrochemical NT sensors have been developed, some challenges still be unresolved. The highly selective, highly sensitive, rapid, portable, small sample volume consumed and simultaneous detection of multi-NTs still need to be realized. In addition, the unsatisfactory stability still limits the practical application of electrochemical sensors, especially those based on enzymes. Besides, although nanostructures have been widely used to enhance the detection performance of electrochemical NT sensors, the detailed mechanism is unclear and accuracy control is hard to achieve. Thus, the theoretical calculation and in-situ techniques should be employed for these sensors to deeply analysis the related mechanism.

References

1. Tavakolian-Ardakani Z, Hosu O, Cristea C, Mazloum-Ardakani M, Marrazza G (2019) Latest trends in electrochemical sensors for neurotransmitters: a review. Sensors 19(9):2037
2. Su Y, Bian S, Sawan M (2020) Real-time in vivo detection techniques for neurotransmitters: a review. Analyst 145(19):6193–6210
3. Madhurantakam S, Karnam JB, Brabazon D, Takai M, Ahad IU, Balaguru Rayappan JB, Krishnan UM (2020) "Nano": an emerging avenue in electrochemical detection of neurotransmitters. ACS Chem Neurosci 11(24):4024–4047
4. Sanghavi BJ, Mobin SM, Mathur P, Lahiri GK, Srivastava AK (2013) Biomimetic sensor for certain catecholamines employing copper(II) complex and silver nanoparticle modified glassy carbon paste electrode. Biosens Bioelectron 39(1):124–132
5. Hasanzadeh M, Shadjou N, de la Guardia M (2017) Current advancement in electrochemical analysis of neurotransmitters in biological fluids. Trac-Trends Anal Chem 86:107–121
6. Luk HN, Chou TY, Huang BH, Lin YS, Li H, Wu RJ (2021) Promotion effect of palladium on BiVO4 sensing material for epinephrine detection. Catalysts 11(9):1083
7. Shankar SS, Shereema RM, Rakhi RB (2018) Electrochemical determination of adrenaline using MXene/Graphite composite paste electrodes. ACS Appl Mater Interfaces 10(50):43343–43351
8. Wierzbicka E, Sulka GD (2016) Fabrication of highly ordered nanoporous thin Au films and their application for electrochemical determination of epinephrine. Sens Actuators B-Chem 222:270–279
9. Yang X, Zhao PC, Xie ZG, Ni MJ, Wang CX, Yang PP, Xie YX, Fei (2021) Selective determination of epinephrine using electrochemical sensor based on ordered mesoporous carbon/nickel oxide nanocomposite. Talanta 233:122545
10. Li HH, Wang HH, Li WT, Fang XX, Guo XC, Zhou WH, Cao X, Kou DX, Zhou ZJ, Wu SX (2016) A novel electrochemical sensor for epinephrine based on three dimensional molecularly imprinted polymer arrays. Sens Actuators B-Chem 222:1127–1133
11. Liu CL, Goel P, Kaeser PS (2021) Spatial and temporal scales of dopamine transmission. Nat Rev Neurosci 22(6):345–358
12. Durairaj S, Sidhureddy B, Cirone J, Chen A (2018) Nanomaterials-based electrochemical sensors for in vitro and in vivo analyses of neurotransmitters. Appl Sci 8(9):1504
13. Sajid M, Nazal MK, Mansha M, Alsharaa A, Jillani SMS, Basheer C (2016) Chemically modified electrodes for electrochemical detection of dopamine in the presence of uric acid and ascorbic acid: a review. Trac-Trends Anal Chem 76:15–29
14. Sajid M, Baig N, Alhooshani K (2019) Chemically modified electrodes for electrochemical detection of dopamine: challenges and opportunities. TrAC Trends Anal Chem 118:368–385
15. Khan AF, Brownson DA, Randviir EP, Smith GC, Banks CE (2016) 2D hexagonal boron nitride (2D-hBN) explored for the electrochemical sensing of dopamine. Anal Chem 88(19):9729–9737
16. Senel M, Dervisevic E, Alhassen S, Dervisevic M, Alachkar A, Cadarso VJ, Voelcker NH (2020) Microfluidic electrochemical sensor for cerebrospinal fluid and blood dopamine detection in a mouse model of Parkinson's disease. Anal Chem 92(18):12347–12355

17. de Queiroz DF, Dadamos TRd, Machado SAS, Martines MAU (2018) Electrochemical determination of norepinephrine by means of modified glassy carbon electrodes with carbon nanotubes and magnetic nanoparticles of cobalt ferrite. Sensors 18(4):1223
18. Moon J-M, Thapliyal N, Hussain KK, Goyal RN, Shim Y-B (2018) Conducting polymer-based electrochemical biosensors for neurotransmitters: a review. Biosens Bioelectron 102:540–552
19. Samdani KJ, Samdani JS, Kim NH, Lee JH (2016) FeMoO4 based, enzyme-free electrochemical biosensor for ultrasensitive detection of norepinephrine. Biosens Bioelectron 81:445–453
20. Samdani KJ, Joh DW, Rath MK, Lee KT (2017) Electrochemical mediatorless detection of norepinephrine based on MoO3 nanowires. Electrochim Acta 252:268–274
21. Chen J, Huang H, Zeng Y, Tang H, Li L (2015) A novel composite of molecularly imprinted polymer-coated PdNPs for electrochemical sensing norepinephrine. Biosens Bioelectron 65:366–374
22. Wang T, Muthuswamy J (2008) Immunosensor for detection of inhibitory neurotransmitter γ-aminobutyric acid using quartz crystal microbalance. Anal Chem 80(22):8576–8582
23. Colombo ML, Sweedler JV, Shen M (2015) Nanopipet-based liquid-liquid interface probes for the electrochemical detection of acetylcholine, tryptamine, and serotonin via ionic transfer. Anal Chem 87(10):5095–5100
24. Alamry KA, Hussein MA, Choi JW, El-Said WA (2020) Non-enzymatic electrochemical sensor to detect gamma-aminobutyric acid with ligand-based on graphene oxide modi fied gold electrode. J Electroanal Chem 879:114789
25. Niwa O, Kurita R, Horiuchi T, Torimitsu K (1998) Small-volume on-line sensor for continuous measurement of gamma-aminobutyric acid. Anal Chem 70(1):89–93
26. Hossain I, Tan C, Doughty PT, Dutta G, Murray TA, Siddiqui S, Iasemidis L, Arumugamr PU (2018) A novel microbiosensor microarray for continuous ex vivo monitoring o gamma-aminobutyric acid in real-time. Front Neurosci 12
27. Iwai NT, Kramaric M, Crabbe D, Wei YY, Chen R, Shen M (2018) GABA Detection with Nano-ITIES pipet electrode: a new mechanism, Water/DCE-Octanoic acid interface. Anal Chem 90(5):3067–3072
28. Reymond F, Fermin D, Lee HJ, Girault HH (2000) Electrochemistry at liquid/liquid interfaces: methodology and potential applications. Electrochim.Acta 45(15–16):2647–2662
29. Khoshnevisan K, Honarvarfard E, Torabi F, Maleki H, Baharifar H, Faridbod F, Larijani B, Khorramizadeh MR (2020) Electrochemical detection of serotonin: a new approach. Clin Chim Acta 501:112–119
30. Dascalescu D, Apetrei C (2021) Nanomaterials based electrochemical sensors for serotonin detection: a review. Chemosensors 9(1):14
31. Sharma S, Singh N, Tomar V, Chandra R (2018) A review on electrochemical detection of serotonin based on surface modified electrodes. Biosens Bioelectron 107:76–93
32. Adumitrachioaie A, Tertis M, Suciu M, Graur F, Cristea C (2019) A novel immunosensing platform for serotonin detection in complex real samples based on graphene oxide and chitosan. Electrochim Acta 311:50–61
33. Mahato K, Purohit B, Bhardwaj K, Jaiswal A, Chandra P (2019) Novel electrochemical biosensor for serotonin detection based on gold nanorattles decorated reduced graphene oxide in biological fluids and in vitro model. Biosens Bioelectron 142:111502
34. Li JX, Liu YX, Yuan L, Zhang BB, Bishop ES, Wang KC, Tang J, Zheng YQ, Xu WH, Niu SM, Beker L, Li TL, Chen G, Diyaolu M, Thomas AL, Mottini V, Tok JBH, Dunn JCY, Cui BX, Pasca SP, Cui Y, Habtezion A, Chen XK, Bao ZN (2022) A tissue-like neurotransmitter sensor for the brain and gut. Nature 606(7912):94–101
35. Nishitani S, Sakata T (2020) Enhancement of signal-to-noise ratio for serotonin detection with well-designed nanofilter-coated potentiometric electrochemical biosensor. ACS Appl Mater Interfaces 12(13):14761–14769
36. Cincotto FH, Canevari TC, Campos AM, Landers R, Machado SA (2014) Simultaneous determination of epinephrine and dopamine by electrochemical reduction on the hybrid material SiO 2/graphene oxide decorated with Ag nanoparticles. Analyst 139(18):4634–4640

37. Kaur H, Siwal SS, Saini RV, Singh N, Thakur VK (2022) Significance of an electrochemical sensor and nanocomposites: toward the electrocatalytic detection of neurotransmitters and their importance within the physiological system. ACS Nanosci Au 3(1):1-27
38. Si B, Song E (2018) Molecularly imprinted polymers for the selective detection of multi-analyte neurotransmitters. Microelectron Eng 187:58–65
39. Ribeiro JA, Fernandes PM, Pereira CM, Silva F (2016) Electrochemical sensors and biosensors for determination of catecholamine neurotransmitters: a review. Talanta 160:653–679
40. Ramirez S, Silva N, Oyarzun MP, Pavez J, Silva JF (2017) Gold nanostructures on self-assembled monolayers activity for epinephrine, noradrenaline and dopamine. J Electroanal Chem 799:349–357
41. Tortolini C, Cass AEG, Pofi R, Lenzi A, Antiochia R (2022) Microneedle-based nanoporous gold electrochemical sensor for real-time catecholamine detection. Microchimica Acta 189(5):180
42. Yin B, Zhai HL, Zhao BQ, Bi KX, Mi JY (2021) Chemometrics-assisted simultaneous voltammetric determination of multiple neurotransmitters in human serum. Bioelectrochemistry 139:107739
43. Aashish A, Sadanandhan NK, Ganesan KP, Saraswathy Hareesh UN, Muthusamy S, Devaki SJ (2018) Flexible electrochemical transducer platform for neurotransmitters. ACS Omega 3(3):3489–3500

Chapter 5
Electrochemical Biosensors for Metabolites Detection

Jiafu Shen, Yu Dai, and Fan Xia

Abstract Metabolites, including intermediate and final products in metabolic pathways, play a crucial role in physiology activities. The detection of metabolites in the blood is very important for preventing disease and understanding health status. The electrochemical biosensors are an effective method to realize highly sensitive and specific detection of metabolites. In this chapter, we focus on the electrochemical detection of metabolites in blood, including glucose, urea, uric acid, bilirubin, cholesterol, lactic acid, creatinine, galactose, triglyceride, and hydrogen peroxide. From the perspective of different metabolites, the innovation of electrochemical biosensors for detecting them is discussed. We hope to provide ideas for the detection of metabolites in the blood through this Chapter and point out the important role of the detection of metabolites in human health.

Keywords Biosensors · Electrochemical detection · Electrodes · Nanomaterials · Blood · Serum · Metabolites · Glucose · Urea · Uric acid · Bilirubin · Cholesterol · Lactic acid · Creatinine · Galactose · Triglyceride · Hydrogen peroxide

5.1 Introduction

Biological fluids such as blood, saliva, tears, urine, and sweat have been used for human health diagnosis. For example, strictly controlling the glucose level in the human body can greatly reduce the incidence rate of diabetes. As a representative biological fluid, blood is widely used because of its safe and reliable characteristics. There are some metabolites in the blood such as glucose, urea, uric acid, bilirubin,

J. Shen
Faculty of Materials Science and Chemistry, China University of Geosciences, Wuhan 430074, China

Y. Dai (✉) · F. Xia
State Key Laboratory of Biogeology and Environmental Geology, Engineering Research Center of Nano-Geomaterials of Ministry of Education, Faculty of Materials Science and Chemistry, China Uni-Versity of Geosciences, Wuhan 430074, China
e-mail: yudai@cug.edu.cn

F. Xia et al. (eds.), *Electrochemical Biosensors for Whole Blood Analysis*,
https://doi.org/10.1007/978-981-99-5644-9_5

Fig. 5.1 Electrochemical biosensors for metabolites detection

cholesterol, lactic acid, creatinine, galactose, triglyceride, and hydrogen peroxide. These metabolites can be used as biomarkers based on the correlation between their content and human physiological status. Therefore, how to detect these metabolites quickly and accurately has attracted widespread attention.

Biosensors can convert biometric signals into physical/chemical signals. They are used to identify biomolecules for health detection. Based on different signal conversion processes, existing biosensors can be divided into mechanical sensors, optical sensors, electrochemical sensors, acoustic sensors, and magnetic sensors [1]. Compared with other sensors, electrochemical biosensors have been widely studied because of their small size, high sensitivity, good specificity, long life, simplicity and convenience [2]. Electrochemical biosensors can be divided into enzyme biosensors and non-enzyme biosensors according to whether they contain specific enzymes. Electrochemical biosensors mainly consist of three parts: mechanical support materials, specific biological receptors that capture target and active materials that convert biological signals into electrical signals [3].

In recent years, conductive polymers, redox media, nanoparticles, and nanocomposites have been used to modify electrodes in electrochemical biosensors to improve detection performance. Among them, nanomaterials are favored because of their controllable morphology, large surface area, and excellent electrocatalytic performance [4]. Some nanomaterials can also directly catalyze biomaterials [5]. The combination of biomaterials and nanoelements can increase the load of biomaterials and make them become specific labels of physiological targets [6]. It can also improve the electronic transfer ability to amplify the electrical signal.

In this Chapter, we focus on the electrochemical detection of metabolites in blood, including glucose, urea, uric acid, bilirubin, cholesterol, lactic acid, creatinine, galactose, triglyceride, and hydrogen peroxide (Fig. 5.1). The background of metabolites in the blood is introduced and the detection of metabolites by electrochemical biosensors is described. It is hoped that this Chapter can bring new insights and thinking to the electrochemical detection of metabolites in blood.

5.1.1 Biosensors for Blood Metabolites Detection

The detection performance of electrochemical biosensors for various blood metabolites mainly depends on electrode materials. Unmodified electrodes typically exhibit low sensitivity and specificity. By introducing nanomaterials with large surface area and good conductivity into electrode materials, their electrical activity and sensitivity can be significantly improved. In addition, all modified electrodes belong to enzyme-based or non-enzymatic electrochemical biosensors. Each of them has its own advantages and disadvantages, such as the relatively high specificity and anti-interference of enzyme-based sensors, and the better sensitivity and stability of non-enzymatic-based sensors. The detection performances of electrochemical biosensors for metabolites in blood have been summarized in Table 5.1.

5.1.2 Glucose Detection

As one of the important metabolites in blood, glucose reflects people's health. The glucose concentration of healthy people ranges from 3.6 to 7.5 mM [17]. The increase in glucose concentration often leads to many diseases, such as stroke, cardiovascular disease, retinal damage, and renal failure [18]. Therefore, early detection of glucose concentration in the human body is of great significance to avoid some complications. Since Clark and Lyons proposed the glucose biosensor in 1962 [19], electrochemical, thermoelectric, colorimetric, and piezoelectric methods have been widely used for glucose detection [20]. Among these methods, electrochemical biosensors have become one of the most popular methods because of their simple manufacturing process, low cost, and outstanding performance. There are two kinds of electrochemical biosensors for glucose detection: enzyme glucose biosensors and non-enzyme glucose biosensors. The development of enzyme glucose biosensors has experienced three generations [21]. For the first and second-generation biosensors, the interference of coexisting electroactive substances (e.g., ascorbic acid, uric acid, acetaminophen) and the lack of oxygen may affect the accuracy of glucose detection. The third-generation enzyme glucose biosensor mainly eliminates the artificial leaching medium through the direct electron transfer of the redox center of the enzyme from glucose to the electrode, and realizes the operation near the redox potential window of the enzyme. Although this biosensor has improved compared with the previous two generations, it is still in the research stage. The main challenge is that the redox center is wrapped by a thick insulating protein shell, which hinders the effective electron transfer to the electrode surface [22]. Although the enzyme glucose biosensors have good selectivity, their repeatability, high cost, and the risk of enzyme inactivation always exist. Non-enzyme glucose biosensors can avoid these problems. In order to manufacture non-enzymatic electrochemical biosensors for glucose detection, researchers have made great efforts to develop new electrode materials. Because the performance of biosensors mainly depends on the shape of electrode materials, various nanomaterials are used to prepare non-enzyme biosensors. It is generally composed of carbon-based components (e.g., carbon paper, carbon fiber, glassy carbon) and nanomaterials (e.g., noble metals, metal oxides, metal hydroxides) [23]. Carbon-based modules provide good conductivity, mechanical elasticity, and biocompatibility, while nanomaterials provide significant catalytic activity and glucose tolerance.

Khosla et al. synthesized CuO nanoleaf by low-temperature hydrothermal method and then used the nanoleaf to prepare an electrochemical non-enzyme glucose biosensor (Fig. 5.2a) [7]. Field emission scanning electron microscopy (FESEM)

Table 5.1 Electrochemical biosensors platform toward human blood metabolite detection

Metabolite	Electrode materials	Type	Electrochemical technique	Sensitivity	Linear range	Detection limit	References
Glucose	Cupric oxide nanoleaves/glassy carbon electrode	Non-enzymatic	Cyclic voltammetry and amperometry	1467.32 $\mu A\ \mu M^{-1}\ cm^{-2}$	5–5890 μM	0.12 μM	[7]
Urea	Silver nanoparticles/ commercial gold palladium electrode	Non-enzymatic	Cyclic voltammetry and electrochemical impedance spectroscopy	9.212 $\mu A\ \mu M^{-1}$	1000–8000 μM	140 μM	[8]
Uric acid	Uricase/gold nanoparticles decorated with reduced graphene oxide/ indium-tin-oxide	Enzymatic	Differential pulse voltammetry	86.62 ± 0.19 $\mu A\ mM^{-1}$	50–800 μM	7.32 ± 0.21 μM	[9]
Bilirubin	Cerium dioxide nanocubes/carbon blacks/ screen-printed electrode	Non-enzymatic	Differential pulse voltammetry	–	0.1–100 μM	0.1 μM	[10]
Cholesterol	Nanoporous gold/ screen-printed electrode	Enzymatic	Cyclic voltammetry	32.68 $\mu A\ mM^{-1}\ cm^{-2}$	50–6000 μM	8.36 μM	[11]

(continued)

Table 5.1 (continued)

Metabolite	Electrode materials	Type	Electrochemical technique	Sensitivity	Linear range	Detection limit	References
Lactic acid	Graphene oxide/ cellulose acetate modification/ screen-printed carbon electrodes	Enzymatic	Cyclic voltammetry	0.14 $\mu A\ mM^{-1}$	250–4000 μM	–	[12]
Creatinine	Carbon dots/ tungstic anhydride/ graphene oxide	Non-enzymatic	Cyclic voltammetry	1.5325 $\mu A\ \mu M^{-1}\ cm^{-2}$	0.0002–0.112 μM	0.0002 μM	[13]
Galactose	Galactose oxidase/ zinc oxide nanorods/nanorods	Enzymatic	Cyclic voltammetry	0.18 $\mu A\ mm^{-2}$	40,000–230,000 μM	–	[14]
Triglyceride	Toluidine blue/ electrochemically reduced graphene oxide/indium tin-oxide	Enzymatic	Cyclic voltammetry	3.2708 $\mu A\ \mu M^{-1}\ cm^{-2}$	4420–26,520 μM	2210 μM	[15]
Hydrogen peroxide	Nile Blue A/ 5-(1,2-dithiolan-3-yl)-N-(4-(4,4,5,5-tetramethyl-1,3,2-dioxaborolan-2-yl)phenyl)pent-anamide/carbon fiber microelectrode	Non-enzymatic	Differential pulse voltammetry	–	0.5–400 μM	0.02 μM	[16]

shows that the morphology of CuO presents a nanoleaf shape and uniform distribution (Fig. 5.2b). X-ray diffraction (XRD) shows that CuO is a monoclinic structure, without diffraction peaks and impurity peaks of other phases (Fig. 5.2c). The characteristic peaks at 421, 521, and 598 cm^{-1} in Fourier transform infrared spectroscopy (FTIR) spectrum further proves the formation of a monoclinic CuO structure (Fig. 5.2d). Bare glassy carbon electrode (GCE) and CuO/GCE do not show redox reaction without glucose (Fig. 5.2e). When glucose is added, CuO/GCE shows an obvious redox peak (Fig. 5.2f). The biosensor exhibits good anti-interference performance without significant changes when adding interfering substances (Fig. 5.2g). Although non-enzyme glucose biosensors have made great progress in the past decade, there are still some problems. The selectivity is poor because the sensing interface cannot selectively identify the substrate. The test environment is often an alkaline solution rather than a physiological condition. The research focuses on the preparation of nanomaterials, and pays less attention to glucose detection.

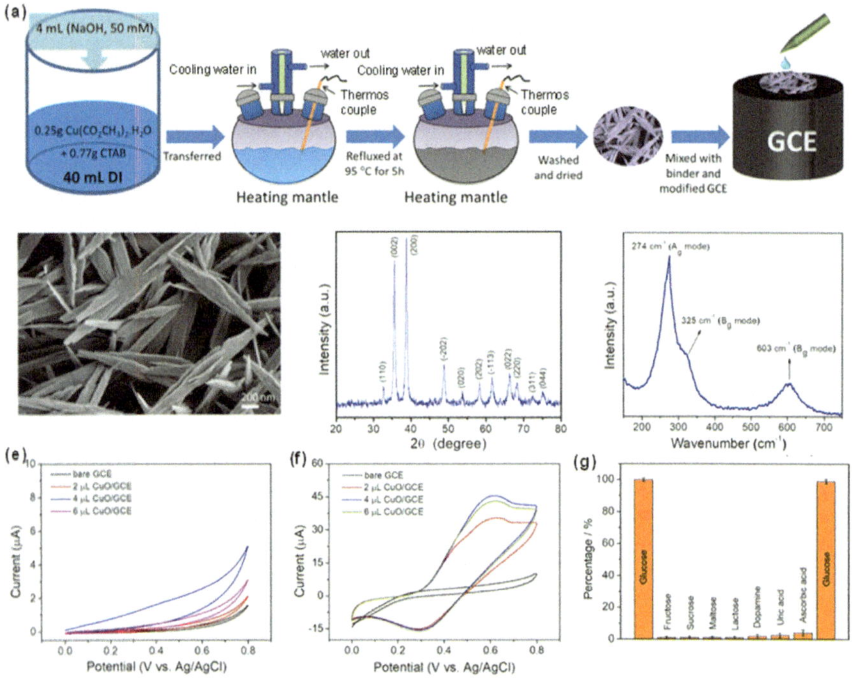

Fig. 5.2 Glucose detection (Reprinted with permission from Ref. [7]. Copyright 2021 The Authors). **a** Schematic diagram of CuO nanoleaf preparation. **b** FESEM image of CuO nanoleaf. **c** XRD spectrum of CuO nanoleaf. **d** FTIR spectrum of nanoleaf. **e** CV diagram of bare GCE and CuO nanoleaf modified GCE in buffer without glucose. **f** CV diagram of bare GCE and CuO nanoleaf modified GCE in buffer with glucose. **g** Anti-interference performance of CuO nanoleaf modified GCE

5.1.3 Urea Detection

Urea is a compound containing nitrogen, which widely exists in organisms. The urea content in human plasma is generally 3.3–6.7 mM [24]. Increased urea content in the body often leads to uremia and some nerve damage [25]. Low levels of urea can also cause some renal and liver failure. Therefore, it is necessary to detect urea quickly, simply and sensitively. Electrochemical, colorimetric, fluorescence, surface plasmon resonance, and chemiluminescence methods can be used to detect urea. Electrochemical biosensors are widely used to detect urea in blood because of their simplicity, economy, and high efficiency.

In the process of electrochemical detection, there will be an interaction between urease and urea, which will cause electron transfer to generate electrochemical signals [26]. In traditional electrochemical biosensors, the instability of enzymes on the electrode often leads to inaccurate detection and short service life. In recent years, some materials such as clay, zeolite, gel, polymer, and nanocomposite have been widely used in urea biosensors [27]. Sarkar et al. reported an electrochemical biosensor whose working electrode was modified by conductive hydrogel [28]. The conductive hydrogel is composed of polyaniline, polyacrylamide, and polyvinyl alcohol. Due to good conductivity and high enzyme loading, the biosensor shows a detection line of 60 nM and a sensitivity of 878 $\mu A\ mM^{-1}\ cm^{-2}$ in urea detection. For urea detection, nanomaterials have great attraction due to their high chemical stability, unique structure, and large specific surface area [29]. Selective biomolecules (e.g., enzymes) can be combined with nanomaterials through adsorption, covalent attachment, or entrapment. Uygun et al. prepared a highly stable urea biosensor by using poly(2-hydroxyethyl methacrylate-glycidyl methacrylate) nanoparticles as the carrier to covalently immobilize the urease [30]. The results show that the biosensor has good performance in detecting urea in serum. This is mainly due to the rapid catalytic oxidation caused by high enzyme loading of nanoparticles.

Due to the change in enzyme activity, the detection of urea by immobilized enzyme will reduce the stability of biosensors to a certain extent, so some non-enzyme urea biosensors show the potential to solve this problem [31]. Wachsmann-Hogiu et al. reported a simple and low-cost urea biosensor by depositing silver nanoparticles (AgNPs) on a commercial glucose test strip [8]. The glucose test strip consists of a plastic substrate with gold and palladium bimetallic channels. Compared with the traditional electrode, it has the advantages of good conductivity and easy access (Fig. 5.3a). Scanning electronic microscopy (SEM) images show that AgNPs are uniformly deposited on the electrode, leading to good electrical properties (Fig. 5.3b). The deposition of AgNPs leads to the increase of electrode resistance, which is mainly attributed to the formation of the Ag_2O layer and the discontinuity of nanoparticles (Fig. 5.3c). The formation of $Ag(OH)_2$ shows that the deposition of AgNPs on the electrode has catalytic activity for the detection of urea in alkaline electrolytes and improves the catalytic efficiency (Fig. 5.3d). After adding the interfering substances, the redox peak of urea only shifts and no other peaks occur, which indicates that the biosensor has good anti-interference ability for urea detection (Fig. 5.3e).

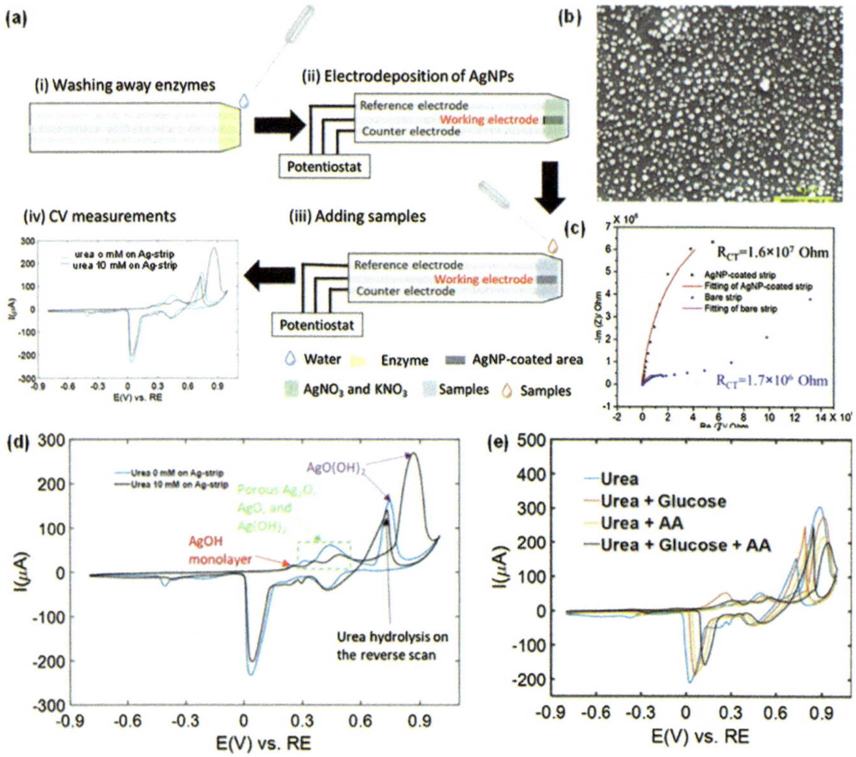

Fig. 5.3 Urea detection (Reprinted with permission from Ref. [8]. Copyright 2020 The Authors). **a** Schematic diagram of sensor workflow. **b** SEM image of AgNPs deposited on the test strip. **c** Electrochemical impedance spectroscopy (EIS) of AgNPs deposited on the test strip. **d** Sensing performance of AgNPs-deposited test strip on urea. **e** Anti-interference performance of AgNP-deposited test strip on urea

5.1.4 Uric Acid Detection

As a purine derivative, uric acid is the final product of human metabolism [32]. The normal level of uric acid in serum is 0.13–0.46 mM. The change in uric acid content is used for early diagnosis of human kidney diseases [33]. It is reported that the high level of uric acid in serum will become a detection indicator of heart-related diseases, and the low level will lead to multiple sclerosis [34, 35].

At present, electrochemical biosensors used for uric acid detection are mainly divided into enzyme biosensors and non-enzyme biosensors. Gupta et al. reported an electrochemical biosensor prepared with a uricase/CuO/Pt/glass electrode for the detection of urea [36]. The introduction of CuO enhances the redox characteristics of the electrode and increases the electronic transmission of the biosensor. The biosensor has a detection range of 0.05–1 mM for uric acid, a response of 2.7 mA mM^{-1}, and a service life of more than 14 weeks. In order to further improve the enzyme activity and electrode kinetics, Singh et al. prepared gold nanoparticles (AuNPs) decorated with reduced graphene oxide (rGO) nanocomposite and covalently immobilized uricase (UOx) enzyme to detect uric acid (Fig. 5.4a) [9]. Transmission electron microscope (TEM) shows that AuNPs are uniformly dispersed on the rGO sheet (Fig. 5.4b). The appearance of an amide-related peak proves that the urease is successfully immobilized on the surface of Au-rGO (Fig. 5.4c). CV curve shows that Au-rGO has good electron transfer performance, which is mainly attributed to the good conductivity of AuNPs and rGO (Fig. 5.4d). The current–time curve shows that UOx/Au-rGO/indium-tin-oxide (ITO) biosensor has the ability to detect uric acid quickly (Fig. 5.4e). The current change of uric acid at different concentrations indicates that uricase plays an important catalytic role in the oxidation of uric acid (Fig. 5.4f). Non-enzyme biosensors can also detect uric acid. Carboxylated cells (CLC) were dissolved in ionic liquid [BMIM]Cl, and then grafted with poly(3,4-ethylene dioxythione) (PEDOT) via atom-free radical polymerization [37]. The interaction between CLC and other components improves the conjugation degree of polymer in the composite, and promotes the formation of fibrous and skin-like composites, thus improving the conductivity and catalytic performance. The detection of uric acid in blood by PEDOT/CLC-based electrochemical biosensor shows a wide linear range of 0.6–620 μM and a detection limit of 5–400 μM.

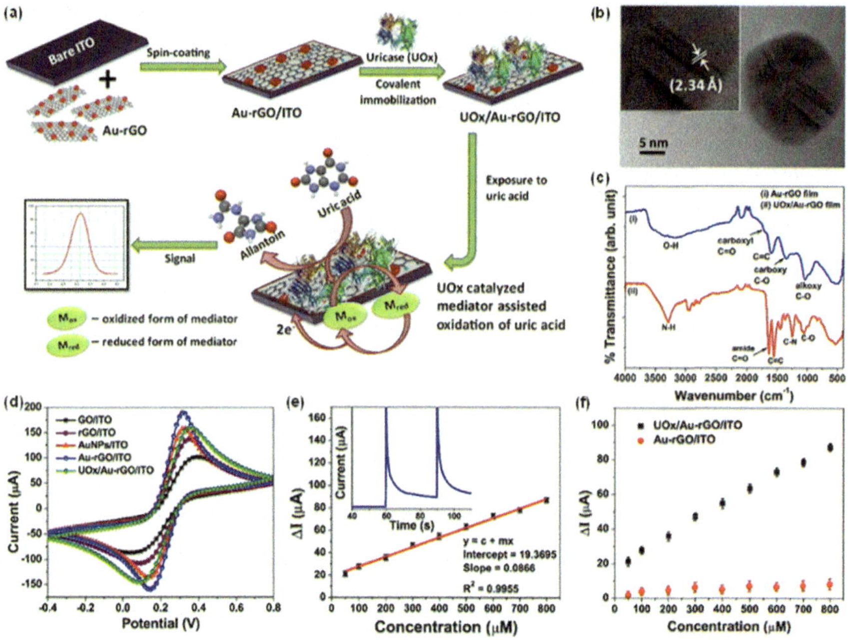

Fig. 5.4 Uric acid detection (Reprinted with permission from Ref. [9]. Copyright 2019 Elsevier B.V.). **a** Schematic diagram of biosensor preparation. **b** TEM photo of AuNPs scattered on rGO. **c** FTIR spectra of films. **d** Electrochemical activity of electrodes. **e** Current–time curve of UOx/Au-rGO/ITO biosensor for uric acid detection. **f** Current change of uric acid at different concentrations

5.1.5 Bilirubin Detection

Bilirubin is the yellow product of hemoglobin decomposition and widely exists in serum. Bilirubin is usually divided into two types, conjugated bilirubin and unconjugated bilirubin. Bilirubin in serum generally exists in an unconjugated form. Excessive bilirubin usually causes hemolytic anemia, sickle cell anemia, and liver infection [38]. Lack of bilirubin can lead to iron deficiency anemia and coronary artery disease. At present, the methods for detecting bilirubin include fluorescence, enzyme analysis, capillary electrophoresis, high-performance liquid chromatography, polarography, and chemiluminescence [39]. In recent years, the detection of bilirubin by electrochemical biosensors has gradually attracted people's attention.

As a common method to detect bilirubin, electrochemical biosensors are also divided into enzyme biosensors and non-enzyme biosensors. Non-enzyme biosensors are mainly divided into two types, one is the direct oxidation of bilirubin on the electrode surface, and the other is the detection of bilirubin through molecularly imprinted polymer. Gold nanoclusters (AuNCs) stabilized by human serum albumin (HSA) were used as biological recognition systems, and then chemically fixed on ITO plate [40]. The biosensor shows high sensitivity (0.34 $\mu A\ \mu M^{-1}$) with a detection limit of 86.32 nM in detecting bilirubin in the blood. Li et al. reported that a non-enzyme bilirubin biosensor prepared by cerium dioxide nanocubes (ceria NCs) and carbon blacks (CBs) modified the screen-printed electrode (Fig. 5.5a) [10]. TEM shows that ceria NCs are uniformly dispersed on the electrode, providing a large number of loading sites for the catalytic bilirubin (Fig. 5.5b). CBs are a good carrier for loading ceria NCs, which can improve the electrochemical performance of the composite (Fig. 5.5c). The ratio of ceria NCs/CBs to 2:1 produces the maximum oxidation current (Fig. 5.5d). The biosensor has good selectivity for bilirubin (Fig. 5.5e).

The enzyme biosensor mainly uses bilirubin oxidase (BOx) for electrochemical detection of bilirubin. Fang et al. reported a composite electrode made of multi-walled carbon nanotubes (CNTs), graphene oxide (GO), and AuNPs with BOx fixed on the electrode by adsorption [41]. The biosensor shows a detection range of 1.33–71.56 mM and a detection limit of 0.34 μM. The combination of conductive polymer and nanomaterials can improve the electrochemical performance of biosensors. BOx was fixed on the polypyrrole (PPy) layer modified by GO nanoparticles, and then electrochemically deposited on F-doped tin oxide (FTO) plate to prepare a nanocomposite electrode [42]. The biosensor assembled by this electrode can reach a detection limit of 0.1 nM and a detection range of 0.01–500 mM.

5.1.6 Cholesterol Detection

As an indispensable component of the cell membrane, cholesterol significantly affects the stability and fluidity of the cell membrane [43]. At the same time, it is also a

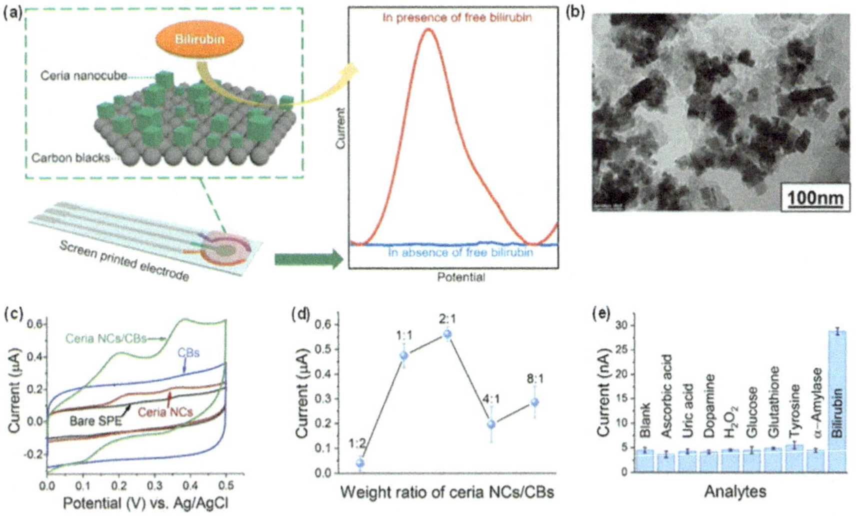

Fig. 5.5 Bilirubin detection (Reprinted with permission from Ref. [10]. Copyright 2020 Elsevier B.V.). **a** Schematic diagram of bilirubin detection in blood. **b** TEM image of ceria NCs dispersed on the electrode. **c** CV curve of bare SPE (black), ceria NCs (red), CBs (blue) and ceria NCs/CBs (green). **d** Current of ceria NCs/CBs. **e** Influence of interfering substances in serum

precursor of hormones, bile acids, and vitamin D in the body. It can also regulate the function of the heart and other organs. The cholesterol level in the blood of healthy people is not higher than 5.17 mM [44]. Increased cholesterol in the blood will lead to hypertension, cerebral thrombosis, and some kidney diseases [45]. Therefore, the detection of cholesterol in blood is particularly important for the prevention of some diseases. There are two main methods to detect cholesterol by electrochemical biosensors. One is the enzymatic reaction of cholesterol, and the other is a non-enzyme method that directly catalyzes the oxidation of cholesterol. In the enzymatic reaction of cholesterol, cholesterol oxidase (ChOx) oxidizes free cholesterol to cholest-3-one, and produces hydrogen peroxide (H_2O_2) by-product. Since there is a certain correlation between H_2O_2 and cholesterol, the content of cholesterol can be calculated by the amount of H_2O_2. Yildiz et al. reported an amperometric cholesterol biosensor in which ChOx was immobilized on 4-(4H-dithienol[3,2-b:2′,3′-d]pyrrole-4)aniline polymer [46]. The introduction of conductive polymer can accelerate electron transfer on the electrode surface. Subsequently, Wang et al. reported a portable paper biosensor that was a screen-printed electrode (SPE) with nanoporous gold (NPG) (Fig. 5.6a) [11]. The catalytic oxidation of cholesterol by ChOx and the reduction mechanism of NPG to H_2O_2 ensure the synergy of this strategy (Fig. 5.6b). The electrode can also generate a reduction current in phosphate buffer saline (PBS) solution containing cholesterol, which proves the synergy (Fig. 5.6c). ChOx oxidizes cholesterol in serum to H_2O_2, and the released H_2O_2 is reduced by NPG (Fig. 5.6d). ChOx has the highest catalytic efficiency for cholesterol at 37 °C (Fig. 5.6e). The

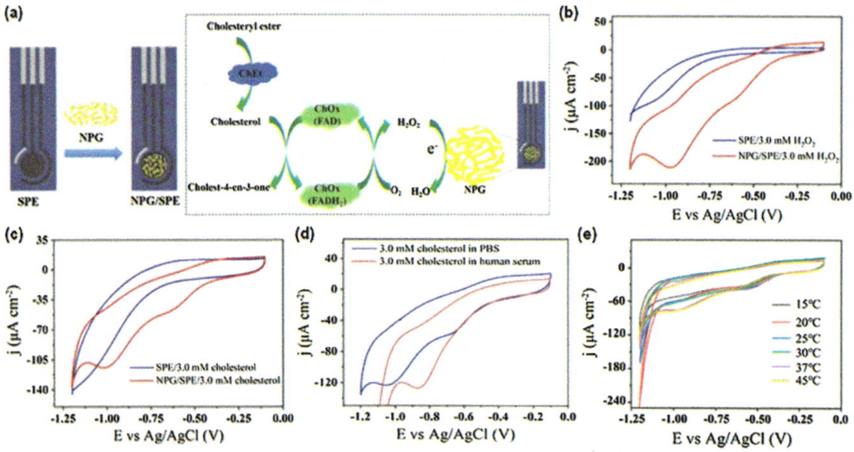

Fig. 5.6 Cholesterol detection (Reprinted with permission from Ref. [11]. Copyright 2021 Elsevier B.V.). **a** Schematic diagram of NPG/SPE biosensor preparation and detection. **b** CV curve of bare SPE and NPG/SPE in PBS solution containing H_2O_2. **c** CV curve of bare SPE and NPG/SPE in PBS solution containing cholesterol. **d** NPG/SPE in PBS solution and serum containing cholesterol, respectively. **e** CV curve of cholesterol oxidation by ChOx at different temperatures

enzyme cholesterol biosensors have some inherent defects, such as high cost and enzyme instability, which limit their application.

Non-enzyme direct catalytic oxidation of cholesterol has attracted much attention because of its high efficiency, stability, and oxygen-free binding. For example, Sun et al. prepared a non-enzyme electrochemical biosensor to improve the conductivity and increase the number of molecular imprints through multi-walled CNTs and AuNPs, with poly-aminothiophenol (PAP) as the sensing core [47]. Copper-based nanomaterials can also be used for non-enzyme electrochemical biosensors. A special rose-like copper sulfide nanostructure (Cu_2S NRS) was synthesized on the copper rod, and then the Cu_2S NRS/Cu rod was used as the electrode for non-enzyme detection of cholesterol [48]. The biosensor shows a wide linear range from 0.01 to 6.8 mmol L^{-1} and a detection limit of 0.1 μmol L^{-1}. However, non-enzyme detection still has some shortcomings, such as poor selectivity, low sensitivity, and electron transfer between active sites.

5.1.7 Lactic Acid Detection

Lactic acid is the main product of glucose decomposition in an anaerobic environment. It exists in cells, and then transported to the liver for oxidation to pyruvate [49]. The level of lactic acid in the blood of healthy people is 0.5–1.5 mM [50]. The lactic acid level in critically ill patients is often greater than 4 mM. The long-term increase of lactic acid in the blood will lead to respiratory failure, hemorrhage, anemia, hypoxemia, and meningitis. The detection of lactic acid is very important for evaluating the health problems related to lactic acid poisoning and anaerobic conditions. Electrochemical biosensors often use two enzymes when detecting lactic acid, one is lactate oxidase (LOx) and the other is lactate dehydrogenase (LDH). Lorenzo et al. prepared a biosensor by coupling carbon dots (CDs) and LOx, and then depositing the coupling compound on the gold electrode [51]. The linear detection range of the biosensor for lactic acid is 3.0–500 μM, the sensitivity is 4.98×10^{-3} μA μM^{-1}, and the detection limit is 0.9 μM. The biosensor prepared by dehydrogenase depends on the permanent presence of the cofactor nicotinamide adenine dinucleotide (NAD^+). This is mainly attributed to the cofactor participating in the electron transfer of dehydrogenase-catalyzed reaction. Schöning et al. prepared a reagent-free electrochemical biosensor by immobilizing LDH, NAD^+, and $Fe(CN)_6{}^{3-}$ to a screen-printed carbon electrode modified with GO and cellulose acetate (CA) (Fig. 5.7a) [12]. The maximum signal increase is observed at the potential of +0.25 V. The generation of this optimal potential is based on the electron transfer mediated by the redox process (Fig. 5.7b). CV curve shows redox characteristics, and the amount of GO added is positively correlated with the oxidation peak current (Fig. 5.7c). With the increase of CA layer, the linear detection range is expanded (Fig. 5.7d). The modification of the electrode with GO and CA increases the immobilization of coenzyme factors on the electrode, thus improving the electrocatalytic performance. GO provides a new process for the preparation of coenzyme-dependent biosensors.

5.1.8 Creatinine Detection

As one of the detection indicators of kidney disease, creatinine is widely used to monitor the progress of kidney disease. As one of the main components of blood and urine, creatinine provides energy for muscle tissue [52]. The traditional detection methods of creatinine include chromatography, spectrophotometry, and colorimetry. Although these methods have good selectivity and sensitivity, the disadvantages of cumbersome pretreatment, time-consuming, and high cost affect their application. Electrochemical biosensors have become favorable candidates because of their simplicity, efficiency, low cost, and high sensitivity.

Wang et al. prepared a creatinine detection biosensor by converting creatinine into 1-methylhydatine through a single enzyme pathway [53]. Due to the denaturation of enzymes, these systems show poor stability, sensitivity, and repeatability. Pundir et al. prepared an electrochemical biosensor for creatinine detection by immobilizing nanoparticles of creatinase, creatinase, and sarcosine oxidase on GCE [54]. The detection range is 0.01–12 μM, and the detection limit is 0.01 μM.

Due to the high cost and variability of enzymes, the development of non-enzyme biosensors is becoming more and more important. Prakash et al. embedded CDs

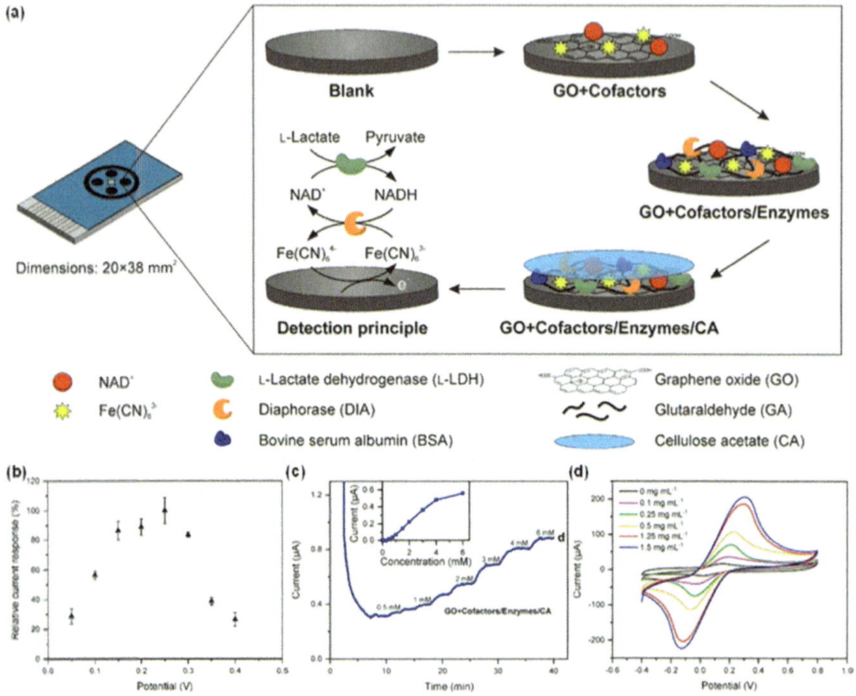

Fig. 5.7 Lactic acid detection (Reprinted with permission from Ref. [12]. Copyright 2019 American Chemical Society). **a** Schematic diagram of biosensor preparation and detection. **b** Effect of applied working potential on the sensor current. **c** Current response of the biosensor. **d** CV curve of the biosensor under different GO loads

doped with tungstic anhydride (WO_3) in GO chip to develop a non-enzyme electrochemical biosensor (Fig. 5.8a) [13]. The embedded structure provides an electrocatalytic basis for the detection of creatinine (Fig. 5.8b). XRD results prove the successful assembly of the nanocomposite (Fig. 5.8c). The nanocomposite modification improves the redox ability of the electrode (Fig. 5.8d). The synergy of CDs and WO_3 provides a good electrocatalytic environment and anti-interference ability (Fig. 5.8e). The electrode shows good selectivity for creatinine detection. The biosensor can realize the detection of creatinine at the level of pimole in blood.

5.1.9 Galactose Detection

The detection of galactose is important for food safety, medicine, fermentation industry, and human health [55]. The imbalance of galactose in blood and urine will lead to symptoms of galactosemia, galactorrhea, and other metabolic disorders

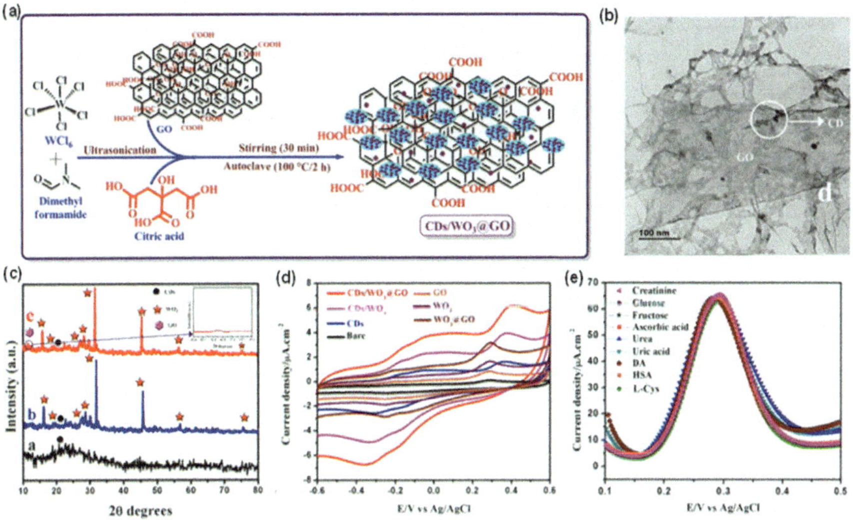

Fig. 5.8 Creatinine detection (Reprinted with permission from Ref. [13]. Copyright 2020 Elsevier B.V.). **a** Schematic diagram of electrode preparation. **b** TEM image of CDs embedded in GO. **c** XRD patterns of CDs, CDs/WO_3, and CDs/WO_3@GO. **d** Oxidation–reduction curves of electrodes modified with different substances. **e** Current density diagram of different concentrations of creatinine

[56]. As a genetic metabolic disease, galactosemia shows the disorder characteristic of weakness. Patients with this disease often lack enzymes that metabolize galactose. When the concentration of galactose in the blood of newborns exceeds 1.1 mM, it will lead to severe galactosemia [57]. This symptom leads to cataracts, kidney problems, and brain damage. Different methods such as chromatography, spectrophotometry, and fluorescence are used to detect galactose. Electrochemical biosensors have attracted much attention because of their fast, simple, and efficient characteristics, which overcome the shortcomings of traditional methods.

As a member of free radical coupled copper oxidase, galactose oxidase (GalOx) can significantly catalyze galactose to galactose acid and H_2O_2 [58]. Cuong et al. reported a simple method to prepare galactose biosensors, that is, grow small ZnO nanorods (NRs) on ZnO NRs, and then immobilize GalOx to form the electrode (GalOx|ZnO NRs/NRs) (Fig. 5.9a) [14]. The growth of small ZnO NRs on ZnO NRs with large surface area leads to different crystal orientations, which proves the successful preparation of the structure (Fig. 5.9b). Because of the high surface volume ratio of ZnO NRs/NRs, the amount of GalOx immobilized on the surface of ZnO NRs/NRs is high (Fig. 5.9c). GalOx|ZnO NRs/NRs electrode has high current density, which indicates that there are many reaction sites on the electrode surface (Fig. 5.9d). The biosensor can detect galactose in blood with a wide detection range of 40–230 mM. In order to make the biosensor simpler and cheaper, a filter paper containing Pt layer and Nafion® membrane is used as the working electrode, and then H_2O_2 produced by oxidation is detected in the presence of GalOx [59]. Nanomaterials

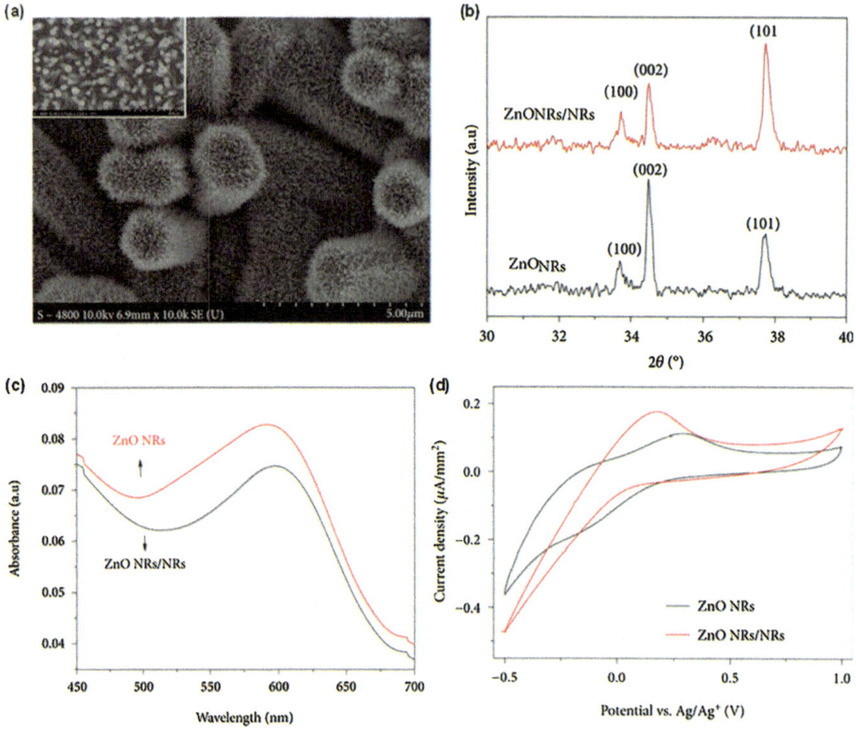

Fig. 5.9 Galactose detection (Reprinted with permission from Ref. [14]. Copyright 2019 The Authors). **a** SEM images of ZnO NRs and ZnO NRs/NRs. **b** XRD spectra of ZnO NRs and ZnO NRs/NRs. **c** Optical density of GalOx immobilized on the surface of ZnO NRs and ZnO NRs/NRs. **d** CV curve of GalOx|ZnO NRs and GalOx|ZnO NR/NR electrodes

with different surface crystal orientations improve the selectivity and sensitivity of non-enzyme galactose biosensors [60].

5.1.10 Triglyceride Detection

Triglyceride is a natural fat with ester bonds [61]. Triglyceride, as low-density lipoprotein, plays an important role in the metabolism [62]. Increased triglyceride in the blood will increase the probability of atherosclerosis, hypolipoproteinemia, and coronary heart disease [63]. The traditional methods can accurately detect triglyceride but require high cost, long time and skilled operators.

Recently, Pundir et al. prepared a sensitive triglyceride biosensor by the covalently immobilizing enzyme (lipase, glycerol kinase, and glycerol 3-phosphate oxidase) on the surface of pencil graphite electrode [64]. The biosensor has excellent detection performance for triglyceride in blood. It is reported that the sensitivity of triglyceride detection based on a single enzyme is sometimes affected by the pH of the medium, resulting in inaccurate detection [65]. Basu et al. prepared a dual-enzyme triglyceride biosensor in which toluidine blue was absorbed by the transparent film of rGO onto the surface of ITO to form an electrode, and then lipase (LIP) and glycerol dehydrogenase (GDH) were immobilized (Fig. 5.10a) [15]. FESEM image shows that the enzymes are immobilized on the surface (Fig. 5.10b). This is mainly attributed to the high specific surface area and the interaction between surface functional groups and enzymes. Raman spectra indicate that there are a large number of sp^2 sites and structural defects (Fig. 5.10c). CV curve shows the reversibility of the dual-enzyme biological electrode (Fig. 5.10d). The increase in tributyl concentration will increase the reaction intensity of the biological electrode (Fig. 5.10e). The biosensor can shield the influence of other interfering substances when detecting triglycerides (Fig. 5.10f). Toluidine blue is introduced into the electrode as a reversible redox medium to replace the traditional irreversible redox pair, which increases the reversibility and extends the service life of the biosensor.

5.1.11 Hydrogen Peroxide Detention

As one of the most stable and abundant reactive oxygen species in nature, hydrogen peroxide (H_2O_2) plays an important role in host defense, immune response, and pathogen invasion mechanisms [66]. The imbalance of H_2O_2 may lead to oxidative stress and protein deterioration, leading to accelerated aging and the occurrence of various diseases [67]. The electrochemical biosensors have the characteristics of simple operation, high sensitivity, fast response, and low cost. The detection of H_2O_2 in the blood is mainly divided into two ways. One is the high sensitivity and specificity of enzyme reactions [68]. The other is based on electron transfer between non-enzymes [69].

Xu et al. reported a switch-type proportional electrochemical biosensor for the detection of H_2O_2 in blood [16]. The biosensor was prepared by co-assembly of 5-(1,2-dithiocyclopentane-3-yl)-N-(4-(4,4,5,5-tetramethyl1,3,2-dioxapron-2-yl)phenyl) pentanediamine and Nile Blue A and then applied to carbon fiber microelectrode (CFME) (Fig. 5.11a). SEM image shows that the Au nanocone on the electrode uniformly covers the CFME surface (Fig. 5.11b). CV curve shows that the biosensor has good recognition ability for H_2O_2 detection (Fig. 5.11c). Due to the high specificity of the reaction between Nile Blue A and H_2O_2, the biosensor has good selectivity and anti-interference ability (Fig. 5.11d). The appearance of the redox peak is attributed to H_2O_2-induced borate ester bond cleavage. Differential pulse voltammetry (DPV) was obtained from three electrodes of blood and serum (Fig. 5.11e). This micro-sensor can rapidly analyze the content of H_2O_2 in trace blood, and provide ideas for detecting other active oxygen species and blood metabolites.

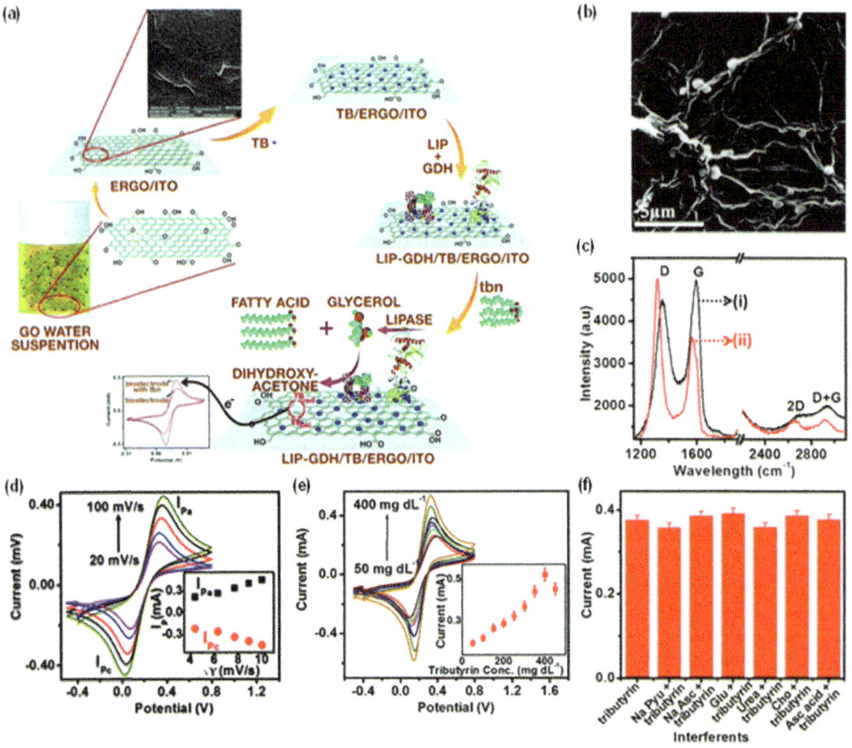

Fig. 5.10 Triglyceride detection (Reprinted with permission from Ref. [15]. Copyright 2019 The Royal Society of Chemistry). **a** Schematic diagram of LIP-GDH/TB/ERGO/ITO electrode preparation and detection. **b** FESEM image of LIP-GDH/TB/ERGO/ITO electrode. **c** Raman spectra of ERGO/ITO and GO/ITO surfaces. **d** CV curve of LIP-GDH/TB/ERGO/ITO electrode with different scan rates. **f** CV curve of LIP-GDH/TB/ERGO/ITO electrode at different triglyceride concentrations. **f** Influence of interfering substances

5.2 Conclusion

Blood is one of the most reliable biological samples in bioinformatics, and its metabolites include glucose, urea, uric acid, bilirubin, cholesterol, lactic acid, creatinine, galactose, triglyceride, and hydrogen peroxide. The level of metabolites in the blood plays a critical role in physiological activities. It is essential to prevent or diagnose diseases by detecting metabolites in blood. Electrochemical biosensors are widely used to detect metabolites in the blood due to their high sensitivity, good reproducibility, simplicity, and low cost. This chapter reviews the development of electrochemical biosensors for the detection of metabolites in blood. Although great progress has been made in this area, there are still some issues that need further discussion.

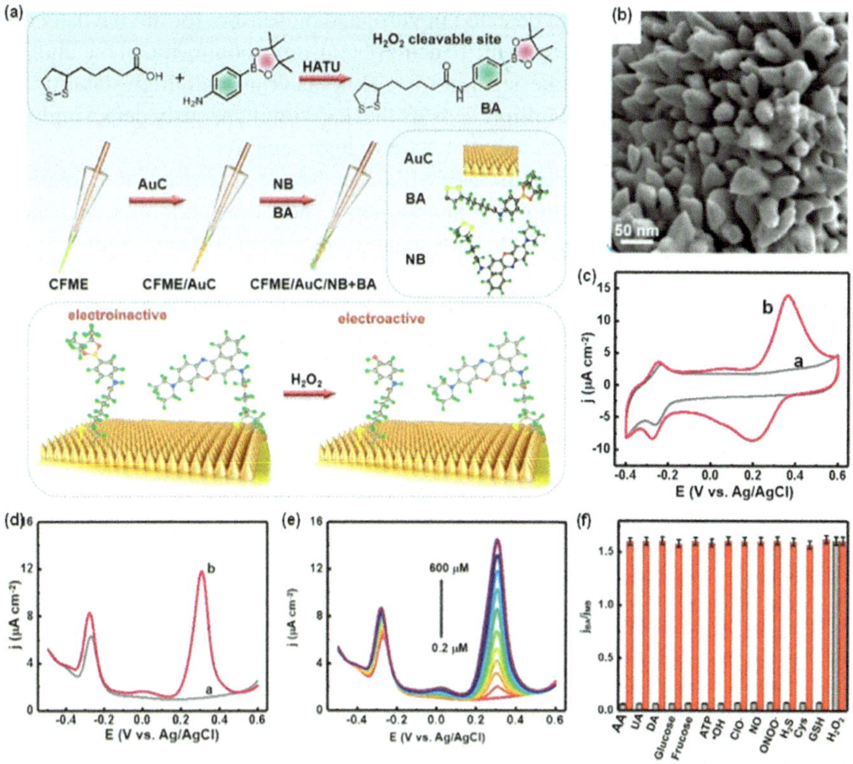

Fig. 5.11 Cholesterol detection (Reprinted with permission from Ref. [16]. Copyright 2020 Elsevier B.V). **a** Schematic diagram of electrode and biosensor preparation. **b** SEM image of gold nanocone. **c** CV curve of electrodes in PBS solution. **d** DPV response of electrodes in PBS solution. **e** DPV response of electrodes at different H_2O_2 concentrations. **f** Influence of interfering substances

Although non-enzyme biosensors have made progress in the detection of blood metabolites, there is still a certain gap in specificity and sensitivity compared with enzyme biosensors. The introduction of some nanomaterials into the biosensor can improve its electron transfer ability and target specific recognition. However, most electrochemical biosensors containing nanomaterials are still in the early stages and cannot be applied to clinical diagnosis. To obtain electrochemical sensors with long-term stability and good biocompatibility, it is necessary to further develop new nanomaterials or other materials.

The detection of metabolites in blood requires immersion injury to the human body. This detection method cannot be used for long-term tracking detection. Although the blood can now be replaced by other body fluids, the sensitivity and specificity of detection cannot reach the level of blood detection. It is necessary to further strengthen the research of electrochemical biosensors capable of incorporating portable, subcutaneous, and micro-implant devices to detect metabolites in blood.

Most blood metabolites are detected in vitro, and biosensors for in vivo detection are important for future development. To achieve real-time monitoring of metabolites in vivo, it is necessary to synthesize non-toxic nanomaterials with high stability and anti-interference capability. Multiple sensors that can simultaneously detect multiple metabolites should also be developed to achieve high sensitivity.

In the future, the detection of diseases will make the use of sensors more frequent, and the ability to process multiple biomarkers in a short time should be increased. In addition, the determination of analytes requires almost no cross-reaction between the target and receptor. Therefore, the rational design of receptors is particularly important. Some advanced automation equipment should be used in routine testing, which can avoid some artificial errors and produce more accurate results.

References

1. Turner APF (2013) Biosensors: sense and sensibility. Chem Soc Rev 42:3184–3196
2. Kimmel DW, LeBlanc G, Meschievitz ME, Cliffel DE (2012) Electrochemical sensors and biosensors. Anal Chem 84:685–707
3. Xu M, Obodo D, Yadavalli VK (2019) The design, fabrication, and applications of flexible biosensing devices. Biosens Bioelectron 124:96–114
4. Zhu CZ, Yang GH, Li H, Du D, Lin YH (2015) Electrochemical sensors and biosensors based on nanomaterials and nanostructures. Anal Chem 87:230–249
5. Kokkinos C (2019) Electrochemical DNA biosensors based on labeling with nanoparticles. Nanomaterials 9:1361
6. Yousefi H, Mahmud A, Chang DR, Das J, Gomis S, Chen JB, Wang HS, Been T, Yip L, Coomes E, Li ZJ, Mubareka S, McGeer A, Christie N, Gray-Owen S, Cochrane A, Rini JM, Sargent EH, Kelley SO (2021) Detection of SARS-CoV-2 viral particles using direct, reagent-free electrochemical sensing. J Am Chem Soc 143:1722–1727
7. Ahmad R, Khan M, Mishra P, Jahan N, Ahsan MA, Ahmad I, Khan MR, Watanabe Y, Syed MA, Furukawa H, Khosla A (2021) Engineered hierarchical CuO nanoleaves based electrochemical nonenzymatic biosensor for glucose detection. J Electrochem Soc 168:017501
8. Liu JJ, Moakhar RS, Perumal AS, Roman HN, Mahshid S, Wachsmann-Hogiu S (2020) An AgNP-deposited commercial electrochemistry test strip as a platform for urea detection. Sci Rep 10:9527
9. Verma S, Choudhary J, Singh KP, Chandra P, Singh SP (2019) Uricase grafted nanoconducting matrix based electrochemical biosensor for ultrafast uric acid detection in human serum samples. Int J Biol Macromol 130:333–341
10. Lu ZJ, Cheng YR, Zhang Y, Wang XF, Xu PC, Yu HT, Li XX (2021) Non-enzymatic free bilirubin electrochemical sensor based on ceria nanocube. Sens Actuator B-Chem 329:129224
11. Wang SJ, Chen SY, Shang KS, Gao XY, Wang X (2021) Sensitive electrochemical detection of cholesterol using a portable paper sensor based on the synergistic effect of cholesterol oxidase and nanoporous gold. Int J Biol Macromol 189:356–362
12. Pilas J, Selmer T, Keusgen M, Schoning MJ (2019) Screen-printed carbon electrodes modified with graphene oxide for the design of a reagent-free NAD^+-dependent biosensor array. Anal Chem 91:15293–15299
13. Ponnaiah SK, Prakash P (2020) Carbon dots doped tungstic anhydride on graphene oxide nanopanels: a new picomolar-range creatinine selective enzymeless electrochemical sensor. Mater Sci Eng C-Mater Biol Appl 113:111010
14. Phan PHL, Tran QT, Dinh DA, Bok KK, Hong CH, Cuong TV (2019) The facile synthesis of novel ZnO nanostructure for galactose biosensor application. J Nanomater 2019:2364327

15. Bhardwaj SK, Chauhan R, Yadav P, Ghosh S, Mahapatro AK, Singh J, Basu T (2019) Bi-enzyme functionalized electro-chemically reduced transparent graphene oxide platform for triglyceride detection. Biomater Sci 7:1598–1606
16. Dong H, Zhou YL, Hao YQ, Zhao L, Sun S, Zhang YT, Ye BX, Xu MT (2020) "Turn-on" ratiometric electrochemical detection of H_2O_2 in one drop of whole blood sample via a novel microelectrode sensor. Biosens Bioelectron 165:112402
17. Ye DX, Liang GH, Li HX, Luo J, Zhang S, Chen H, Kong JL (2013) A novel nonenzymatic sensor based on CuO nanoneedle/graphene/carbon nanofiber modified electrode for probing glucose in saliva. Talanta 116:223–230
18. Zuo MH, Tang J, Xiang MM, Long Q, Dai JP, Yu GD, Zhang HG, Hu H (2019) Clinical observation of the reduced glutathione in the treatment of diabetic chronic kidney disease. J Cell Biochem 120:8483–8491
19. Valentini F, Fernandez LG, Tamburri E, Palleschi G (2013) Single walled carbon nanotubes/polypyrrole-GOx composite films to modify gold microelectrodes for glucose biosensors: study of the extended linearity. Biosens Bioelectron 43:75–78
20. Akhtar MA, Batool R, Hayat A, Han DX, Riaz S, Khan SU, Nasir M, Nawaz MH, Niu L (2019) Functionalized graphene oxide bridging between enzyme and Au-sputtered screen-printed interface for glucose detection. ACS Appl Nano Mater 2:1589–1596
21. Sehit E, Altintas Z (2020) Significance of nanomaterials in electrochemical glucose sensors: an updated review (2016–2020). Biosens Bioelectron 159:112165
22. Guo CX, Li CM (2010) Direct electron transfer of glucose oxidase and biosensing of glucose on hollow sphere-nanostructured conducting polymer/metal oxide composite. Phys Chem Chem Phys 12:12153–12159
23. Adeel M, Rahman MM, Caligiuri I, Canzonieri V, Rizzolio F, Daniele S (2020) Recent advances of electrochemical and optical enzyme-free glucose sensors operating at physiological conditions. Biosens Bioelectron 165:112331
24. Ray H, Perreault F, Boyer TH (2019) Urea recovery from fresh human urine by forward osmosis and membrane distillation (FO-MD). Environ Sci-Wat Res Technol 5:1993–2003
25. Raghavan R, Eknoyan G (2018) Uremia: a historical reappraisal of what happened. Clin Nephrol 89:305–313
26. Karazehir T, Ates M, Sarac AS (2016) Covalent immobilization of urease on poly(pyrrole-3-carboxylic acid): electrochemical impedance and Mott Schottky study. J Electrochem Soc 163:B435–B444
27. Pundir CS, Jakhar S, Narwal V (2019) Determination of urea with special emphasis on biosensors: a review. Biosens Bioelectron 123:36–50
28. Das J, Sarkar P (2016) Enzymatic electrochemical biosensor for urea with a polyaniline grafted conducting hydrogel composite modified electrode. RSC Adv 6:92520–92533
29. Capek I (2009) Dispersions, novel nanomaterial sensors and nanoconjugates based on carbon nanotubes. Adv Colloid Interface Sci 150:63–89
30. Ondes B, Akpinar F, Uygun M, Muti M, Uygun DA (2021) High stability potentiometric urea biosensor based on enzyme attached nanoparticles. Microchem J 160:105667
31. Kumar THV, Sundramoorthy AK (2018) Non-enzymatic electrochemical detection of urea on silver nanoparticles anchored nitrogen-doped single-walled carbon nanotube modified electrode. J Electrochem Soc 165:B3006–B3016
32. Martinez-Perez D, Ferrer ML, Mateo CR (2003) A reagent less fluorescent sol-gel biosensor for uric acid detection in biological fluids. Anal Biochem 322:238–242
33. Raj CR, Ohsaka T (2003) Voltammetric detection of uric acid in the presence of ascorbic acid at a gold electrode modified with a self-assembled monolayer of heteroaromatic thiol. J Electroanal Chem 540:69–77
34. Alderman M, Aiyer KJV (2004) Uric acid: role in cardiovascular disease and effects of losartan. Curr Med Res Opin 20:369–379
35. Misra N, Kumar V, Borde L, Varshney L (2013) Localized surface plasmon resonance-optical sensors based on radiolytically synthesized silver nanoparticles for estimation of uric acid. Sens Actuator B-Chem 178:371–378

36. Jindal K, Tomar M, Gupta V (2012) CuO thin film based uric acid biosensor with enhanced response characteristics. Biosens Bioelectron 38:11–18
37. Liu HY, Jamal R, Abdiryim T, Simayi R, Liu LL, Liu YJ (2021) Carboxylated cellulose as a soft template combined with PEDOT derivatives in BMIM Cl: a competent biosensor for detection of guanine and Uric Acid in the blood. ACS Sustain Chem Eng 9:5860–5871
38. Noh HB, Won MS, Shim YB (2014) Selective nonenzymatic bilirubin detection in blood samples using a Nafion/Mn-Cu sensor. Biosens Bioelectron 61:554–561
39. Rawal R, Kharangarh PR, Dawra S, Tomar M, Gupta V, Pundir CS (2020) A comprehensive review of bilirubin determination methods with special emphasis on biosensors. Process Biochem 89:165–174
40. Santhosh M, Chinnadayyala SR, Singh NK, Goswami P (2016) Human serum albumin-stabilized gold nanoclusters act as an electron transfer bridge supporting specific electrocatalysis of bilirubin useful for biosensing applications. Bioelectrochemistry 111:7–14
41. Feng QL, Du YL, Zhang C, Zheng ZX, Hu FD, Wang ZH, Wang CM (2013) Synthesis of the multi-walled carbon nanotubes-COOH/graphene/gold nanoparticles nanocomposite for simple determination of Bilirubin in human blood serum. Sens Actuator B-Chem 185:337–344
42. Chauhan N, Rawal R, Hooda V, Jain U (2016) Electrochemical biosensor with graphene oxide nanoparticles and polypyrrole interface for the detection of bilirubin. RSC Adv 6:63624–63633
43. Kumar B, Poddar S, Sinha SK (2022) Electrochemical cholesterol sensors based on nanostructured metal oxides: current progress and future perspectives. J Iran Chem Soc 19:4093–4116
44. Alagappan M, Immanuel S, Sivasubramanian R, Kandaswamy A (2020) Development of cholesterol biosensor using Au nanoparticles decorated f-MWCNT covered with polypyrrole network. Arab J Chem 13:2001–2010
45. Ahmad M, Nisar A, Sun HY (2022) Emerging trends in non-enzymatic cholesterol biosensors: challenges and advancements. Biosensors-Basel 12:955
46. Cevik E, Cerit A, Gazel N, Yildiz HB (2018) Construction of an amperometric cholesterol biosensor based on DTP(aryl)aniline conducting polymer bound cholesterol oxidase. Electroanalysis 30:2445–2453
47. Ji J, Zhou ZH, Zhao XL, Sun JD, Sun XL (2015) Electrochemical sensor based on molecularly imprinted film at Au nanoparticles-carbon nanotubes modified electrode for determination of cholesterol. Biosens Bioelectron 66:590–595
48. Ji R, Wang LL, Wang GF, Zhang XJ (2014) Synthesize thickness copper(I) sulfide nanoplates on copper rod and It's application as nonenzymatic cholesterol sensor. Electrochim Acta 130:239–244
49. Alam F, RoyChoudhury S, Jalal AH, Umasankar Y, Forouzanfar S, Akter N, Bhansali S, Pala N (2018) Lactate biosensing: the emerging point-of-care and personal health monitoring. Biosens Bioelectron 117:818–829
50. Kost GJ (1993) New whole-blood analyzers and their impact on cardiac and critical care. Crit Rev Clin Lab Sci 30:153–202
51. Bravo I, Gutierrez-Sanchez C, Garcia-Mendiola T, Revenga-Parra M, Pariente F, Lorenzo E (2019) Enhanced performance of reagent-less carbon nanodots based enzyme electrochemical biosensors. Sensors 19:5576
52. Gonzalez-Gallardo CL, Arjona N, Alvarez-Contreras L, Guerra-Balcazar M (2022) Electrochemical creatinine detection for advanced point-of-care sensing devices: a review. RSC Adv 12:30785–30802
53. Dasgupta P, Kumar V, Krishnaswamy PR, Bhat N (2020) Serum creatinine electrochemical biosensor on printed electrodes using monoenzymatic pathway to 1-methylhydantoin detection. ACS Omega 5:22459–22464
54. Kumar P, Jaiwal R, Pundir CS (2017) An improved amperometric creatinine biosensor based on nanoparticles of creatininase, creatinase and sarcosine oxidase. Anal Biochem 537:41–49
55. Yu KM, Yang P, Huang TY, Shen TYS, Lau JYN, Hu OYP (2022) A novel galactose electrochemical biosensor intended for point-of-care measurement of quantitative liver function using galactose single-point test. Anal Bioanal Chem 414:4067–4077

56. Khun K, Ibupoto ZH, Nur O, Willander M (2012) Development of galactose biosensor based on functionalized ZnO nanorods with galactose oxidase. J Sens 2012:696247
57. Kanyong P, Pemberton RM, Jackson SK, Hart JP (2013) Development of an amperometric screen-printed galactose biosensor for serum analysis. Anal Biochem 435:114–119
58. Sharma M, Yadav P, Sharma M (2019) Novel electrochemical sensing of galactose using GalOxNPs/CHIT modified pencil graphite electrode. Carbohydr Res 483:107749
59. Bouri M, Zuaznabar-Gardona JC, Novell M, Blondeau P, Andrade FJ (2021) Paper-based potentiometric biosensor for monitoring galactose in whole blood. Electroanalysis 33:81–89
60. Mie Y, Katagai S, Ikegami M (2020) Electrochemical oxidation of monosaccharides at nanoporous gold with controlled atomic surface orientation and non-enzymatic galactose sensing. Sensors 20:5632
61. Hooda V, Gahlaut A, Gothwal A, Hooda V (2018) Recent trends and perspectives in enzyme based biosensor development for the screening of triglycerides: a comprehensive review. Artif Cell Nanomed Biotechnol 46:626–635
62. Skulas-Ray AC, Kris-Etherton PM, Harris WS, Heuvel JPV, Wagner PR, West SG (2011) Dose-response effects of omega-3 fatty acids on triglycerides, inflammation, and endothelial function in healthy persons with moderate hypertriglyceridemia. Am J Clin Nutr 93:243–252
63. Okazaki M, Komoriya N, Tomoike H, Inoue N, Usui S, Itoh S, Hosaki S (1998) Quantitative detection method of triglycerides in serum lipoproteins and serum-free glycerol by high-performance liquid chromatography. J Chromatogr B 709:179–187
64. Narwal V, Pundir CS (2017) An improved amperometric triglyceride biosensor based on co-immobilization of nanoparticles of lipase, glycerol kinase and glycerol 3-phosphate oxidase onto pencil graphite electrode. Enzyme Microb Technol 100:11–16
65. Bhardwaj SK, Yadav P, Ghosh S, Basu T, Mahapatro AK (2016) Biosensing test-bed using electrochemically deposited reduced graphene oxide. ACS Appl Mater Interfaces 8:24350–24360
66. Cai LP, Deng LY, Huang XY, Ren JC (2018) Catalytic chemiluminescence polymer dots for ultrasensitive In vivo imaging of intrinsic reactive oxygen species in mice. Anal Chem 90:6929–6935
67. Abdesselem M, Ramodiharilafy R, Devys L, Gacoin T, Alexandrou A, Bouzigues CI (2017) Fast quantitative ROS detection based on dual-color single rare-earth nanoparticle imaging reveals signaling pathway kinetics in living cells. Nanoscale 9:656–665
68. Zhou J, Liao CN, Zhang LM, Wang QG, Tian Y (2014) Molecular hydrogel-stabilized enzyme with facilitated electron transfer for determination of H_2O_2 released from live cells. Anal Chem 86:4395–4401
69. Dai HX, Chen YL, Niu XY, Pan CJ, Chen HL, Chen XG (2018) High-performance electrochemical biosensor for nonenzymatic H_2O_2 sensing based on Au@C-Co_3O_4 heterostructures. Biosens Bioelectron 118:36–43

Chapter 6
Electrochemical Biosensors for Drugs Detection

Ruiqi Ma, Rui Liu, and Fan Xia

Abstract Drugs detection could help to monitor the treatment of diseases in order to adjust drug dosage and usage, to ensure maximum therapeutic effect while reducing unnecessary side effects and danger. The electrochemical biosensors are an effective method to realize highly sensitive and specific detection of drugs. In this Chapter, we focus on the electrochemical detection of drugs in blood, including nanomaterials and bio-enzyme. From the perspective of different drugs, the innovation of electrochemical biosensors for detecting them is discussed. We hope to provide ideas for the detection of drugs in the blood through this Chapter and point out the important role of the detection of drugs in human health.

Keywords Biosensors · Electrochemical detection · Drug detection · Nanomaterials · Blood

6.1 Introduction

Drugs detection is primarily used to monitor the feedback of drugs during the development process and the monitoring of the metabolism, absorption, distribution, and excretion in the human body [1, 2]. In the process of actual diagnosis and treatment, drug testing can help doctors adjust drug dosage and usage plans to ensure maximum therapeutic effect while reducing unnecessary side effects and danger. In the past decades, the technology of drug testing has made fruitful achievements, expanding from traditional drug concentration analysis to the exploration of fields

---

R. Ma
Faculty of Materials Science and Chemistry, China University of Geosciences, Wuhan 430074, China

R. Liu (✉) · F. Xia
State Key Laboratory of Biogeology and Environmental Geology, Engineering Research Center of Nano-Geomaterials of Ministry of Education, Faculty of Materials Science and Chemistry, China University of Geosciences, Wuhan 430074, China
e-mail: ruiliu@cug.edu.cn

F. Xia et al. (eds.), *Electrochemical Biosensors for Whole Blood Analysis*,
https://doi.org/10.1007/978-981-99-5644-9_6

such as molecular diagnosis and personalized medicine, and the scope of drug testing continues to expand [3–5].

Traditional drug-detection methods include mass spectrometry, magnetic resonance imaging, gas chromatography, high-performance liquid chromatography, etc. [6]. These methods have high precision and reliability, but generally require complex sample preparation and expensive equipment. The operations are generally complex, time and economic costing, and difficult to real-time monitoring. In recent years, drug detection based on electrochemical sensor technology has attracted more and more attention. Biosensing technology is a technology that combines biomolecules with signal conversion devices to detect biomolecules by means of electrochemistry, optics, and acoustics [7]. With high sensitivity, selectivity, and real-time performance, biosensor technology does not require expensive equipment or complex pretreatment, which has been widely considered and applied.

Electrochemical biosensors, a kind of signal conversion device, could detect biomolecules selectively by immobilizing biorecognition elements on electrochemical transducers. The biological recognition elements of the sensor are usually biological molecules, such as an enzyme, antibody, nucleic acid, or cell, which could specifically bind to the target and release an electric signal [7, 8]. Electrochemical biosensors have received widespread attention in cancer drug testing.

Different detection methods and bio-recognition elements are employed in electrochemical biosensors for different drugs. Nanomaterials, like nanoparticles/ nanotubes/ nanosheets, have good electrochemical properties and biocompatibility, and are often chosen to improve the selectivity of the sensor, so as to achieve rapid and accurate detection of drugs [9]. Nanoparticles have a high surface volume ratio and good biocompatibility. Such as, gold nanoparticles were used to modify screen-printed electrodes (AuNP-SPE) to combine with electrochemical sensors to achieve the detection of the anticancer drug Fluorouracil (5-FU) [10]. Carbon nanotubes have excellent conductivity and biocompatibility. Carbon nanotubes are used to modify the electrodes of electrochemical biosensors and to detect the concentration of anti-tumor drug cisplatin [11]. Multi-walled carbon nanotubes (MWCNTs) are also excellent supporting material, which have low resistivity, high stability, and have a high surface volume ratio. Researchers have prepared a new DNA-free electrochemical biosensor for the detection of mitoxantrone anticancer drugs by using silver nanoparticles and polythiophene (PT) loaded on multi-walled carbon nanotubes [12]. Enzymatic electrochemical biosensors are also a common type of biosensors that can be used to detect various substances, such as glucose, lactate, and cholesterol, among others. Enzyme-catalyzed reactions are used to convert the target substance into a measurable electrical signal, thereby enabling detection of the target substance. A typical enzyme electrochemical sensor consists of an electrode, enzyme, mediator, and target molecule. When the target molecule presents, it reacts with the enzyme, triggering an electrochemical peak. Based on the size and shape of the electrochemical peak, the concentration of the target molecule could be determined [13]. For example, a glucose sensor takes glucose oxidase (GOD) to convert glucose into glucose acid, generating electrons. These electrons can be transmitted through the electrode, forming an electrochemical signal. By continuously improving and optimizing the construction materials and mediators of the sensor, the sensitivity and selectivity of the sensor can be further improved, enabling more accurate and reliable detection results [14].

In drug testing in the human body, samples including blood, urine, and saliva are used, among which blood testing is widely used in clinical medicine mainly because blood is one of the easiest biological samples to obtain. In blood samples, whole blood refers to blood that has not been centrifuged to separate plasma or serum. Plasma refers to the liquid component separated under the action of an anticoagulant, including plasma proteins, clotting factors, etc., while serum refers to the liquid component after centrifugation that has removed clotting factors. In in vivo drug testing, whole blood and plasma samples can be used to detect the concentration of drugs in the body, while serum samples are mainly used to detect drug metabolites.

In this chapter, the application of electrochemical biosensors mainly using biological samples such as whole blood and serum in drug detection is reviewed. A summarized performance standard of the published electrochemical biosensors can be seen in Table 6.1.

6.2 Nanomaterials-Based Electrochemical Biosensors

6.2.1 Zero-Dimensional Nanoparticles-Based Electrochemical Biosensors

With particle sizes ranging from 1 to −100 nm, nanoparticles have a high surface-to-volume ratio and good biocompatibility. Nowadays, hundreds of nanomaterials have been discovered with peroxidase-like catalytic activities. Most of the electrochemical biosensors consist of peroxidase mimics, the peroxidase-like nanozymes comes first.

Methotrexate (MTX) is a folic acid antagonist broadly used in different cancers curage. However, it also exhibits high toxicity, which could make a big difference in the growth and proliferation of healthy cells. Traditional MTX detection methods require complex and expensive equipment and often require multiple-sample processing steps, increasing the cost and time of detection. In order to overcome these limitations, Fe_3O_4/indium tin oxide (ITO) electrode modification technology based on the solvothermal method by El-Said W. A. et al. They used the solvothermal method to grow Fe_3O_4 nanoparticles directly on the surface of ITO, and prepared magnetite Fe_3O_4 nanoparticles/ITO electrode. The modified electrode was highly sensitive to MTX. The experimental results show that this electrode is highly sensitive to MTX drugs and can be met in a wide range of concentrations. The electrode can also effectively detect MTX in human serum and clinical samples. Compared to the traditional MTX detection method, the electrochemical biosensor proposed in this chapter has many advantages. First, the process of preparing the electrode is very simple, and only Fe_3O_4 nanoparticles can be directly grown on the ITO surface by solvothermal method. Secondly, the electrode has high sensitivity

Table 6.1 Electrochemical biosensors for drug detection

Analyte	Specimen	Biosensor construction	Linear range	Detection limit	References
Mitoxantrone anticancer drug	Urine and blood serum	MWCNT-Ag- polythiophene (PT)	0.05–100.0 μM	0.013 μM	[12]
Methotrexate (MTX)	Blood serum	Magnetite Fe_3O_4 nanoparticles/indium tin oxide (ITO) electrode	10^{-8}–10.0 μM	0.4×10^{-9} μM	[15]
Imipramine (IMI)	Urine and whole blood	CPE/Sr $(VO_3)_2$-Phytic acid (PA)	0.02–1000 and 0.2–1000 ng/mL	0.001 and 0.01 ng/mL	[16]
Ethanol	Human blood, beer, red wine, and Chinese liquor	Alcohol dehydrogenase (ADH)-chloranil (TCBQ)-lipidic cubic phases (LCPs)/Single-walled carbon nanotubes (SWCNTs)/GCE	200.0–1.3×10^4 μM	50.0 μM	[17]
Epirubicin (EPI)	Human serum	Gold nanoparticles (Au NPs)/graphite screen-printed electrodes (GSPCEs)	0.04–0.80 μM 0.8–20.0 μM	0.01 μM	[18]
Rutin	Utin tablets serum	Chitosan (Chi)/hydroxyl fullerenes (HFs)-laccase (Lac)/ MWCNTs-CTAB/GCE	0.1 μM–2 μM 2 μM–11 μM	0.03 μM 9.55×10^{-2} μM	[19]
MTX	Whole blood	Phosphide (WP)/nitrogen-doped carbon nanotubes (N-CNT)	0.01–540 μM	0.045 μM	[20]
5-amino salicylic acid	Human blood serum and human urine	Ni-ZrO_2/MWCNT	0.001–500 μM	0.0029 μM	[21]
MTX	Human blood serum and diluted urine	DNA/graphene oxide (GO)/GCE	0.055–2.2 μM	7.6×10^{-3} μM	[22]
Amyloid β (42) oligomers	Physiological buffer human serum	Chemical vapor deposition (CVD)/ carbon cloth (CC)/ SnS_2	10^{-4}–10^{-3} ng/mL	2.389×10^{-4} ng/mL 5.69×10^{-5} ng/mL	[26]

(continued)

Table 6.1 (continued)

Analyte	Specimen	Biosensor construction	Linear range	Detection limit	References
Paracetamol dopamine	Human serum Saliva	Polystyrene microspheres (PSs)/TiO_2/glassy carbon electrode (GCE)	0.3–50 μM 0.3–50 μM	0.05 μM 0.04 μM	[27]
RR-ETB SS-ETB	Blood, urine, and pharmaceutical	CD-CuMOF-CNF-GCE	0.1–100.0 μM 0.5–250.0 μM	3.10×10^{-2} μM 8.52×10^{-2} μM	[28]
Propofol	Human serum	Pencil Graphite Electrode (PGE)/Pt electrode	30–240 μM	7.2 ± 3.0 μM	[29]
Ciprofloxacin (CIP)	Urine and blood serum	Co-Fe-PBA@CN/GCE	0.005–300 and 325–741 μM	0.7389×10^{-3} and 1.0313×10^{-3} μM	[30]
Glucocorticoid receptor alpha (GRα)	Hippocampus and blood cells	Amino-ion graphene oxide (IL-rGO) and amino acid-coated gold nanoparticles (AA-AuNPs)	0.001 ng/mL–500 ng/mL	2.83×10^{-4} ng/mL	[31]
SARS-CoV-2	Saliva and whole blood	AuNP-peptide-DNA combined nanoprobe/ gold electrode	1×10^{-3} ng/mL–1×10^{4} ng/mL ($I = 1.63 + 4.44\ lgC$)	2.718×10^{-5} μM	[33]
Doxorubicin (DOX)	Whole blood	Gold electrode / DNA tetrahedron	0.03–0.11 μM	0.024 μM	[34]
Topotecan (TPT)	Human blood serum and urine	Graphene pastes electrode (GPE)	0.7–90.0 μM	0.37 μM	[36]
Etoposide	Human blood plasma, serum, and urine	ZnAl/layered double hydroxide modified cobalt ferrite-graphene oxide nanocomposite (GO/$CoFe_2O_4$/ZnAl-LDH)	0.2–10 μM	0.001 μM	[37]
H_2O_2	Whole blood	Pentanamide-Carbon fiber microelectrode (CFME)/Au	0.5–400 μM	0.02 μM	[38]

and selectivity, and can detect MTX drugs in a low concentration range. Most importantly, the cost of MTX detection using this sensor is low, so it can provide more convenient and economical detection services for the general population [15].

Imipramine (IMI) is a drug commonly used to treat severe endogenous depression, anxiety, hyperactivity disorder, attention deficit and enuresis. In order to maximize its therapeutic effect and reduce side effects, an appropriate IMI dose must be determined. Recently, Mofidi, Z. et al. developed an ultra-trace detection system for IMI in whole blood by combining electromembrane extraction with fast Fourier transform square wave voltammetry. The system uses nanoparticles-modified carbon paste electrodes and composite polymers to improve the sensitivity of electrochemical measurement. The researchers also used nano Sr $(VO_3)_2$ doped with phytic acid to modify CPE for determination of IMI. By using 2-nitrophenyloctyl ether as a support liquid membrane and extracting IMI molecules from a phosphate buffer solution, this method can be directly used for ultra-trace detection of IMI in biological complexes without any pretreatment. Because EME-FFTSWV technology and the electrochemical biosensor have simple operation, cost-effectiveness, and shortened analysis time, it has great practical application potential in the medical scientific analysis [16].

6.2.2 One-Dimensional Nanotubes-Based Electrochemical Biosensors

Nanotubes, especially carbon-based nanotubes show excellent mechanical, electrical, and thermal properties, and had been applicated in analysis and biomedical measurement. The integration of enzymes and their nanotube electronic media onto a single electrode surface to prepare enzyme-based bio-electrochemical devices is a research hotspot.

Wang S. Z. et al. proposed a simple and efficient immobilization method using liquid crystal lipid cubic phase (LCPs) coated single-walled carbon nanotubes (SWCNT) film as substrate. LCPs are a kind of biomimetic membrane with a three-dimensional ordered water channel structure, which effectively carries alcohol dehydrogenase's activity. Since SWCNTs have a special electronic structure and high conductivity, once coated on traditional glassy carbon electrode, they could promote the transfer of electrons. The immobilization of ethanol dehydrogenase (ADH) and NADH oxidation catalyst tetrachloro-1,4-benzoquinone (TCBQ) on LCPs formed on CNTs was demonstrated for NADH detection and ethanol biosensor. TCBQ exhibited excellent electrocatalytic activity for the electrooxidation of NADH at an overpotential. LCPs-TCBQ-SWCNTs composites exhibit good NADH electrocatalytic oxidation at extremely low anode potential, fast reaction, and wide linear range. Under the optimized conditions, the integrated enzyme electrode has a fast reaction, high sensitivity, and stable for ethanol oxidation. The ethanol biosensor developed has been used for the detection of ethanol content in whole blood. The results are

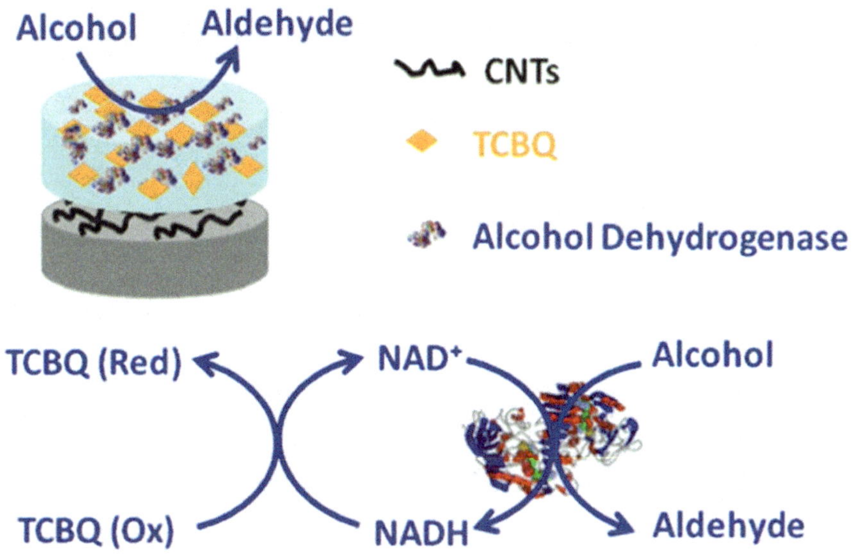

Scheme 1 Schematic illustration of ethanol biosensor based on an enzymatic electrode integrated with alcohol dehydrogenase and chloranil in liquid-crystalline cubic phase on carbon nanotubes [17]. Copyright (2019) Journal of The Electrochemical Society

satisfactory. It provides a simple, but effective method for the preparation of enzyme biosensors and bioelectrochemical devices such as biofuel cells, and also shows its application in analysis and biomedical measurement [17] (Scheme 1).

Abedi et al. prepared an unmarked electrochemical DNA hybridization biosensor, which uses gold nanotubes to modify graphite, to detect the interaction between epirubicin and short-length DNA of prostate cancer. It was validated for the detection of epirubicin in human serum and was used as a stable and reproducible EPI determination device. Further studies have shown that the developed biosensor has less disturbance in the presence of other drugs, and has been checked can detect EPI in human serum easily [18] (Scheme 2).

Multi-walled carbon nanotubes (MWCNTs) are also an excellent supporting material. Rutin is mainly transported through the blood circulation system in the human body. To monitor rutin's blood concentration is important in the fields of pharmacology and pharmacokinetics. Song, X. Y. et al. developed a biosensor by modifying the MWCNTs to glassy carbon electrode. The dispersed of CTAB to MWCNTs improved the hydrophilicity and biocompatibility. Then by the use of Lac, the oxidation of rutin enhanced, thereby significantly improving the sensitivity and selectivity. The biosensor can be used for the detection of rutin tablets and serum [19].

Nanocomposites with multiple active sites are also commonly used to prepare electrodes for electrochemical sensors, which are also used to monitor the concentration of MTX in blood. Mofidi, Z. et al. proposed a portable and reusable electrochemical detection platform based on nitrogen-doped carbon nanotubes. They

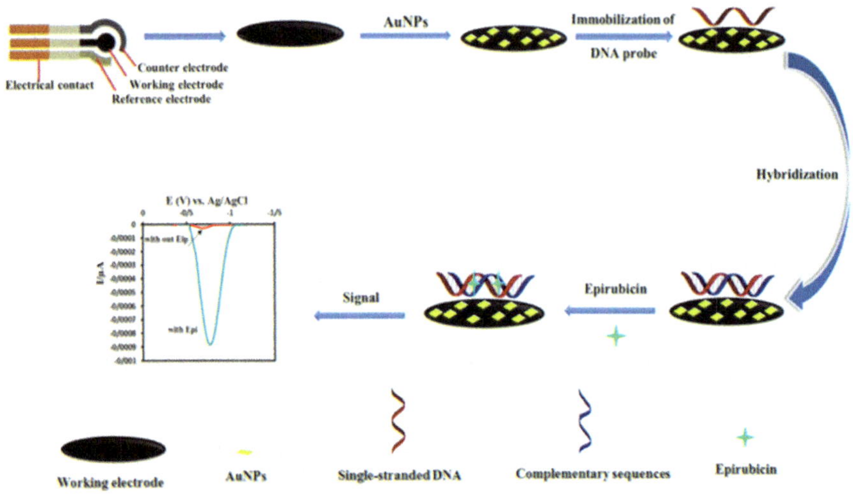

Scheme 2 The schematic illustration of the geno sensor fabrication [18]. Copyright (2021) Elsevier

successfully prepared a new type of electrode with multiple active sites. The electrode turned out to be good at conductivity and electrocatalytic activity. The dispersed active sites are connected through the electron transport channel N-CNT effectively. This sensor provides a low-cost, portable, and instant detection platform for cancer treatment. Although anticancer drugs cannot be monitored in real time in vivo, this research shows that detection in in vitro samples could be realizable. Future research will address these issues by developing new micro/nanoelectrodes [20].

5-Aminosalicylic acid (5-ASA) is a non-steroidal anti-inflammatory drug, which has been widely used in the treatment of inflammatory bowel diseases, such as Crohn 's disease, ulcerative colitis. However, long-term overdose of these drugs can cause lots of side effects. It may cause dizziness, rash, stomach cramps or vomiting. To test the content of 5-ASA is important for early diagnosis. In recent years, researchers have successfully developed a Ni-ZrO_2/MWCNT nanocomposite as a highly sensitive 5-ASA electrochemical detection electrode material by modifying MWCNT with Ni-doped ZrO_2 nanoparticles (NPs). The electrode material has excellent catalytic performance and high specificity for 5-ASA detection, even in different interfering species. In addition, it shows a satisfactory recovery rate in complex biological samples and has broad practical application prospects [21].

6.2.3 Two-Dimensional Nanosheets-Based Electrochemical Biosensors

Two-dimensional nanosheets are another kind of material to make flexible and cost-effective electrochemical biosensors. Chen et al. also designed a DNA/ GO/ GCE sensor for the detection of MTX. They immobilized DNA on a graphene oxide modified glassy carbon electrode (GCE). The sensor has a better adsorption effect on DNA than the bare GCE sensor. It can be applied to the determination of methotrexate in real samples such as human serum, and shows good recovery. This study provides an effective method for in vivo monitoring of methotrexate [22].

As an incurable, irreversible, and progressive brain disease, Alzheimer 's disease (AD) takes more than 10% of over-65-year-old humans to darkness [23, 24]. The content of brain amyloid β (Aβ) is an important indicator to determine whether patients with Alzheimer's disease [25]. Recent electrochemical studies reduced the detection limits to 100 pg/mL and 0.638 fg/mL in human serum/plasma. Murti, B. T. et al. constructed an ultrasensitive electrochemical biosensor for most neurotoxic AD biomarkers (i.e., Aβ (42) oligomers) in human serum by photoelectric properties, stability and reactivity of SnS_2 through SnS_2 nanosheets. They synthesized nanosheets with SnS_2 vertically aligned on a flexible conductive carbon cloth substrate by chemical vapor deposition. The designed sensing platform was covalently immobilized on 3-mercaptopropyl-trimethoxysilane-SnS_2 by Aβ (42)-specific capture DNA aptamer, which bound to AβOs in the sample. The AβO concentration in the serum of AD mice was distinguished by the developed biosensor. The as-designed SnS_2 -aptasensor provides a design idea for the design of an electrochemical sensing platform based on nanomaterials, and paves the road for AD's early prediction and diagnosis [26].

6.2.4 Three-Dimensional Nanomaterials-Based Electrochemical Biosensors

Paracetamol (PA) is commonly used as an antipyretic analgesic drug, which often coexists with the neurotransmitter dopamine (DA). When they coexist in the body, they interfere with each other and are difficult to detect, so the simultaneous determination of their content is an important difficulty. Yang, H. et al. designed and synthesized H-C/N@TiO_2 by simple alcohol method and hydrothermal method. At the same time, H-TiO_2 or H-C/N were also prepared by the same method. H-C/N@TiO_2 has good charge separation, and stable current response for the detection of PA and DA in human serum and saliva [27] (Scheme 3).

The detection of chiral molecules has always been a challenging problem because their stereochemical configurations and enantiomers can undergo racemization. Upadhyay S. S. et al. developed a new type of enantioselective electrochemical biomimetic sensor. In this sensor, square wave voltammetry was used to identify

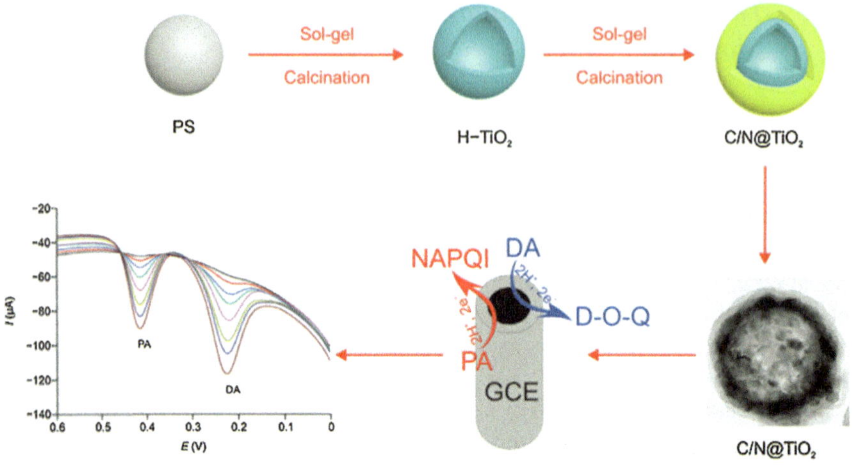

Scheme 3 Schematic diagrams of 3-dimensional nitrogen-doped carbon@TiO_2 double-shelled hollow sphere synthesis and electrochemical sensing [27]. Copyright (2022) Journal of Pharmaceutical Analysis

ethambutol isomers. The sensor used a chiral host-cyclodextrin-based copper metal–organic framework (CD-CuMOF) to chelate the ETB isomer and combines it with a carbon nanofiber (CNF) composite to chemically modify a GCE to construct a sensor. CD-CuMOF has excellent peak potential difference when using SWV to measure ETB isomers in blood, urine, or drug samples, showing obvious differences in racemic mixtures. CD-CuMOF not only has low cost, but also has better stability than enzymes, which provides a new idea for future sensing and catalytic applications [28].

Accurate measurement of drug concentration during anesthesia is a challenging task, and electrochemical biosensor technology provides a feasible solution. By monitoring the drug concentration, the sensor can dynamically adjust the anesthesia infusion to meet individual needs, thereby improving the success rate of the anesthesia process. Traditional therapeutic drug monitoring (TDM) methods usually use blood spot sampling, but this method is not suitable for continuous monitoring in the operating room. To overcome this limitation, Aiassa, S. et al. developed a sensor to measure anesthetics in the patient's vein, directly. And the use of excellent performance of cyclic voltammetry (CV) for drug online detection. In CV, the voltage step acts on the three-electrode electrochemical cell and triggers the redox on the cell surface. The pencil graphite electrode (PGE) is suitable for propofol monitoring to ensure compensation for fouling. The new low-cost and disposable electrode is realized by using a sub-millimeter-sized needle-like PGE pencil lead electrode, which has robustness and stability. The sensor could be used to develop into a complete portable electronic system for anesthesia monitoring [29].

Ciprofloxacin (CIP) is commonly used for the cure of respiratory tract infections and urinary tract infections. Due to the widespread use of CIP, high levels of CIP have been detected in many public places. Therefore, it is necessary to construct an electrochemical sensor with a fast response, strong selectivity, and sensitive detection for the CIP level. Umesh N. M. et al. proposed a new Co-Fe-PBA@CN modified carbon electrode for the detection of antibiotic drug CIP. The electrode has nice reusability and stability, and the electrode and the CIP electrode not only have specific ion adsorption and electrostatic repulsion but also exhibit excellent selectivity. In addition, the effect of the sensor was tested by analyzing real samples of human serum containing ciprofloxacin tablets. It shows that the electrode can be well used for the analysis and detection of ciprofloxacin. It also has a certain reference value for the detection of other drugs, pesticides, neurotransmitters, and electrochemical research [30].

Sometimes, different types of materials can also be combined to make more suitable electrochemical biosensors.

Depression is a mental disease that is difficult to judge by objective methods. Yang, S. N. et al. developed an electrochemical biosensor that takes the key biomarker detection of trace glucocorticoid receptor α (GRα) in the hippocampus and blood cells. The electrochemical signal is significantly amplified by the integration of amino-ion graphene oxide and amino acid coated gold nanoparticles, in which amino acids play important roles as a stabilizer and bridging agent. In addition, it is not only particularly suitable for detecting GRα changes in the blood cells of depressed samples, but also for assessing the level of GRα changes after treatment. Therefore, this unique biosensor provides an integrated solution for depression diagnosis and evaluation of new drugs [31].

6.3 Enzymatic Electrochemical Biosensors

The detection strategies of enzymatic electrochemical biosensors were mainly two categories. One was nanozymes acting as catalytic unit to combine with the biological recognition elements. The other one was that nanozymes acted as electrode modifiers.

The combination of electrochemical biosensors with nucleic acids is becoming more and more popular. By fixing the nucleic acid recognition layer on the electrochemical sensor, the detection of DNA-targeted drugs can be realized. The detection of MTX also utilizes the complementary pairing relationship between MTX and nucleic acid to fix the nucleic acid recognition layer on the electrochemical sensor and successfully detect its content. This method is easy to operate, with high sensitivity and high selectivity, which could detect relevant drug concentrations quickly and accurately [22].

Rapid detection of diseases that are difficult to detect is a development direction of electrochemical biosensors. Detection of blood galactose levels can diagnose several clinical conditions, including liver dysfunction and galactose deficiency. However, it is difficult to detect potential liver damage, since the liver is a silent organ, once damaged, it can only be detected by tools in the late stage. But it is always too late. A novel point-of-care (POC) electrochemical biosensor can rapidly and accurately quantify the galactose content in the blood using fingertip capillary blood for routine assessment of liver dysfunction and other potential diseases. The detection system employs enzyme electrochemical sensing technology to coat galactose dehydrogenase on the test strip to catalyze chemical reactions and sends signals to the instrument. Then, the electrical signal is quantized into digital data. The detection method is very simple to operate, can be performed by professional medical personnel and non-professional users, and is suitable for various clinical environments, as well as families. This new POC diagnostic equipment is not only efficient and accurate, but also very affordable, truly improving the popularity and efficiency of medical care [32].

The novel coronavirus (SARS-CoV-2), which once raging globally, caused lots of loss. In order to curb the spread of such a virus, rapid detection of the virus, early diagnosis, and targeted treatment. At the same time, the selection and development of virus inhibitors during treatment are also crucial. Liang, Q. Z. et al. prepared an electrochemical biosensor using papain-like cysteine protease (Plpro) as a biomarker, which can be used not only for the detection of SARS-CoV-2 but also for the screening of antiviral inhibitors. The biosensor innovatively uses the specific enzyme activity of viral functional proteins. In screening and verification of target protease inhibitors, it was found that Sepantronium Bromide (YM155) had an obviously inhibitory effect on the activity of SARS-CoV-2 PLpro. The specificity has also been verified in whole blood. The electrochemical biosensor proposed by the author has great practical application potential in SARS-CoV-2 detection or inhibitor screening, which can show some enlightenment for disease detection and antiviral drug development [33] (Scheme 4).

Liu et al. developed an electrochemical biosensor with excellent selectivity and sensitivity to target tumor cells. The sensing platform of the sensor is built on antifouling bilayer lipids and fluid DNA tetrahedrons, which can resist non-specific protein adsorption in blood and achieve sensitive selective target binding in the presence of rare target cells. The development of this biological analysis technology will provide a new and feasible method for the detection of circulating tumor cells, and has broad application prospects in biological analysis and clinical diagnosis [34] (Scheme 5).

Chaum et al. developed a biosensor to detect and quantify the concentration of propofol in blood continuously, accurately and stably. The sensor is based on electrochemical methods and has been verified in vitro to be able to detect and quantify propofol at sub-micromolar concentrations. The sensor uses a special coating material that can quickly and stably detect drugs without interference from biological pollution. In addition, the study also used known drug pharmacokinetics, using published algorithms, to demonstrate in a modeled 'patient', demonstrating that the platform can automatically deliver and maintain propofol levels. This will provide a safer and more accurate monitoring method for the use of propofol [35].

Topotecan (TPT) is a major anticancer drug that acts by inhibiting the mechanism of parasitic topoisomerase I. When topoisomerase I is inhibited, DNA strand breaks, leading to cancer cell death. At present, the study of the interaction between cells and DNA is an important research direction, especially for new DNA-targeted drug

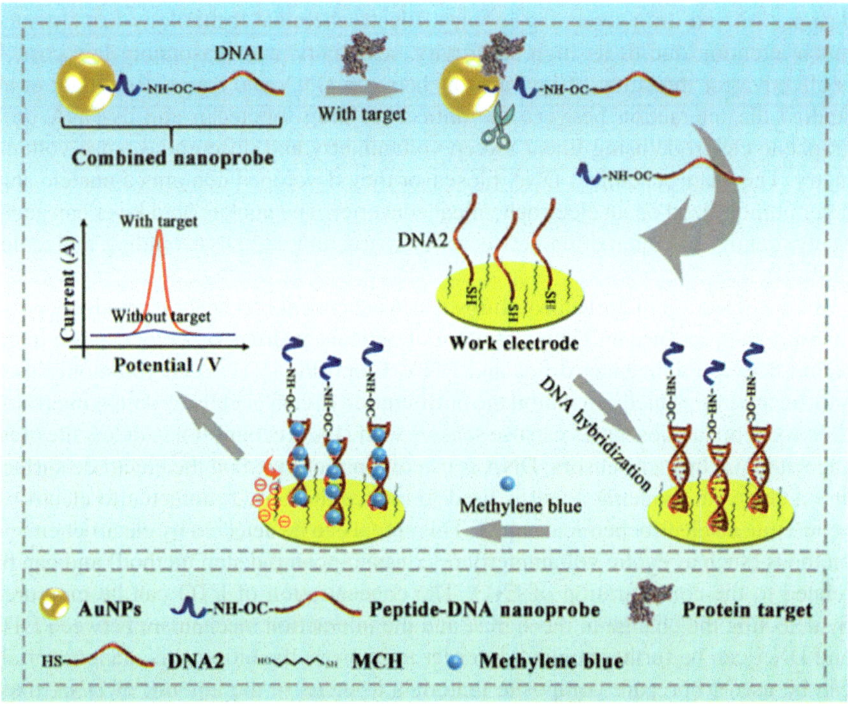

Scheme 4 Schematic diagram of the electrochemical biosensor to detect SARS-CoV-2 [33]. Copyright (2023) Elsevier

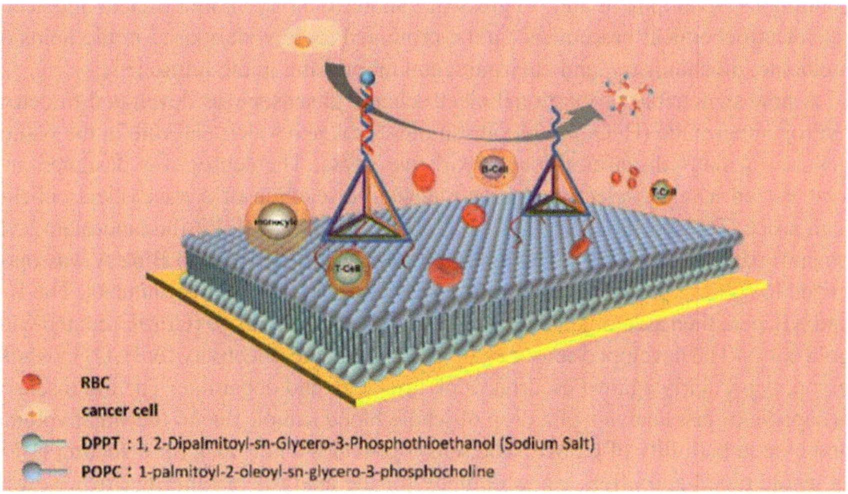

Scheme 5 Schematic illustration of the preparation of a fluid electrochemical adaptor sensor [34]. Copyright (2020) Elsevier

designation and drug screening in vitro. Electrochemical technology has attracted much attention due to its high sensitivity, selectivity, and reasonably low cost. It is effective for the study of interactions between DNA and drugs. Beitollahi et al. studied the interaction between the anticancer drug topotecan and ds-DNA on a graphene electrode using linear sweep voltammetry and differential pulse voltammetry. The electrochemical DNA biosensor they developed contains a nucleic acid layer immobilized on an electrochemical converter. The nucleic acid layer can accurately identify the interaction between DNA structure and DNA binding molecules [36].

In the detection of anticancer drugs, electrochemical biosensors can identify and measure the presence and concentration of anticancer drugs by detecting the interaction between anticancer drugs and DNA. Etoposide (ETO) is a commonly used anticancer drug, which can inhibit the proliferation of cancer cells by damaging DNA. Electrochemical biosensors can use sensors with DNA recognition to detect the presence of ETO. In such sensors, DNA is usually immobilized on the electrode surface and a chemical reducing agent is used to reduce the Eto O molecule to electrons, generating an electrochemical signal. This signal can be detected by electrochemical methods (such as cyclic voltammetry or constant potential step method) and can be related to the concentration of ETO. The concentration of ETO can be measured by detecting the change in the signal, and the interaction mechanism between ETO and DNA can be further studied. In order to improve the biosensors, nanomaterials can be used to prepare composite materials to increase the amount of DNA fixed and improve the stability of the signal. The electrochemical biosensor detects the presence and concentration of anticancer drugs (such as ETO) and can provide an important reference for the research and development of anticancer drugs. With the continuous development and progress of nanotechnology, more efficient and accurate electrochemical biosensors can be promoted to be widely used in the fields of medicine, biochemistry, and environmental monitoring in the future [37].

A new switchable proportional electrochemical sensor was developed to detect hydrogen peroxide (H_2O_2) in blood samples, as it plays a crucial role in the pathogenesis of many diseases through oxidative stress. The sensor was designed and synthesized using a selective one-step acylation reaction of a novel electrochemical probe, 5-(1,2-dithiol-3-yl)-N-(4-(4,4,5,5-tetramethyl-1,3,2-dioxaborolan-2-yl) phenyl) pentanamide (BA). The internal reference molecule Nile Blue A was optimized to achieve precise quantification of H_2O_2 in complex environments. The BA and NB were then assembled onto a carbon fiber microelectrode (CFME) coated with gold cones. The developed sensor exhibited remarkable selectivity for H_2O_2 in whole blood, particularly against ascorbic acid, uric acid, and dopamine. The microsensor is capable of fast analysis of a drop of whole blood sample due to the small volume and biocompatibility of CFME. The sensor provides a new idea for detecting H_2O_2 in whole blood and offers a new strategy for the analysis of other reactive oxygen species (ROS), which can have a profound impact on the early diagnosis, pathological mechanisms, drug discovery, and treatment of ROS-related diseases [38] (Scheme 6).

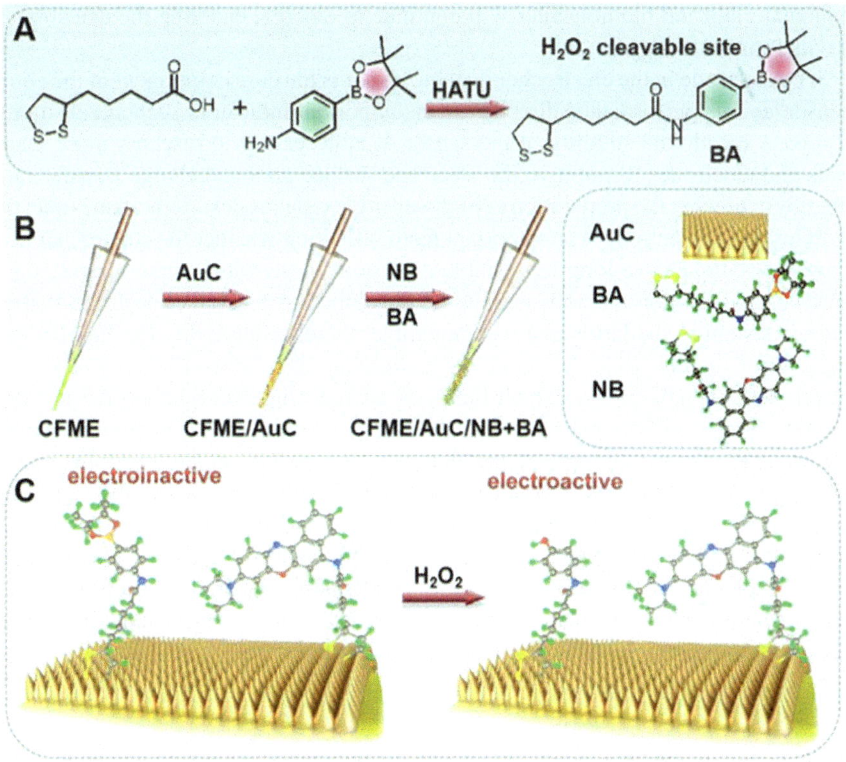

Scheme 6 **A** Synthesis route for specific recognition molecule of H_2O_2. **B** Preparation procedures for the modified electrodes. **C** Working principle for the "turn-on" ratiometric electrochemical monitoring of H_2O_2 [38]. Copyright (2020) Elsevier

6.4 Conclusion and Outlook

This chapter reviews the research progress on electrochemical biosensors for drug detection using whole blood and serum samples, both domestically and abroad, and summarizes the performance of the developed sensors. Most biosensors for drug detection rely on linear changes in a certain indicator in the blood to determine drug content, while others directly detect the reaction of the drug with a substance modified on the electrode for indirect detection of drug content. Currently, electrochemical biosensors are not limited to detecting a single indicator in the blood, but can also detect multiple components in the blood. The diversity and comprehensiveness of the detected substances are important aspects of their development. Combined with the advantages of biomolecular-specific recognition and electrochemical signal sensing, electrochemical sensors are very advantageous for the convenient and sensitive detection of drug biomarkers in whole blood. In addition, new types of sensors based on nanomaterials and biomolecules will continue to emerge, enabling the detection of a wider range of drugs.

The electrode in the electrochemical biosensor is the core component of the entire sensor, and its performance directly affects important indicators such as sensitivity, response speed, and stability of the sensor. It is necessary to research more materials and electrodes to improve the electrode surface area and charge transfer rate, thereby achieving the preparation of high-sensitivity electrodes. At the same time, the stability of the electrode is an important factor affecting whether the electrochemical biosensor can achieve long-term stable detection. In the future, new materials and surface modification methods need to be developed to make the electrode surface more resistant to oxidation and biocompatible, thereby improving the stability and lifespan of the electrode. Moreover, in traditional electrochemical biosensors, the electrode is usually planar, which limits the increase in electrode surface area and the rate of electrochemical reactions. Research on new types of three-dimensional electrode structures can increase the electrode surface area, and improve the rate of electrochemical reactions, and sensitivity. Finally, with the advent of the intelligent era, future electrodes can be integrated with intelligent chips and other technologies to achieve higher-level signal processing and data analysis, thereby improving the intelligence and reliability of the sensor. In conclusion, the application prospects of electrochemical biosensors in the field of drug detection are broad and will become one of the important research directions in the future of drug detection.

References

1. Garzon V et al (2019) Optical biosensors for therapeutic drug monitoring. Biosensors-Basel 9(4):132
2. Vlasceanu GM et al (2018) Versatile graphene biosensors for enhancing human cell therapy. Biosens Bioelectron 117:283–302
3. Cicha I et al (2022) Biosensor-integrated drug delivery systems as new materials for biomedical applications. Biomolecules 12(9):1198
4. Wei SG et al (2021) Recent development of graphene based electrochemical sensor for detecting hematological malignancies-associated biomarkers: a mini-review. Front Chem 9:1–6
5. Pirzada M, Altintas Z (2019) Nanomaterials for healthcare biosensing applications. Sensors 19(23):5311
6. Gupta AK, Dubey RS, Malik JK (2013) Application of modern electroanalytical techniques: recent trend in pharmaceutical and drug analysis. Int J Pharm Sci Res 4(7):2450–2458
7. Sanvicens N et al (2011) Biosensors for pharmaceuticals based on novel technology. TrAC Trends Anal Chem 30(3):541–553
8. Burcu Bahadir E, Kemal Sezginturk M (2015) Applications of electrochemical immunosensors for early clinical diagnostics. Talanta 132:162–174
9. Kour R et al (2020) Review—recent advances in carbon nanomaterials as electrochemical biosensors. J Electrochem Soc 167(3):037555
10. Wang S, Fu SS, Ding H (2012) Determination of 5-fluorouracil using disposable gold nanoparticles modified screen-printed electrode. Sens Lett 10(3–4):974–978
11. Ye LQ et al (2014) Electrochemical determination of cisplatin in serum at graphene oxide/multi-walled carbon nanotubes modified glassy carbon electrode. Int J Electrochem Sci 9(3):1537–1546

12. Saljooqi A, Shamspur T, Mostafavi A (2019) The MWCNT-Ag-PT GCE electrochemical sensor functionalized with dsDNA for mitoxantrone sensing in biological media. IEEE Sens J 19(12):4364–4368
13. Wang X, Dong S, Wei H (2022) Recent advances on nanozyme-based electrochemical biosensors. Electroanalysis 35(1):38–49
14. Song YH et al (2018) Ratiometric electrochemical glucose biosensor based on GOD/AuNPs/Cu-BTC MOFs/macroporous carbon integrated electrode. Sens Actuators B-Chem 257:792–799
15. El-Said WA et al (2019) Electrochemical monitoring of methotrexate anticancer drug in human blood serum by using in situ solvothermal synthesized Fe3O4/ITO electrode. Electroanalysis 31(5):829–837
16. Mofidi Z et al (2018) Ultra-trace determination of imipramine using a Sr(VO3)2 doped phytic acid carbon paste electrode after preconcentration by electromembrane extraction coupled with FFT square wave voltammetry. J Electrochem Soc 165(5):H205–H212
17. Wang SZ et al (2019) An enzymatic electrode integrated with alcohol dehydrogenase and chloranil in liquid-crystalline cubic phases on carbon nanotubes for sensitive amperometric detection of NADH and ethanol. J Electrochem Soc 166(10):G116–G121
18. Abedi R et al (2021) Highly sensitive and label-free electrochemical biosensor based on gold nanostructures for studying the interaction of prostate cancer gene sequence with epirubicin anti-cancer drug. Microchem J 170:106668
19. Song XY et al (2022) MWCNTs-CTAB and HFs-Lac nanocomposite-modified glassy carbon electrode for rutin determination. Biosensors-Basel 12(8):632
20. Zhou HF et al (2018) Novel tungsten phosphide embedded nitrogen-doped carbon nanotubes: a portable and renewable monitoring platform for anticancer drug in whole blood. Biosens Bioelectron 105:226–235
21. Nataraj N et al (2021) Ni-Doped ZrO2 nanoparticles decorated MW-CNT nanocomposite for the highly sensitive electrochemical detection of 5-amino salicylic acid. Analyst 146(2):664–673
22. Chen JC et al (2018) A graphene oxide-DNA electrochemical sensor based on glassy carbon electrode for sensitive determination of methotrexate. Electroanalysis 30(2):288–295
23. DeTure MA, Dickson DW (2019) The neuropathological diagnosis of Alzheimer's disease. Mol Neurodegener 14(1):18
24. De La Torre JC (2010) Alzheimer's disease is incurable but preventable. J Alzheimers Dis 20(3):861–870
25. Sevigny J et al (2016) The antibody aducanumab reduces a beta plaques in Alzheimer's disease. Nature 537(7618):50–56
26. Murti BT et al (2023) Free-standing vertically aligned tin disulfide nanosheets for ultrasensitive aptasensor design toward Alzheimer's diagnosis applications. Chem Eng J 452:139394
27. Yang H et al (2022) Nitrogen-doped Carbon@TiO2 double-shelled hollow spheres as an electrochemical sensor for simultaneous determination of dopamine and paracetamol in human serum and saliva. J Pharm Anal 12(3):436–445
28. Upadhyay SS, Gadhari NS, Srivastava AK (2020) Biomimetic sensor for ethambutol employing beta-cyclodextrin mediated chiral copper metal organic framework and carbon nanofibers modified glassy carbon electrode. Biosens Bioelectron 165:112397
29. Aiassa S et al (2020) Pencil graphite needle-shaped biosensor for anaesthetic monitoring in human serum. In: IEEE sensors conference. Electr network
30. Umesh NM, Jesila JAA, Wang SF (2022) Amperometric detection of antibiotic drug ciprofloxacin using cobalt-iron Prussian blue analogs capped on carbon nitride. Microchim Acta 189(1):31
31. Yang SN et al (2022) Multifunctional amino acids empowering bifunctional biosensing platform for depression study. Biosens Bioelectron 201:113972
32. Yu KM et al (2022) A novel galactose electrochemical biosensor intended for point-of-care measurement of quantitative liver function using galactose single-point test. Anal Bioanal Chem 414(14):4067–4077

33. Liang QZ et al (2023) An electrochemical biosensor for SARS-CoV-2 detection via its papain-like cysteine protease and the protease inhibitor screening. Chem Eng J 452:139646
34. Liu YB et al (2020) Mobile DNA tetrahedron on ultra-low adsorption lipid membrane for directional control of cell sensing. Sens Actuators B-Chem 307:127570
35. Chaum E, Lindner E (2021) A "smart" biosensor-enabled intravascular catheter and platform for dynamic delivery of propofol to "close the loop" for total intravenous anesthesia. Mil Med 186:370–377
36. Beitollahi H et al (2017) A sensitive electrochemical DNA biosensor for anticancer drug topotecan based on graphene carbon paste electrode. J Electrochem Soc 164(12):II812–II817
37. Vajedi FS, Dehghani H (2020) A high-sensitive electrochemical DNA biosensor based on a novel ZnAl/layered double hydroxide modified cobalt ferrite-graphene oxide nanocomposite electrophoretically deposited onto FTO substrate for electroanalytical studies of etoposide. Talanta 208:120444
38. Dong H et al (2020) "Turn-on" ratiometric electrochemical detection of H2O2 in one drop of whole blood sample via a novel microelectrode sensor. Biosens Bioelectron 165:112402

Chapter 7
Electrochemical Biosensors for Amino Acids Detection

Yanyan Wang, Li Dai, Cihui Luo, Yidan Tu, Xugang Wang, and Yu Huang

Abstract Amino acids are important compounds in life and are the basic units that makeup proteins. The detection of amino acids and their metabolites is of significant importance for the diagnosis of diseases and the progress of life activities. Various approaches have been established for the detection of amino acids, among which electrochemical detection methods are highly sensitive, fast simple to operate, and very attractive. This chapter focuses on the applications of electrochemical biosensors for amino acid detection, including qualitative and quantitative analysis of amino acids, as well as the identification of chiral amino acids, with identical molecular structures. The powerful capabilities of electrochemical biosensors for amino acid detection are demonstrated.

Keywords Amino acids · Electrochemical biosensors · Non-derivatization analysis · High sensitivity · Specificity

7.1 Introduction

7.1.1 Amino Acids

Amino acids are the building blocks of life because they have two essential roles: they form the structural units of proteins, which are macromolecules that perform various biological functions, and they participate in metabolic pathways that regulate cellular processes [1, 2]. Each amino acid molecule contains at least one carboxyl and one amino group, and the side chains in the molecule determine their physicochemical properties. Among them, the most common is the α-amino acid, which contains a primary amine and a carboxylic acid attached to the same carbon atom, facilitating the formation of an amide or peptide bond (Fig. 7.1) [3].

Y. Wang · L. Dai · C. Luo · Y. Tu · X. Wang · Y. Huang (✉)
Faculty of Materials Science and Chemistry, China University of Geosciences, Wuhan 430074, China
e-mail: yuhuang@cug.edu.cn

F. Xia et al. (eds.), *Electrochemical Biosensors for Whole Blood Analysis*,
https://doi.org/10.1007/978-981-99-5644-9_7

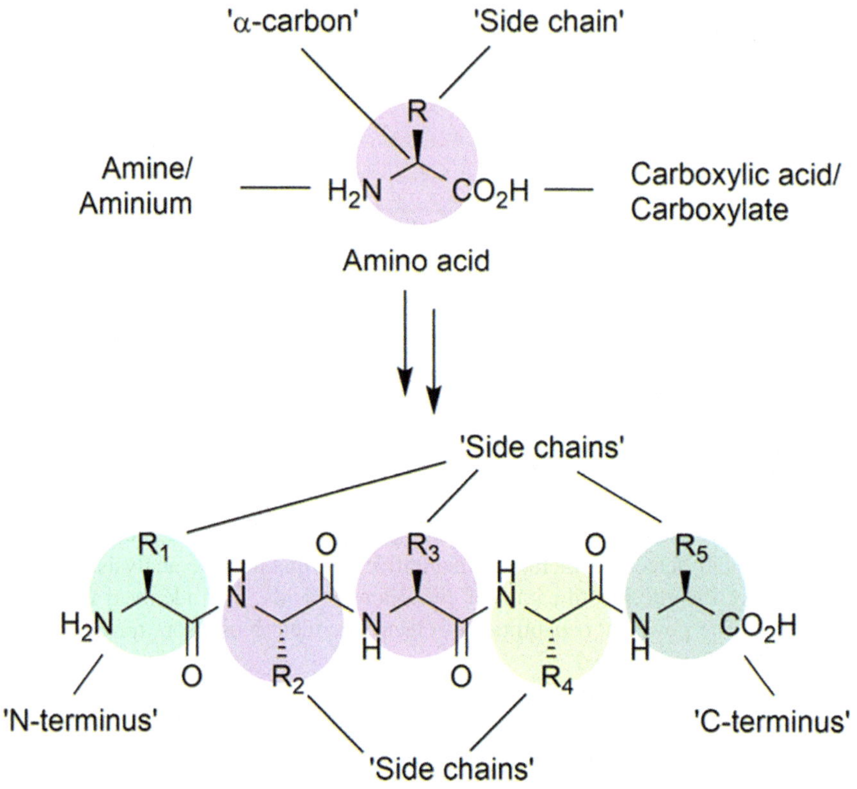

Fig. 7.1 The naming of amino acids. Each amino acid molecule contains at least one carboxyl and one amino group, and the side chains in the molecule determine their physicochemical properties. (Reproduced with permission from Ref. [3]. Copyright 2021 Royal Society of Chemistry)

Amino acids can be classified into two categories, namely essential and non-essential amino acids. Essential amino acids are indispensable as they cannot be manufactured by the body and must be provided through dietary sources or amino acid supplements. Conversely, non-essential amino acids can be synthesized naturally by the body. There are 20 common amino acids, which are Glycine (Gly), Alanine (Ala), Valine (Val), Leucine (Leu), Isoleucine (Ile), Phenylalanine (Phe), Tryptophan (Trp), Tyrosine (Tyr), Aspartate (Asp), Histidine (His), Asparagine (Asp), Glutamate (Glu), Lysine (Lys), Glutamine (Gln), Methionine (Met), Arginine (Arg), Serine (Ser), Threonine (Thr), Cysteine (Cys), Proline (Pro) [4, 5].

7.1.2 Significance of Amino Acid Detection

Amino acids are involved in various diseases such as metabolism, tumor, immunity, cardiovascular, nervous system, kidney disease diabetes, and various health aspects of human growth and development, nutritional health, musculoskeletal growth, and hormone secretion. In addition to revealing the state of amino acid metabolism in the human body, warning of physiological disorders, and early warning of various diseases, amino acid testing and analysis can also indicate the state of nutritional health of the human body and provide guidance and reference for health maintenance, health care, and treatment [6–8].

Recently, it has been shown that amino acids can act as important molecules with a significant impact on signaling pathways and metabolic regulation. Amino acid levels are combined and synchronized at different pathways for the organism, and these ways work together to ensure overall metabolic homeostasis (Fig. 7.2) [9]. In a study by Mohamad et al., patients with chronic renal failure showed abnormal levels of several amino acids. The mean serum L-arginine level was lower in 61.9% of the patients compared to the control group. Patients who had never experienced thromboembolism had significantly lower glutamate levels and significantly higher phenylalanine levels than patients who had a history of thromboembolism [10]. Thalacker-Mercer et al. examined the relationship between glucose disposal rate (GDR) and amino acids by direct measurement. The relationship between GDR and serum amino acids was found by measuring it in all subjects including insulin insensitivity, insulin resistance, and type 2 diabetes mellitus (T2DM) patients. Among them, glycine (Gly) had the strongest positive correlation with GDR, followed by leucine/isoleucine (Leu/Ile) with a negative correlation. The correlation between Leu/Ile and GDR was stronger in male patients with T2DM than in female patients [11].

7.1.3 Common Methods for Amino Acid Detection

Amino acid analysis is widely used in biochemistry, proteomics, biomedicine, food science, and other fields. The detection of amino acids has become an important tool for health diagnosis and disease screening. Currently, amino acid determination methods can be divided into derivatization and non-derivatization methods.

7.1.3.1 Derivatization Methods of Amino Acid Detection

There are many derivatization methods available for the detection of amino acids in complex mixtures, but only for a few amino acids with strong chromophores, fluorophores, or electroactive groups. Currently, there are three main methods

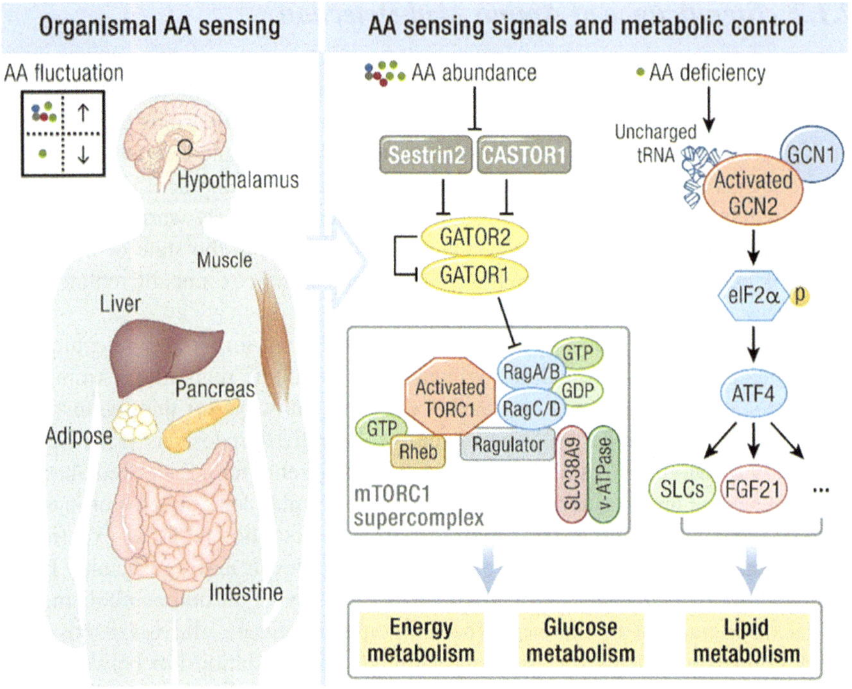

Fig. 7.2 Amino acid sensing signal integration coordinates organismal levels to maintain overall metabolic homeostasis. (Reproduced with permission from Ref. [9]. Copyright 2020 Oxford University Press)

for performing derivatization processes: post-column derivatization, pre-column derivatization, and capillary derivatization methods [12].

The post-column derivatization methods are used only to increase the detection capacity of amino acids by absorption fluorescence spectroscopy, without affecting the previous separation. Positively charged amino acids can be separated on a cation exchange resin with a progressively higher pH of the mobile phase. When the isoelectric point (pI) of the amino acid under study is reached, the analyte is eluted and separated. Post-column derivatization is a step that happens after the sample leaves the column and before it reaches the detector. It saves time by avoiding pre-column derivatization and does not affect the chromatographic performance [13].

Gas chromatography analytical methods often use pre-column derivatization, which can be done with one or two substances. In two-step derivatization, the amino acids are first esterified with alcohols and hydrochloric acid. Then, the non-volatile intermediates are evaporated and acylated to produce different esters of N-acyl amino acids. The derivatization can also be done with one substance that reacts with both the amino and carboxylic acid groups. Pre-column derivatization is considered to be the most flexible derivatization strategy, because it can be combined with the need for many extraction or washing steps [14].

Capillary derivatization is particularly suitable for very small sample volumes. In most cases, capillary derivatization is achieved by the difference in mobility between analyte and reagent [15].

7.1.3.2 Non-derivatization Methods of Amino Acid Detection

Direct methods for the analysis of amino acids without derivatization have always been sought by researchers. Non-derivatization analysis is simple and flexible, and it also avoids problems such as instability of derivatization and interference of reagents. The main non-derivatization methods are UV, fluorescence, conductivity, evaporative light scattering (ELS), mass spectrometry (MS), electrochemistry, and voltammetry.

For amino acids containing aromatic substituents, there is UV absorption at 254 nm. In contrast, the method is less sensitive and has an unstable baseline because most of its carboxyl groups can only be detected at low UV wavelengths (190–210 nm). This method can detect UV-absorbing amino acids, histidine dipeptides, and creatinine in urine, but not all amino acids. It is useful for the chemical diagnosis of inherited metabolic diseases. Fluorescence detection, similar to UV detection, can only be used for amino acids with fluorescence properties. The detection limits range from 12 to 58 ng/mL, depending on the amino acid and how long it stays in the column. Conventional contact conductivity detectors are versatile for the detection of charged ions and can be used to detect ions that are not sensitive to UV. However, contact conductivity detectors are prone to electrode contamination, difficult to clean, and difficult to detect glutamate. The non-contact conductivity detector applies the AC voltage signal to the detection cell through capacitive coupling, which is more conducive to the fabrication and use of the device.

ELS detection is widely considered to be the widespread detection technique for liquid chromatography. It is more widely used for the determination of non-volatile and non-absorbing compounds above 210 nm because their response is independent of their optical properties. The ELS detector measures the light scattered by the analyte particles when the solvent evaporates as the solution passes through the beam. The detector can detect all solutes that are less volatile than the mobile phase. This makes it suitable for the analysis of amino acids without derivatization. In recent years, MS detection techniques have provided new conditions for the detection of natural amino acids. The current mass spectra utilized for the analysis of amino acids in complex biological matrices are coupled with various detection tools, and methods for the direct analysis of amino acids include ion-pair liquid chromatography-tandem mass spectrometry (IP-LC-MS), hydrophilic interaction chromatography-mass spectrometry (HILIC-MS) and capillary electrophoresis-mass spectrometry (CE-MS). It can be seen that although MS has the advantage of high accuracy for the detection of amino acids, its practical application requires the use of multiple methods in tandem.

Electrochemical detection has the characteristics of straightforward instrumentation, inherent sensitivity, and very high selectivity for high-performance liquid chromatography. However, the constant potential current method on traditional carbon-based electrodes can only distinguish a limited number of amino acids. The sensing selectivity of electrochemical high-potential detection is not very high. To this end, EC baseline drift was minimized by using three tandem detectors (dual UV detector, fluorescent detector, and several electrochemical detectors equipped with dual electrodes) and by using microfiber electrodes.

Simple electroanalytical techniques such as voltammetry are mainly used for stationary solutions. Three amino acids, namely tryptophan, tyrosine, and cysteine (and homocysteine), which are readily oxidized on most electrode materials, can be used for voltammetric analysis. Carbon-based materials are widely used for electrodes, such as glassy carbon, graphite, carbon nanotubes, and boron-doped diamond. Copper and platinum electrodes are also common. These electrodes can be used directly or revised with numerous forms of redox media or adsorbents to improve their catalytic activity. Modified electrodes can also show selectivity in the assay, while detection on unmodified electrodes (e.g., carbon nanotubes) is achieved by their specific properties or by using multivariate analysis methods.

7.2 Application of Electrochemical Biosensors for Amino Acid Detection

Amino acids are essential compounds in human life. Any abnormal fluctuations in the physiological level of amino acids are usually manifested in familiar metabolic disarrays, severe neurological diseases, and cardiovascular diseases. Therefore, the design and determination of amino acids is significant for effective clinical diagnosis. Currently, several amino acid assays have been developed, among which electrochemical assays are highly sensitive, fast, and simple to operate, and have good detection of both essential and non-essential amino acids for the human.

7.2.1 Electrochemical Biosensors for Essential Amino Acid Detection

It is generally believed that there are eight essential amino acids in the human body, including Met, Val, Lys, Ile, Phe, Leu, Trp, Thr. In recent years researchers have reported some work on the use of electrochemical sensors for the detection of essential amino acids.

Methionine is an essential amino acid, which is the basis of tissue functionality such as cell metabolism and immune cell production. The lack of methionine can result in diseases AIDS, Parkinson's disease, and other diseases, so the clinical and physiological research on methionine is of great significance. Hun et al. developed an electrochemical biosensor for the discovery of Met using MoS_2-modified electrode. The means are simple, rapid, selective, sensitive, and low cost. The reliability and application potential of this method for the detection of methionine were evaluated, and the concentration of methionine in human serum samples and other practical samples was determined. The results show that the recovery rate is 95.0–100.4% and the relative standard deviation is 3.7–4.4%. Experiments show that the biosensor can effectively detect and quantify methionine in real samples [16].

Tuma et al. determined plasma levels of Val, Ile, Leu, ala-9, and Glu by capillary electrophoresis and contactless conductivity. A novel covalently covered capillary on account of polyacrylamide and (3-acrylamidopropyl) trimethylammonium chloride was used. Baseline separation of all five amino acids from other plasma mixtures was attained by separation of amino acids in counter-current condition, where analyte migration under counter-current electroosmotic flow was achieved. The results showed detection limits of treated and untreated samples were 0.13–0.14 and 0.50–0.58 μmol/L, respectively. This means can be used to test venoarterial amino acid variances in hyper-insulinemia cachectic patients [17].

Chauhan et al. immobilized lysine oxidase on core–shell magnetic nanoparticles (core–shell MNPs) / multi-walled carbon nanotubes (MWCNT) films on gold electrodes by carbodiimide bonds to prepare an amperometric lysine biosensor. The linear detection range of the electrochemical biosensor is 0.05–700 μM, and LOD is 0.05 μM. This showed that there was a great connection between the lysine level determined by this method and the serum lysine level determined by standard high-performance liquid chromatography ($r = 0.98$), and the other substances in the serum were almost undisturbed. This biosensor will be used to detect lysine in the food and pharmaceutical industries [18].

Singh et al. developed a new gene-encoded nanosensor that uses the periplasmic binding protein of E. coli sandwiched between two fluorescent proteins, ECFP and Venus, to measure isoleucine levels in living cells. The biosensor name possesses pH-stable and isoleucine-specific binding properties. The biosensor successfully monitored isoleucine in bacteria and yeast cells in real-time, thus establishing its biocompatibility for the regulation of isoleucine in living cells. As well as its applicability in understanding the potential function of isoleucine receptors in metabolic monitoring, it can also be employed in the metabolic production of bacteria to improve the manufacture of isoleucine in the animal fodder industry [19].

With the progress of flexible electronics and micro/nanofabrication technology, wearable devices with the preponderance of great sensitivity, great integration, and lightweight have attracted wide attention [20]. Due to good fit and high sensitivity, the wearable devices can conduct all-round, multi-angle, and multi-level collection for special patients. By analyzing the signals collected by the wearable devices, the patient's indicators can be monitored in real time. Patients with Phenylketonuria (PKU) lack the capability to metabolize L-phenylalanine (PHE), which accumulates in large quantities in the brain and leads to varying degrees of psychiatric illness. A wearable screen-printing electrode has been developed by Wael's team to quantitatively detect PHE in saliva and serum samples [21]. The electrode surface was modified with Na 1,2-naphthoquinone-4-sulphonate (NQS) and Nafion layer. PHE reacted with NQS to produce an electroactive product, the electroactive product was oxidized on the superficial of the carbon electrode to produce specific electrochemical signals, and the electronegative NQS achieved satisfactory results against charged interferents. The PHE concentration was measured by the characteristic electrochemical curve of the electrode, and the device showed a range of 20–1000 μM in the buffer, with a LOD of 3.0 μM. The screen-printed electrode can also directly detect PHE in undiluted saliva and undiluted serum without specimen preprocessing (Fig. 7.3a). With the popularity of smart devices, the use of smartphone-assisted field detection has aroused wide attention. As Fig. 7.3b schemed, Arduini et al. developed an electrochemical device that can be used to detect L-Tyrosine (Tyr) in serum [22]. The device is composed of screen-printed electrodes and smartphone potentiostat, which has the characteristics of low price, easy to operate, and convenient to carry. The screen-printed electrode is based on graphene and coated with carbon black (CB). CB can improve the electron transfer capability of redox probe $Fe(CN)_6]^{4-}/^{3-}$ couple to improve the sensitivity of detection. After connecting the portable potentiostat EmStat3 Blue to a smartphone, then potentiostat is used to detect electrodes loaded with test samples, and the concentration of Try can be judged by the size of the electric signal. The electrode had a detection range of 30–500 μM, with a LOD of 4.4 μM in Try standard solution, and was equally reliable in real serum, with results comparable to those obtained by conventional HPLC.

Liu et al. developed a new electrochemiluminescence (ECL) sensor for the first time, which immobilized $Ru(bpy)_3{}^{2+}$ using dendritic palladium nanoparticles as support. The electrochemical and electro-chemiluminescent activities of immobilized $Ru(bpy)_3{}^{2+}$ were explored. Leucine level detection was achieved by the revised electrode exhibiting an enhanced electrocatalytic reply to the oxidation of derived leucine generating sensitized ECL signals. In addition, the present ECL sensor shows long-term stability and specificity only for leucine [23].

7.2.2 Electrochemical Biosensors for Non-essential Amino Acid Detection

Non-essential amino acids do not mean that the body does not need these amino acids, but that the body can obtain them through its own synthesis or conversion from other amino acids, not necessarily from food. The intake of a few non-essential amino acids can also impact the requirement of essential amino acids.

Glutamate is the chief excitative neurotransmitter in the central nervous system of vertebrates, excessive Glu can result in excitotoxicity and neurodevelopmental disorders. Therefore, it's significant to accurately detect and monitor the level of glutamate in biofluids. At present, electrochemical biosensors can provide immediate, low-cost, and constant monitoring of Glu in biofluids [24]. Windmiller et al. designed a platinum/platinum bimetallic oxide microneedle array biosensor, which can be used for electropolymerization interception of Glu oxidase in PoPD film [25]. The quantification of Glu in undiluted human serum specimens was evaluated. The experimental results showed high sensitivity ($S_x = 8.077$ μM) and low deviation (RSD = 6.53%). And high-fidelity Glu measurement with LOD as low as 10 μM

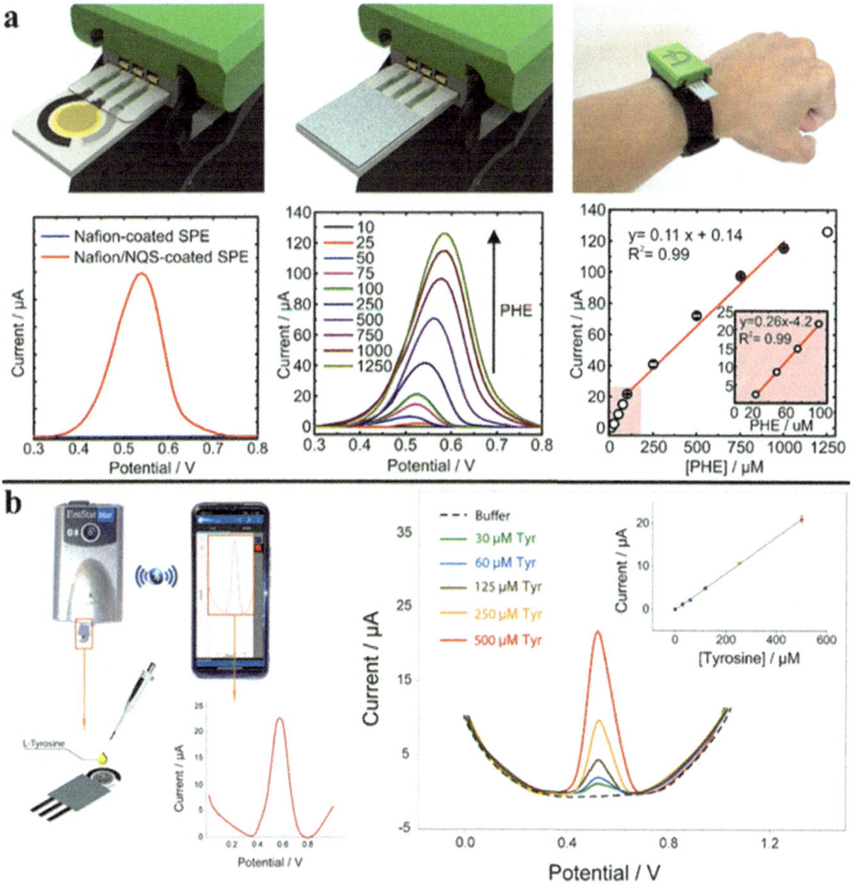

Fig. 7.3 The wearable and device portable for detecting amino acids. **a** The schematic graph of the paper-based sampling with the impregnated paper on the wearable device, and the response curve of an electrical signal and the concentration of PHE. **b** Smartphone-assisted electrochemical device, and the detection of various Try concentrations based on this device. (Reproduced with permission from Ref. [21, 22]. Copyright 2022 Elsevier)

was obtained. The existence of PoPD film achieves high sensitivity, stability, and anti-interference. Ma et al. established a non-enzymatic electrochemical biosensor for the determination of Glu using chitosan-based foam carbon as an electrode [26]. The sensitivity is 1.9×10^4 μA/mM·cm^2, the linear detection range is 0.001–1000 μM, and LOD is 0.001 μM., 200 μM Glu solution (150–300 μM Glu in whole blood) was used to simulate the environment of Glu in blood. The meddling experiments were carried out when different interfering substances were added at the same time with a high concentration of 50 μM (DA, AA, UA, etc.) (Fig. 7.4). The results show that the sensor has ideal anti-interference ability, long-lasting stability and even

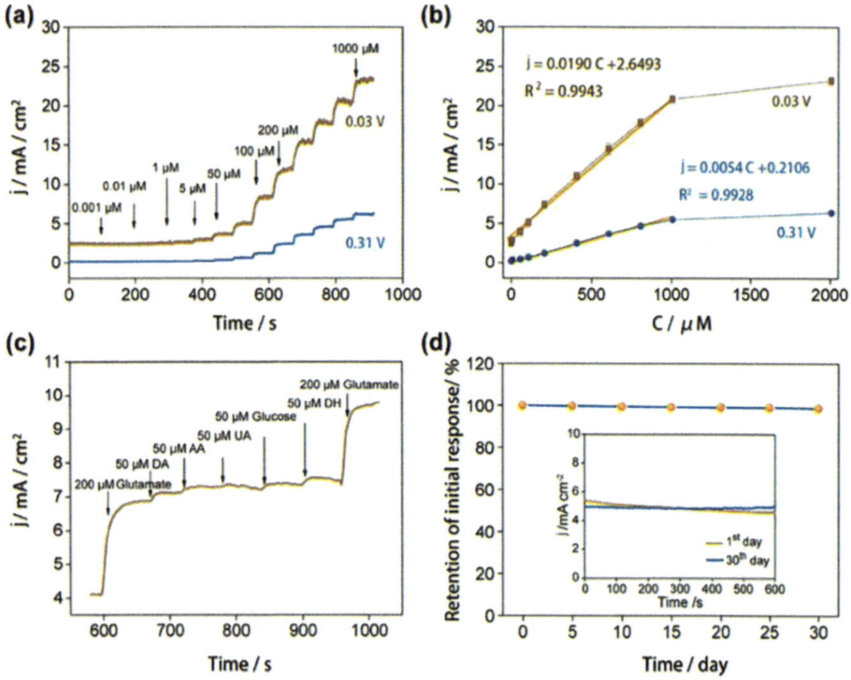

Fig. 7.4 Anti-interference and selective detection of glutamate in blood. (Reproduced with permission from Ref. [26]. Copyright 2022 Multidisciplinary Digital Publishing Institute)

meet the detection requirements of plasma, serum, whole blood, and other clinical environments.

Asparagine is a sign of brain impairment during severe stroke, and its anomalous concentration in blood is usually related to Parkinson's disease. So asparagine monitoring in biofluids is a brook no delay problem. Hermans et al. demonstrated a strategy of combining the biological nanopore cytolysin A (Cly A) with portable electronic devices, and developed a new electrochemical sensor that can quantify metabolites in real time and accurately from nanoliters of blood [27]. The method is easy to use, no sample preparation, and can continuously monitor the concentration of asparagine. In the experience, SBD1 was selected as the substrate (only asparagine was recognized but not glucose). When asparagine was added to the reverse side of the nanopores, it led to an extra current level corresponding to the closed formation of SBD1, and the concentration of asparagine in serum can be calculated. Compared with high-performance liquid chromatography combined with fluorescent detection, it was found that the two results were similar (Fig. 7.5). This confirms the accuracy of the technology.

Glutamine is a significant energy in the human. It supplies 60% of the amino acids in muscles, accounting for 20% of the body's total amino acid cycle. The level of glutamine in body fluid is closely associated with the state of many diseases, so the

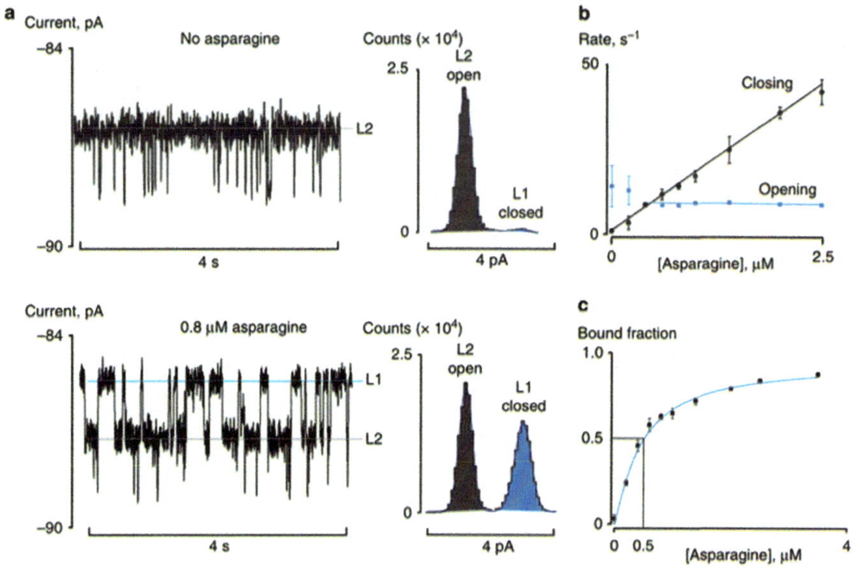

Fig. 7.5 Binding of asparagine to substrate SBD1 (recognition only of asparagine). (Reproduced with permission from Ref. [27]. Copyright 2018 Springer Nature)

monitoring of glutamine is very necessary. Karaku et al. prepared a novel glutamine electrochemical biosensor using self-made zinc oxide nano-rods and chitosan as carriers [28]. In the experiment, the sensor was immersed in a phosphate buffer solution containing a serum sample, adding glutamine solution, and the Lineweaver–Burk curve of potential value and an additional amount of glutamine solution was obtained. The linear detection range of the biosensor is 1×10^{-5}–1×10^{-7} M. It has the advantages of long life, good stability, and can be easily applied to the detection of glutamine.

As Fig. 7.6a described, an extremely selective and reproducible electroanalytical approach for Histidine (His) was established by Wang and colleagues [29]. They used Benzenetricarboxylic acid (BTC) as a ligand precursor to fabricate copper metalorganic frameworks (Cu MOFs), and the fabricated Cu - BTC MOFs were modified on the electrode to detect His. In the electrochemical test, Cu0/CuCl will undergo a redox reaction. After the addition of His, CuCl and His will bind specifically to form Cu-His complex, solid CuCl will have a sharp and extremely stable oxidation peak at −0.020 V. As a result, the concentration of his can be known by testing CuCl oxidation current. This strategy can have a wide detection range in His buffer solution and blood. The LOD of His buffer solution was about 0.025 mM, and the LOD of blood was about 0.125 mM. Li et al. combined Polydimethylsiloxane (PDMS) and Zeolite imidazole framework (ZIF) and used them to modify glassy carbon electrode to prepare an anti-interference self-cleaning electrode [30]. When combined with a simple ratio measurement method, Adrenaline (Adr), Serotonin (5-HT), and Tryptophan (Trp) possessed unique redox potential at the electrode, thus the device could

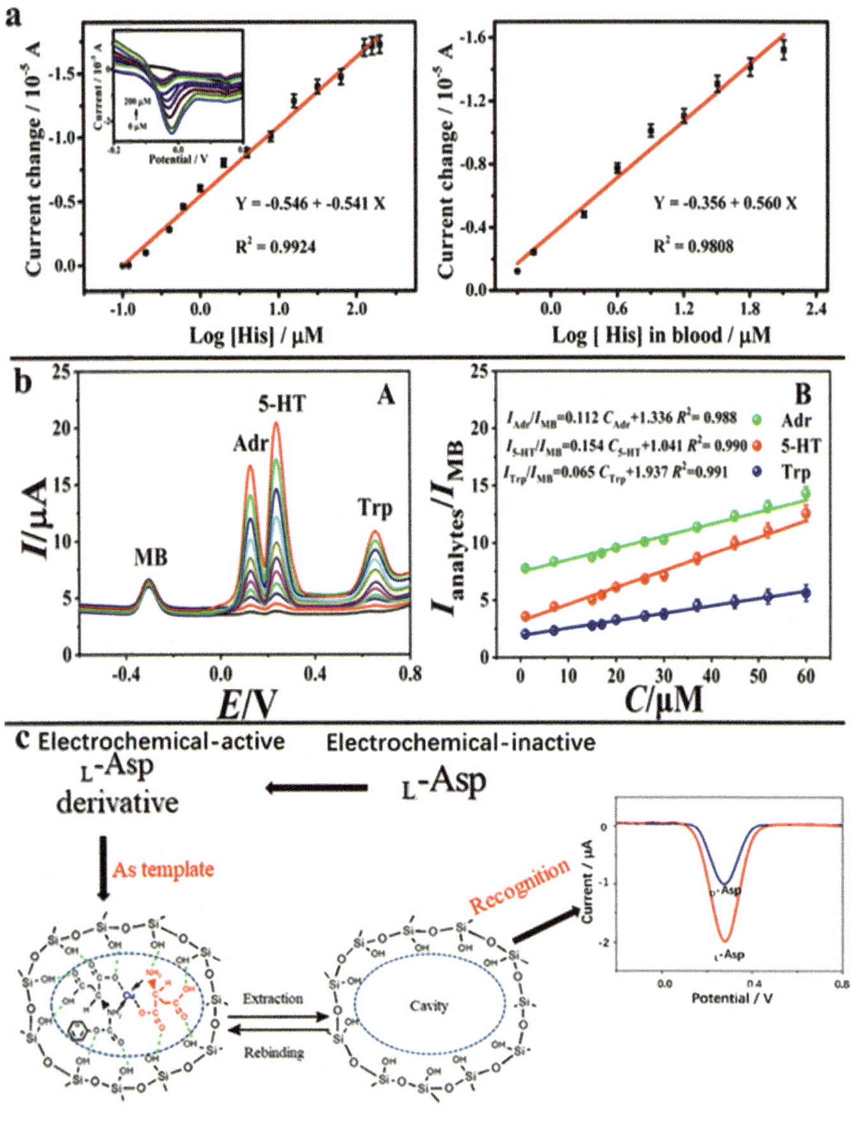

Fig. 7.6 The electronic sensors for detecting amino acids. **a** Cu-BTC MOFs modified electrode in the buffer and blood generated electrical response signal and His concentration calibration curve. **b** The DPV voltammograms for concurrent detection of Adr, 5-HT, and Trp based on fabricated electrodes. **c** The preparation and recognition process of MIS template for L-Asp identification. (Reproduced with permission from Ref. [29]. Copyright 2019 Royal Society of Chemistry, Reproduced with permission from Ref. [30]. Copyright 2019 American Chemical Society, Reproduced with permission from Ref. [32]. Copyright 2019 Elsevier)

detect three substances simultaneously. As shown in Fig. 7.6b, the corresponding detection ranges of Adr, 5-HT, and Trp were 1.0–60 μM, and LODs were 0.13 μM, 0.03 μM, and 0.5 μM, respectively. The strategy was also reliable in rat blood samples. Recognition of chiral amino acids is of great significance to basic research, many researches have focused on chiral recognition of amino acids in recent years [31]. Chen et al. fabricated a molecularly imprinted sol–gel (MIS) membrane with an imprinting cavity by compounding sol–gel technology with molecular imprinting technique, and an electrochemical chiral sensor for aspartic acid (Asp) was obtained by modifying the surface of a glassy carbon electrode (GCE) with MIS membrane [32]. Due to the reduction of copper ions on the electrode, the electrode can produce specific electrical signals for specific pairs of molecules. The electrochemical test results obtained by square-wave stripping voltammetry (SWSV) showed that the sensor showed satisfactory adsorption capacity for L-Asp derivatives, and the recognition efficiency was 2.1. At the same time, the MIS sensor realized the quantitative detection of L-Asp, with a LOD of 1.77 mM (Fig. 7.6c).

Wang et al. introduce a novel electrochemical sensor that can measure the amount of glycine in various biological fluids. The sensor is based on a new quinone protein that specifically catalyzes the oxidation of glycine. The catalytic process involves the redox conversion of Prussian blue when hydrogen peroxide is produced by an enzyme reaction. The study was carried out by efficiently encapsulating the quinone protein in a chitosan matrix and adding an outer Nafion layer of organisms behind it to achieve optimal tailoring of the sensor and thereby suppressing matrix interference. In addition, the rapid response (<7 s), satisfactory reversibility, reproducibility, and stability (<6% variation) and wide linear response range (25–500 μM) of this biosensor enable the monitoring of healthy (and even unhealthy) physiological levels of glycine in blood, serum, urine, and sweat. In six real sample assays, validated by commercially available fluorescent kits, the variation was less than 9% for all samples. The unique analytical performance of this new glycine biosensor will be used in a wide range of clinical applications [33].

A microneedle (MN)-based biosensor was used to detect intradermal glycine (Gly) in interstitial fluid (ISF). The sensor detects Gly levels through the formation of hydrogen peroxide in the reaction between an externally tailored MN tip and a quinone protein-based Gly oxidase. This MN biosensor has a rapid responsive time (<7 s); good reversibility, reproducibility, and stability; and extensive dynamic ranges (25–600 μM) of analytical performance. Moreover, the MN biosensor is highly selective for Gly, and the response is unaffected by temperature, pH, or skin insertion. Authenticated intradermal examinations of Gly were gained at the in vitro (with rat skin slices), in vitro (in vivo tests in euthanized rats) and in vivo (in vitro measurements in anesthetized rats) levels, enabling a dependable, immediate strategy for intradermal Gly measurement in ISF through a slightly invasive method (Fig. 7.7) [34].

A simple and precise magneto electrochemical sensor capable of ultrasensitive distinction of chiral D-amino acids (D-AAs) has been developed. Controllably prepared Fe_3O_4@Au@Ag@Cu_xO nanoparticles (NP) ($x = 1, 2$) were used as electrochemical beacons. The sensor could detect D-alanine (D-Ala) enantiomers in the

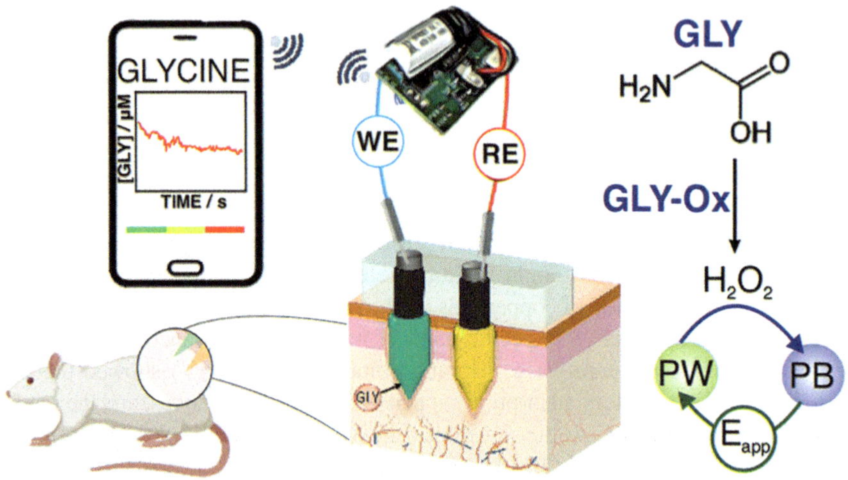

Fig. 7.7 Schematic of intradermal glycine measurement with a wearable microneedle biosensor. (Reproduced with permission from Ref. [34]. Copyright 2022 American Chemical Society)

range of 100 pM–10 μM by using plasma metal NPs that enhanced the electron transfer efficiency. Besides, the limit of detection (LOD) is 52 pM (Fig. 7.8) [35].

Wen et al. constructed a photoelectrochemical biosensor based on a fiber optic probe that enables distant measurement of cysteine. In the miniaturized device, the analyte reacts with the fiber optic probe that serves as a working electrode. The probe also guides the light accurately to the target location. The diffusion-controlled electrochemical reaction of cysteine at the working electrode surface enables its quantitative detection. The fiber-optic-based cysteine biosensor showed a linear response in the detection range of 0.01–1 μM with a regression coefficient (R^2) of 0.9943 at a

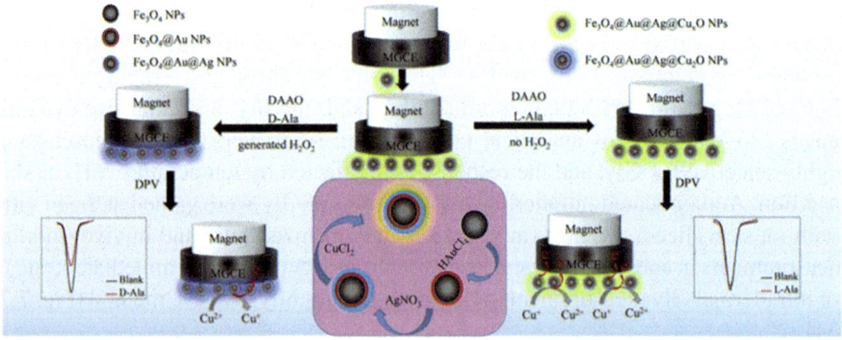

Fig. 7.8 Schematic diagram of Fe_3O_4@Au@Ag@Cu_xO NPs magneto electrochemical chiral sensor designed to detect D-Ala. (Reproduced with permission from Ref. [35]. Copyright 2020 Elsevier)

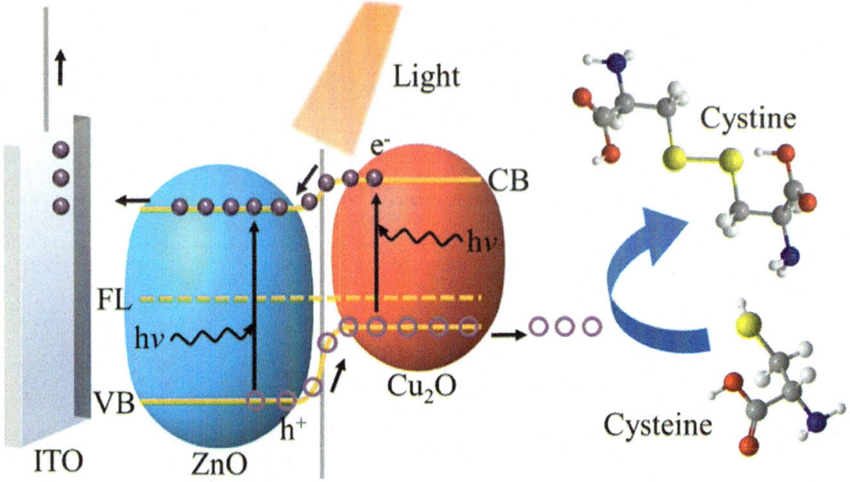

Fig. 7.9 Schematic diagram of the probable mechanism of the developed PEC biosensor. (Reproduced with permission from Ref. [36]. Copyright 2023 Elsevier)

bias voltage of 0 V. The measurement of cysteine in spiked artificial urine was also achieved (Fig. 7.9) [36]. The examination of different amino acids is summarized in Table 7.1.

7.3 Conclusions

Amino acids, known as the cornerstone of life, are of great significance in their detection. The electrochemical biosensor has good selectivity, high sensitivity, and fast and simple operation, which is attractive in amino acid detection. In this chapter, we describe the information about amino acids and the application of electrochemical biosensors for amino acid detection. It has been shown that electrochemical biosensors are effective for both essential and non-essential amino acids, enabling qualitative and quantitative analysis of amino acids. Moreover, accurate amino acid detection can still be achieved in complex matrices such as blood and plasma. In addition, it has been shown that real-time and remote detection can be achieved using electrochemical biosensors, which provides a basis for the wide application of electrochemical biosensors. However, most of the current research is still limited to the basic laboratory research stage, and there is still a certain distance from the application to clinical testing. In the process of detecting complex biomarkers, their sensitivity and stability are not up to the expected effect, and it is not yet adaptable to various complex biological environments. Therefore, it is still a challenge to further reduce the detection limit of amino acid detection, improve the sensitivity of sensors,

Table 7.1 The summary of the different amino acids detection

Sample	Classification	Detection strategy	Linearity range (μM)	Limit of detection {μM)	Literature resources
Methionine	Essential amino acid	Electrochemistry	100-1	0.00003	Ref. [16]
Lysine	Essential amino acid	Electrochemistry	0.05-700	0.05	Ref. [18]
Isoleucine	Essential amino acid	Electrochemistry	0,096-1493	/	Ref. [19]
Phenylalanine	Essential amino acid	Electrochemistry	20-1000	3	Ref. [21]
Tyrosine	Essential amino acid	Electrochemistry	30-500	4.4	Ref. [22]
Leucine	Essential amino acid	Electrochemistry	3.0-182	1	Ref. [23]
Glutamate	Non-essential amino acid	Electrochemistry	0.001-1000	0.001	Ref. [26]
Glutaminc	Non-cssential amino acid	Electrochemistry	0.1-10	/	Ref. [28]
Hislidine	Non-essential amino acid	Electrochcmistry	0.5-128	125	Ref. [29]
Tryptophan	Non-essential amino acid	Electrochemistry	1-60	0.5	Ref. [30]
Asparagine	Non-essential amino acid	Electrochemistry	10-1000	1770	Ref. [32]
Glycine	Non-essential amino acid	Electrochemistry	25-600	7.9	Ref. [34]
Alanine	Non-essential amino acid	Electrochemistry	0.0001-10	0.000052	Ref. [35]
Cysteine	Non-essential amino acid	Electrochemistry	0.01-1	0.00317	Ref. [36]

and achieve stable detection in complex biological environments. With the continuous improvement of related technologies and the unremitting efforts of scientific and technological workers, it is believed that these problems will be gradually solved in the near future, and electrochemical biosensors will also play a crucial part in the practical application of biomedicine, diagnosis and so on.

References

1. Paola C, Lucia T, Andrea P et al (2011) Drug discovery targeting amino acid racemases. Chem Rev 111:6919–6946
2. Xue YP, Cao CH, Zheng YG (2018) Enzymatic asymmetric synthesis of chiral amino acids. Chem Soc Rev 47:1516–1561
3. Thomas AK, Jiyan MK, Stephen JW et al (2021) Photocatalytic methods for amino acid modification. Chem Soc Rev 50:39–57
4. Yoko T, Philip L, Marina K (2019) Amino acid metabolism in hematologic malignancies and the era of targeted therapy. Blood 134:1014–1023
5. Hou YQ, Wu GY (2018) Nutritionally essential amino acids. Adv Nutr 9:849–851
6. Brandon JG, Jamie EB, Esther WL et al (2021) Extracellular serine and glycine are required for mouse and human skeletal muscle stem and progenitor cell function. Mol Metab 43:101106–101121
7. Handzlik MK, Gengatharan JM, Frizzi KE et al (2023) Insulin-regulated serine and lipid metabolism drive peripheral neuropathy. Nature 614:118–124
8. Restrepo-Pérez L, Joo C, Dekker C (2018) Paving the way to single-molecule protein sequencing. Nat Nanotech 13:786–796
9. Hu XM, Guo F (2021) Amino acid sensing in metabolic homeostasis and health. Endocr Rev 42:56–76
10. Mohamad AES, Mona MZ, Ihab ZE et al (2012) Serum amino acid abnormalities in pediatric patients with chronic renal failure with and without history of thromboembolic manifestations. Egypt J Med Human Genet 13:73–80
11. Anna E, Thalacker M, Katherine HI, Guo FJ et al (2014) BMI, RQ, diabetes, and sex affect the relationships between amino acids and clamp measures of insulin action in humans. Diabetes 63:791–800
12. Sabrina F, Víctor G, Davy G et al (2019) Analytical strategies for the determination of amino acids: past, present and future trends. J Chromatogr B 1132:121819–121841
13. Kaspar H, Dettmer K, Gronwald W et al (2009) Advances in amino acid analysis. Anal Bioanal Chem 393:445–452
14. Ruiz-Matute AI, Hernández-Hernández O, Rodríguez-Sánchez S et al (2011) Derivatization of carbohydrates for GC and GC-MS analyses. J Chromatogr B 879:1226–1240
15. Heleen AB, Joop CM, Henk L et al (1997) Pre-, on- and post-column derivatization in capillary electrophoresis. Electrophoresis 18:2214–2227
16. Li Y, Mei S, Liu S et al (2019) A photoelectrochemical sensing strategy based on single-layer MoS_2 modified electrode for methionine detection. J Pharm Biomed Anal 165:94–100
17. Tuma P, Gojda J, Som-merova B et al (2020) Measuring venous-arterial differences of valine, isoleucine, leucine, alanine and glutamine in skeletal muscles using counter-current electrophoresis with contactless conductivity detection. J Electroanal Chem 857:113772
18. Narang J, Jain U, Malhotra N et al (2015) Development of lysine biosensor based on core shell magnetic nanoparticle and multiwalled carbon nanotube composite. Adv Mater Lett 6:407–413
19. Singh S, Sharma MP, Alqarawi A et al (2019) Real-time optical detection of isoleucine in living cells through a genetically-encoded nanosensor. Sensors 20:146
20. Wang QY, Molinero-Fernandez A, Casanova A et al (2022) Intradermal gly-cine detection with a wearable microneedle biosensor: the first in vivo assay. Anal Chem 94:11856–11864
21. Parrilla M, Vanhooydonck A, Watts R et al (2022) Wearable wristband-based electrochemical sensor for the detection of phenylalanine in biofluids. Biosens Bioelectron 197:113764
22. Fiore L, De Lellis B, Mazzaracchio V et al (2022) Smartphone-assisted electro-chemical sensor for reliable detection of tyrosine in serum. Talanta 237:122869
23. Liu L, Bao JC, Min F et al (2009) Electrogenerated chemiluminescence for the sensitive detection of leucine using $Ru(bpy)_3{}^{2+}$ immobilized on dendritic Pd nanoparticle. Sensor Actua B Chem 139:527–531
24. Schultz J, Uddin Z, Singh G et al (2020) Glutamate sensing in biofluids: recent advances and research challenges of electrochemical sensors. Analyst 145:321–347

25. Windmiller JR, Valdés-Ramírez G, Zhou N et al (2011) Bicomponent microneedle array biosensor for minimally-invasive glutamate monitoring. Electroanalysis 23:2302–2309
26. Ma Y, Han J, Tong Z et al (2022) Porous carbon boosted non-enzymatic glutamate detection with ultra-high sensitivity in broad range using Cu ions. Nanomaterials (Basel) 12:1987
27. Galenkamp NS, Soskine M, Hermans J et al (2018) Direct electrical quantification of glucose and asparagine from bodily fluids using nanopores. Nat Commun 9:4085
28. Albayrak D, Karakuş E (2016) A novel glutamine biosensor based on zinc oxide nanorod and glutaminase enzyme from Hypocria jecorina. Artif Cells Nanomed Biotechnol 44(1):92–97
29. Hua Y, Lv X, Cai Y et al (2019) Highly selective and reproducible electroanalysis for histidine in blood with turn-on responses at a potential approaching zero using tetrahedral copper metal organic frameworks. Chem Commun 55:1271–1274
30. Zhang J, Wang DY, Li YC (2019) Ratiometric electrochemical sensors as-sociated with self-cleaning electrodes for simultaneous detection of adrenaline, serotonin, and tryptophan. ACS Appl Mater Interfaces 11:13557–13563
31. Liu ZH, Li X, Masai H et al (2021) A single-molecule electrical approach for amino acid detection and chirality recognition. Sci Adv 7:eabe4
32. Chen X, Zhang S, Shan X et al (2019) Derivative chiral copper (II) complexes as template of an electrochemical molecular imprinting sol-gel sensor for enantiorecognition of aspartic acid. Anal Chem Acta 1072:54–60
33. Wang Q, Liu Y, Campillo-Brocal JC et al (2021) Electrochemical biosensor for glycine detection in biological fluids. Biosens Bioelectron 3:113154
34. Wang Q, Molinero-Fernandez A, Casanova A et al (2022) Intradermal glycine detection with a wearable microneedle biosensor: the first in vivo assay. Anal Chem 94:11856–11864
35. Liu H, Shao J, Shi L et al (2020) Electroactive NPs and D-amino acids oxidase engineered electrochemical chiral sensor for D-alanine detection. Sensor Actua B Chem 304:127333.1–127333.9
36. Wen X, Chang X, Li A et al (2023) ZnO/Cu2O heterojunction integrated fiber-optic biosensor for remote detection of cysteine. Biosens Bioelectron 223:115021

Chapter 8
Electrochemical Biosensors for Nucleic Acids Detection

Wenjing Chu, Mengyu Yang, Meihua Lin, and Fan Xia

Abstract The highly sensitive and selective detection of nucleic acids in clinical samples is of great significance to the diagnosis and prognosis of diseases. Electrochemical biosensors have been widely used in nucleic acid detection due to their advantages of fast response, simple operation, and low cost. In this chapter, we summarize the progress of electrochemical biosensors using one-dimensional, two-dimensional, and three-dimensional probes for nucleic acids detection, aiming to provide general guidance for the design of electrochemical detection methods for nucleic acids in complex matrices, such as whole blood. We also discuss the limitations and challenges in each kind of probe in terms of sensitivity, selectivity, and anti-fouling, providing ideas for further development of electrochemical biosensors for nucleic acid detection in complex matrices.

Keywords Electrochemical biosensor · Nucleic acid · DNA nanostructure · Complex matrices

8.1 Introduction

With the development of genomics, transcriptomics, proteomics, and molecular pathology, more and more nucleic acids related to the occurrence and development of tumors have been discovered [1]. Tumor nucleic acid markers can be used for tumor

W. Chu · M. Yang
Faculty of Materials Science and Chemistry, China University of Geosciences, Wuhan 430074, China
e-mail: 18803762254@163.com

M. Lin (✉) · F. Xia
State Key Laboratory of Biogeology and Environmental Geology, Engineering Research Center of Nano-Geomaterials of Ministry of Education, Faculty of Materials Science and Chemistry, China University of Geosciences, Wuhan 430074, China
e-mail: linmh@cug.edu.cn

F. Xia
e-mail: xiafan@cug.edu.cn

F. Xia et al. (eds.), *Electrochemical Biosensors for Whole Blood Analysis*,
https://doi.org/10.1007/978-981-99-5644-9_8

classification, tumor progression monitoring, and prognosis evaluation. Moreover, pathogens and viruses that cause infectious diseases can be quickly and specifically identified by their nucleic acids, and molecular-level information on cancer and other diseases can be collected by sequence-level analysis [2]. Thus, the application of nucleic acid detection, as a kind of promising biomarkers, in disease analysis and diagnosis is of vital significance. However, it is challenging to specifically detect nucleic acids in clinical samples, due to the low concentration of target nucleic acids in the complex matrices (Fig. 8.1). For example, as a promising liquid biopsy, circulating tumor DNA (ctDNA) is released from tumors into circulation, and the concentration is as low as 0.01% of the total cell-free DNA (1–1 000 ng/ml in plasma) at the early stage of cancer [3]. And the sequences of microRNAs (miRNAs) involved in the origin and development of diseases are highly similar to their family members, which brings difficulties to their detection [4]. Therefore, the methods for the detection of nucleic acids must be extremely sensitive and specific in order to apply for early diagnosis of the disease [5].

Polymerase chain reaction (PCR) is the gold standard for gene expression quantification and the most classic method used to detect nucleic acids in clinical medicine. PCR can directly amplify any fragment of DNA sequence, while PCR for the detection of RNA sequence often requires the process of reverse transcription before exponential amplification [6]. Although PCR test is sensitive and specific, it still has some drawbacks, such as high cost, complex primer design, long detection time, and the need for laborious procedure which makes the PCR technique unable to detect nucleic acids at home or at a point-of-care location. Therefore, researchers have made great efforts to develop simple and sensitive nucleic acid detection platforms [7].

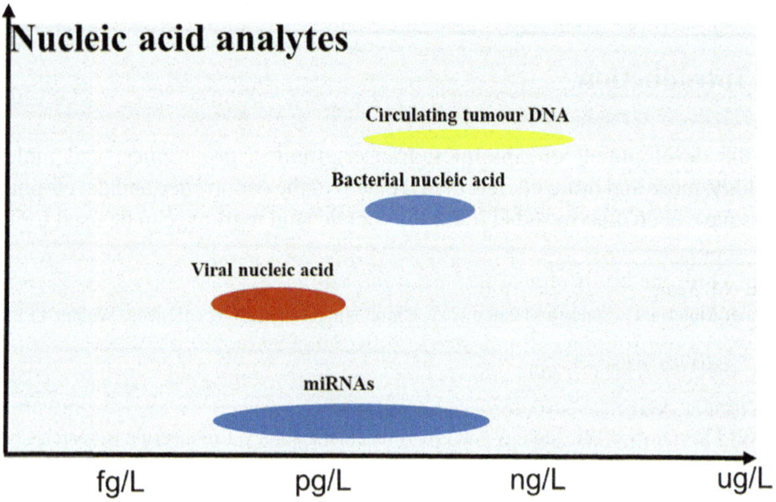

Fig. 8.1 Reference ranges for clinically relevant nucleic acids. (Reprinted with the permission from Ref. [5] Copyright 2016 American Chemical Society)

Electrochemical biosensors are recognized as promising candidates for nucleic acid detection due to their simple operation, fast response, high sensitivity, low cost, miniaturization, and intelligence [8]. Due to the low abundance of target nucleic acids and the high number of proteins and blood cells in clinical samples, the development of electrochemical biosensors must not only achieve clinically relevant sensitivity and specificity, but also be resistant or insensitive to biofouling [9]. The electrochemical biosensing interfaces play a vital role in the interactions between recognition probes and target nucleic acids. Thus, many strategies for engineering the interfaces have been proposed for improved target nucleic acids hybridization and decreased nonspecific adsorption. As an appealing probe, DNA nanostructures with various shapes have been constructed and applied for electrode surface engineering to improve electrochemical biosensor performance [10]. Herein, in this chapter, we will highlight DNA nanostructures from one dimension to three dimensions for engineering the electrochemical interfaces to detect nucleic acids in complicated matrices, which have been published in the past decade. It is well known that there are various DNA nanostructures have been employed in electrochemical biosensors for the detection of nucleic acids, such as one-dimensional (1D) probes, two-dimensional (2D) probes [11], and three-dimensional (3D) probes [12], as shown in (Fig. 8.2).

8.2 Design Principles of Electrochemical Biosensors for Nucleic Acid Detection

In general, electrochemical biosensors consist of two parts. One part is the sensor element consisting of the recognition probe and the interface. The other is an electronic system consisting of signal amplifiers and displays that can output signals in different modes, such as volt-ampere/current type, conductance type, and impedance type [13]. In nucleic acid electrochemical biosensors, biological receptors are usually DNA probes with different structures, which are often immobilized on substrates that are capable of conducting electrical signals, such as electrodes, nanoparticles, and nanowire arrays. Based on how the target nucleic acids interact with the capture probes and the reporting probes, nucleic acid electrochemical biosensors can be divided into three strategies: direct type, competitive type, and sandwich type (Fig. 8.3). In direct mode, there are two kinds of detection methods. One is label-free detection, in which capture probes without labeled redox are immobilized on the electrode surface via covalent or non-covalent interaction. After target nucleic acids hybridization, both the surface steric hindrance and negative charges are increased, which can be detected directly by the impedance technique, or the changed signal can be generated by adding electroactive indicators that can bind to DNA. The common electroactive indicators are methylene blue (MB), ferrocene (Fc), toluidine blue, or some metal complexes like hexaammineruthenium (III) chloride (RuHeX) [14]. The other is labeled redox electrochemical assay, in which capture probes labeled with a redox are also fixed on the electrode surface. Due to the flexibility of single-stranded

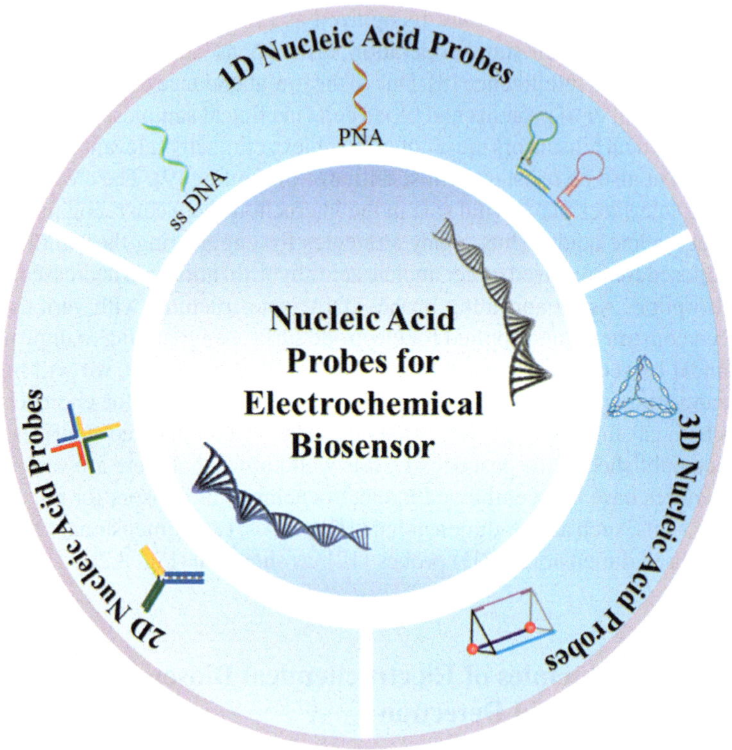

Fig. 8.2 Schematic of DNA nanostructure probes with different dimensions for electrochemical interface engineering

DNA (ss DNA), the collision chance between the electrode and the terminal redox label is great, resulting in a relatively large signal. While the double-stranded DNA (ds DNA) is formed after targeting nucleic acid hybridization, which is rigid and brings the redox moiety far away from the electrode surface, thereby reducing the signal. Compared to label-free detection, the labeled-redox electrochemical assay is more sensitive and often used for real-time measurements in complicated matrices, even in vivo.

The competition pattern and the sandwich pattern are two other complex recognition patterns that require the introduction of a third sequence in addition to the capture probe and target nucleic acid. In the competitive mode, capture probes prehybridize with partly complementary DNA which is often modified with a redox-active reporter, producing an electrochemical signal. In the presence of target nucleic acids, capture probes will hybridize with target sequences and release the labeled sequences through displacement reaction, causing a decreased signal. Thus, competitive assays are common "signal-off", which is prone to false positives in a complicated environment. In the sandwich mode, target nucleic acid acts as a bridge to hybridize with a

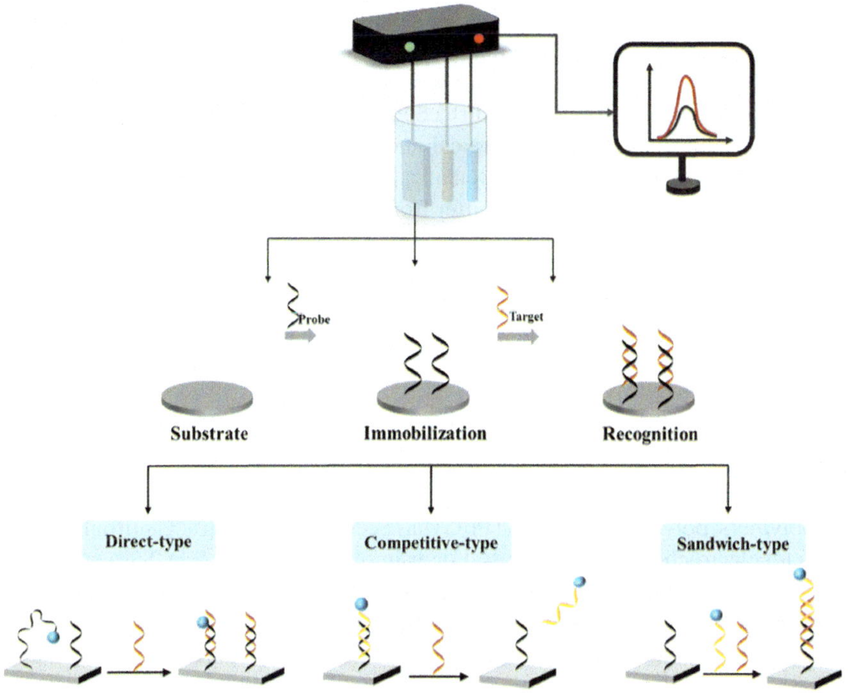

Fig. 8.3 Principle of electrochemical biosensor and three recognition strategies for target nucleic acids

capture probe and a signaling probe. The signal DNA probe can be used as an amplification template to achieve signal amplification, further improving the sensitivity of biosensors [13].

8.3 Biosensors with DNA Nanostructures Probes

8.3.1 1D DNA Probes

As the simplest recognition probe, the ss DNA probe has been widely immobilized on the electrode surface for analytes detection. Ideally, DNA probes should adopt an upright orientation and have a tunable distance between the neighbor probes on the surface, which is critical for target recognition. However, due to the flexibility of ss DNA, the strong interstrand entanglement and the non-specific interactions between bases and the substrate, such as nitrogen atoms and the gold surface, often occur, which largely restricts the performance of DNA sensors [15]. Many molecules have been introduced into the surface to reduce the nonspecific interactions, and improve the stability of the sensors and the resistance to biological contamination, such as 6-mercapto-1-hexanol (MCH), phorsphorycholine (PC), sulfopropylmethacrylate potassium (SP), and bovine serum albumin (BSA). MCH that can be self-assembled on a gold electrode surface formed as a monolayer is one of the most used block molecules at electrochemical nucleic acid sensors [16]. Although MCH can displace the nonspecific adsorption of capture probes to increase the accessibility of target nucleic acids, the sensors suffer from severe baseline drift when challenged in complex matrices [17]. Thus, Li et al. have reported a cell membrane-mimicking monolayer on the gold electrode surface by PC head groups to resist protein and cell fouling, which can reduce the baseline drift and improve the stability of sensors in real-time measurement in whole blood [18].

8.3.1.1 ss DNA Probes

Polymer antifouling is a rapidly growing chemical antifouling strategy [19]. As one of the polymers, polyethylene glycol (PEG) has been used to construct antifouling interfaces for electrochemical biosensors. For example, as shown in (Fig. 8.4A), Luo et al. developed antifouling nanofibers by grafting PEG polymer onto polyaniline (PANI) nanofibers detecting breast cancer susceptibility gene (BRCA1). They showed that the electrochemical sensors could detect target DNA in BSA, human serum albumin (HAS), immunoglobulin G, and hemoglobin matrices. And this biosensor based on the PEGylated PANI nanofibers supported the quantification of BRCA1 in human serum, indicating great potential application in clinical samples [20]. Based on the antifouling properties of PEG, the research group also developed glycosaminoglycan and PEG-coated magnetic nanoparticles for the detection of human papillomavirus (HPV). The magnetic nanoparticle-based electrochemical biosensor has excellent antifouling performance in 10 mg/mL BSA, bacteriophage, and 100% milk (about 30 mg/mL protein). And it has been successfully used for rapid and direct detection of HPV virus DNA in 100% serum [21]. Thus, the backfilling of molecules on the electrochemical DNA sensors can improve the sensing performance, especially in complex matrices.

Physical antifouling is another antifouling strategy, which can change the roughness and wettability of the electrode surface. There are three methods to achieve physical antifouling, including engineering design to prevent the adsorption of pollutants, using directional flow to effectively reduce pollutants reaching the electrode surfaces [22], and constructing nanoporous structures to implement size limitation and filtration to minimize electrode pollution [23]. Among them, the construction of nano-porous structures is a simple and effective method that is easy to implement in batch. Seker et al. reported a novel approach using a multifunctional matrix, nanoporous gold for nucleic acid detection (Fig. 8.4B). The nanoporous gold electrodes modified with DNA probes can sensitively and selectively detect target DNA in fetal bovine serum (FBS) containing interfering DNA fragments [24]. Although the adding molecules and physical design can improve the stability and the resistance to non-specific adsorption, the process is complicated and the monolayer of

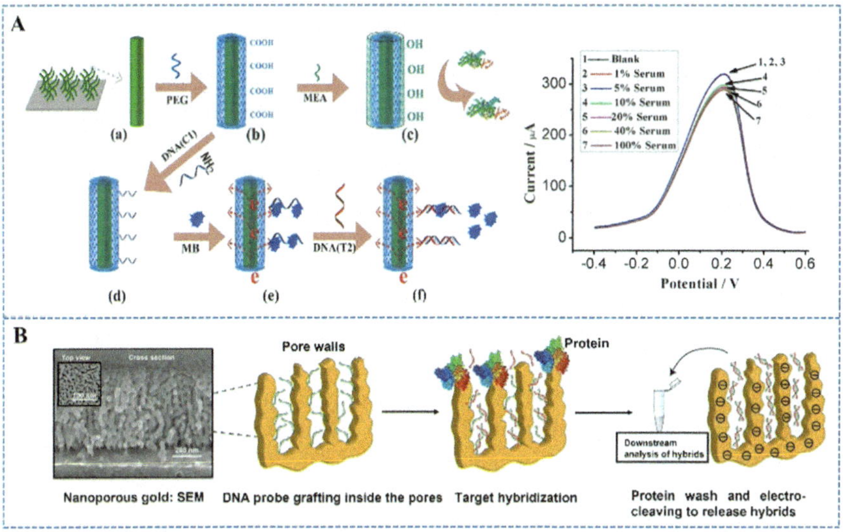

Fig. 8.4 Typical 1D ssDNA probe electrochemical biosensor for detection of nucleic acids in complex matrices. **A** Introducing polymer antifouling strategy for the detection of BRCA1 in human serum. (Reprinted with the permission from Ref. [20] Copyright 2017 American Chemical Society) **B** Using a physical antifouling strategy for the detection of complementary DNA in FBS. (Reprinted with the permission from Ref. [24] Copyright 2016 American Chemical Society)

the capture probe on the electrode surface is heterogeneous. To solve these concerns, Liu's group has reported electrochemical biosensors for bacterial 16S rRNA detection based on the intrinsic affinity of consecutive adenine nucleotides (polyA) for gold surfaces [25–27]. All adenines in polyA, independent of the length, can block nearly all of the adsorption sites on the gold surface, thus preventing nonspecific binding of the recognition probes to the surface [28, 29]. The recognition probes can adopt an upright orientation that is favorable for target hybridization. And the fully covered surface can resist the nonspecific adsorption of protein and blood cells. Thus, polyA-mediated electrochemical biosensors are suitable for the detection of nucleic acids in complex matrices.

8.3.1.2 Hairpin DNA Probes

Typically, hairpin probes are ss DNA with a stem-loop structure. Compared to ss DNA, hairpin DNA is more rigid and can prevent interstrand entanglement. The step-loop structure can switch into ds DNA after hybridization with complementary target nucleic acids [30]. Based on this, Fan et al. developed the first electrochemical DNA (E-DNA) sensor in 2003 [31]. The hairpin DNA was dually modified with a thiol group (for immobilization at the gold electrode surface) and a redox moiety (for signaling). At the initial state, the redox moiety was closed to the electrode surface, resulting in a relatively large signal. After hybridization, the conformational change made the redox moiety far away from the surface, reducing the signal. The first E-DNA sensor can detect as low as ~30 pM DNA. After that, we have developed the photoresponsive E-DNA biosensors with tunable dynamic range by using only one photocleavable capture probe [32]. Although our sensors can span the dynamic range by UV irradiation, the low detection limit is 1 pM, which may be not sensitive enough for the detection of some nucleic acid biomarkers. Therefore, hairpin DNA probes can combine with many isothermal amplification methods for improving sensitivity, such as catalytic hairpin assembly (CHA) [33], hybrid chain reaction (HCR) [34], and rolling circle amplification (RCA) [35].

CHA strategies have been explored for the application of electrochemical biosensors to achieve highly sensitive detection of miRNA. Dong et al. established a new ratiometric electrochemical biosensing platform for ultrasensitive miRNA detection based on CHA amplification (Fig. 8.5A). The detection limit of the proposed biosensor is as low as 50 aM, which also exhibits excellent detection performance in complex serum environments and tumor cell lysates [36].

HCR is an enzyme-free self-assembly strategy that produces long ds DNA structures. The formed ds DNA structure can be loaded with lots of electrically active substances to generate amplified electrical signals. For example, Gao et al. combined a hairpin probe with HCR for the detection of exosomal miRNA (Fig. 8.5B). In this method, target miRNA can trigger HCR to load more RuHex to increase the electrochemical signal. Using this approach, the sensitivity of the electrochemical assay is greatly improved to 53 aM for miRNA extracted from MCF 7 cells [37].

RCA is also an isothermal amplification method with the assistance of DNA polymerase. The product of RCA is a long and repetitive ss DNA which is complementary to circular template DNA. Thus, due to the programmability of DNA sequence, RCA products can be customized for subsequent signal amplification [38]. Gao et al. combined hairpin DNA and RCA products for sensitive detection of miRNA. RCA triggered by target miRNA and produced tandem periodic oligonucleotides that could form DNAzyme with hairpin probes. The DNAzyme catalysis reaction rate and efficiency could be improved due to the multivalent effect between RCA products and hairpin probes, thus achieving the sensitive detection of target miRNA in a complex practical biological matrix. Thus, it is convenient to combine hairpin probes on the electrode surface with amplification techniques for the sensitive detection of biomarkers, however, the process is complicated and time-consuming.

8.3.2 PNA Probes

As a synthetic DNA analog, peptide nucleic acid (PNA) is electrically neutral because the sugar-phosphate backbone of DNA is replaced by a pseudo-peptide polymer [39]. Based on this, PNA can hybridize with DNA/RNA with higher stability and higher specificity [40]. Thus, PNA probes have been widely used in electrochemical biosensors for improving the sensitivity and specificity in complex matrices. Kelley's

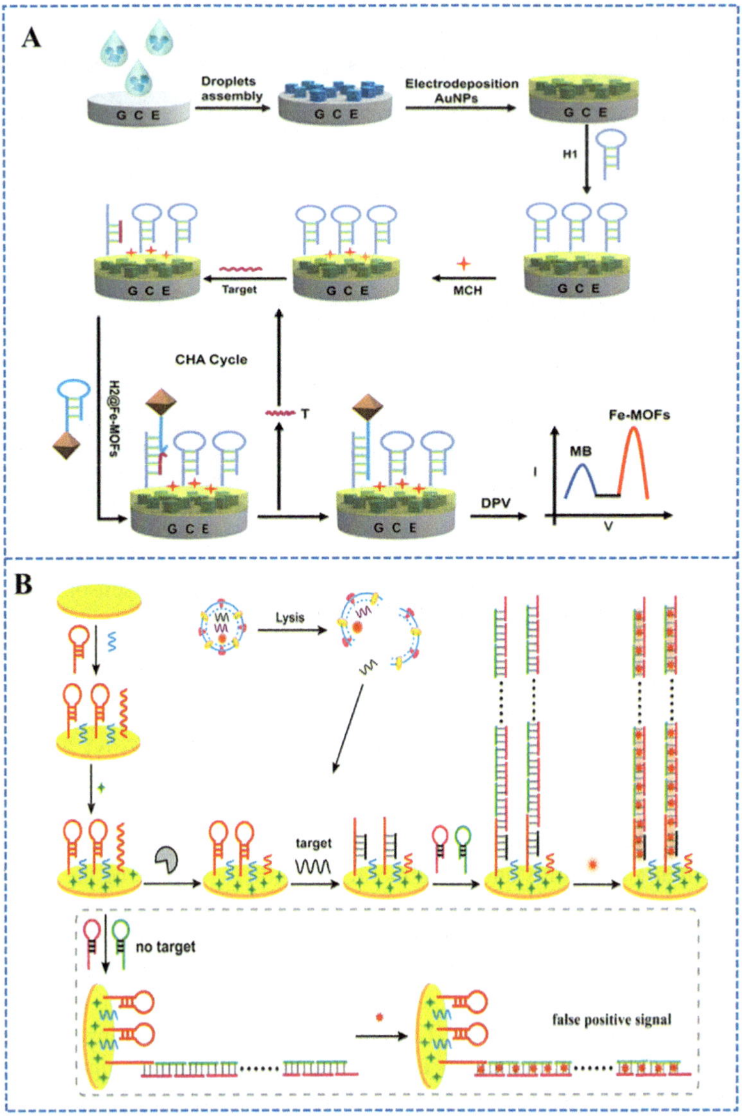

Fig. 8.5 Typical electrochemical biosensors by combining hairpin probes with **A** CHA (Reprinted with permission from Ref. [36] Copyright 2022 American Chemical Society) and **B** HCR for detection of miRNA. (Reprinted with the permission from Ref. [37] Copyright 2020 American Chemical Society)

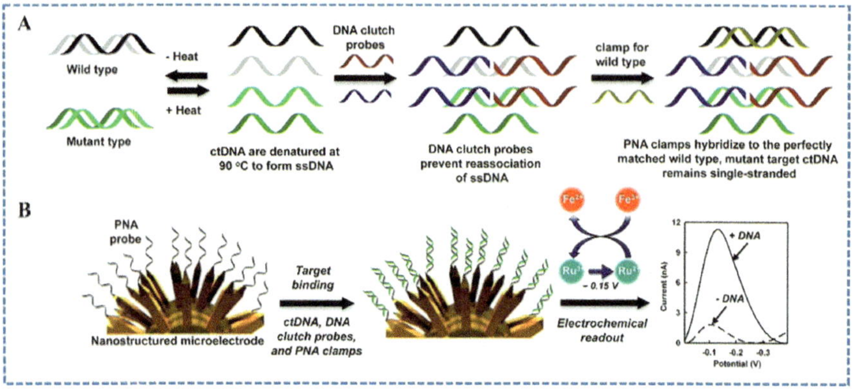

Fig. 8.6 Typical PNA probes electrochemical biosensor for detection of ds ctDNA. **A** Get a mutated ss DNA from ds ctDNA with the aid of DNA clutch and PNA clamp. **B** The microelectrodes are modified with PNA probes for detection of the mutated ctDNA with label-free mode. (Reprinted with the permission from Ref. [44] Copyright 2016 American Chemical Society)

group has developed a series of PNA-based electrochemical biosensors for various molecules detections [41–43]. They modified PNA probes on the nanostructured microelectrodes for the detection of circulating RNAs and ctDNA with label-free mode. When detection of the mutated DNA sequence from wild-type ds ctDNA, DNA clutch probes were designed and added into the samples to prevent reassociation of denatured ssDNA, releasing ssDNA [44]. As shown in (Fig. 8.6A), PNA clamps were designed to hybridize with the perfectly matched wild-type DNA, remaining the mutated ctDNA which would be captured on the microelectrode surface by PNA probes. Due to the uncharged PNA, PNA could rapidly hybridize with target ctDNA with higher affinity and not attract RuHeX, thereby reducing the background current and improving the sensitivity and specificity (0.01% mutated DNA can be detected). Moreover, this PNA-based electrochemical biosensor can directly detect serum samples to profile the mutational spectrum of a tumor. Zhang et al. also reported an electrochemical point-of-care testing (POCT) platform based on PNA-modified carbon-fiber microelectrode [45]. The platform shows a linear range of 10 fM to 1 μM with a detection limit as low as 3.29 fM. PNA probes can improve the sensitivity and specificity of electrochemical biosensors, but the synthesis cost of PNA is high and the length is usually limited to 24 residues.

8.3.3 2D DNA Probes

Although single-point fixed 1D DNA probe possesses low steric hindrance, some inherent defects still exist, such as poor stability and uneven distribution on the electrode surface. Therefore, lots of 2D DNA nanostructures were designed and applied to the construction of electrochemical sensors. These biosensors not only realize the effective detection of target molecules, but also have good anti-protein interference ability. Compared to 1D probes, 2D nanostructures show stronger rigidity and better stability for construction of electrochemical biosensors. For example, Zhang et al. have designed a novel 2D DNA nanostructure as an electrochemical platform for miRNA detection (Fig. 8.7A). The Fe-labeled bipedal 2D nanostructures showed low steric hindrance and strong stability, which significantly improved the space utilization rate. Meanwhile, combining 2D nanostructures with toehold-mediated strand-displacement reactions can recycle target miRNA and achieve signal amplification. The biosensor showed a linear relationship with the target concentration in the range of 1.0 fM–10 nM. Although the detection limit is as low as 0.31 fM, the mechanism of the biosensor is signal-off that is prone to false positives in complicated environments [46]. To improve reliability and precision, Zhou et al. developed a signal-on electrochemical biosensor using Y-type DNA nanostructures (Fig. 8.7B). They combined Y-type DNA as a capture probe with a nonlinear HCR for miRNA analysis. This biosensor has a linear response range to miRNA from 1 fM to 10 pM, and the detection limit is 0.33 fM. It is worth mentioning that the biosensor showed high specificity for the detection of target miRNA, and also enabled label-free detection in serum sample [47]. Therefore, the electrochemical biosensors based on 2D DNA nanostructures have high sensitivity and specificity for nucleic acid detection, but the structures in complex matrices are not robust enough to keep their stability.

8.3.4 3D DNA Probes

During the past two decades, we have witnessed rapid progress in the development of DNA nanotechnology, which provides unprecedented opportunities for developing electrochemical biosensors [48]. Based on the unique Watson–Crick base-pairing rules, various DNA nanostructures with well-defined size, shape, and geometry have been created in a bottom-up manner [49]. In particular, 3D DNA nanostructures, such as DNA polyhedral, have been readily assembled, which can be fabricated on the electrode surface to construct electrochemical biosensors. Due to structural stability, programmability, and natural biocompatibility, 3D DNA nanostructures engineering the biosensing interface not only effectively control probe orientation, but also have great potential to resist biological contamination. Therefore, the rational design of 3D DNA nanostructures is of great significance for improving the sensing performance of electrochemical biosensors in complex matrices.

As a typical 3D nanostructure, tetrahedral DNA nanostructure (TDN) was first employed for electrochemical biosensors in 2010 by Fan's group [50]. TDN was synthesized from four ss DNA with high yield by one step of thermal denaturation. TDN can be self-assembled on the gold electrode surface by three thiol groups at three vertices of TDN, resulting in more stable immobilization. And a capture probe can be appended at the fourth vertex of TDN. Due to the high mechanical rigidity of TDN, the capture probes can adopt an ordered upright orientation, which can improve the accessibility of target nucleic acids and then improve the detection sensitivity.

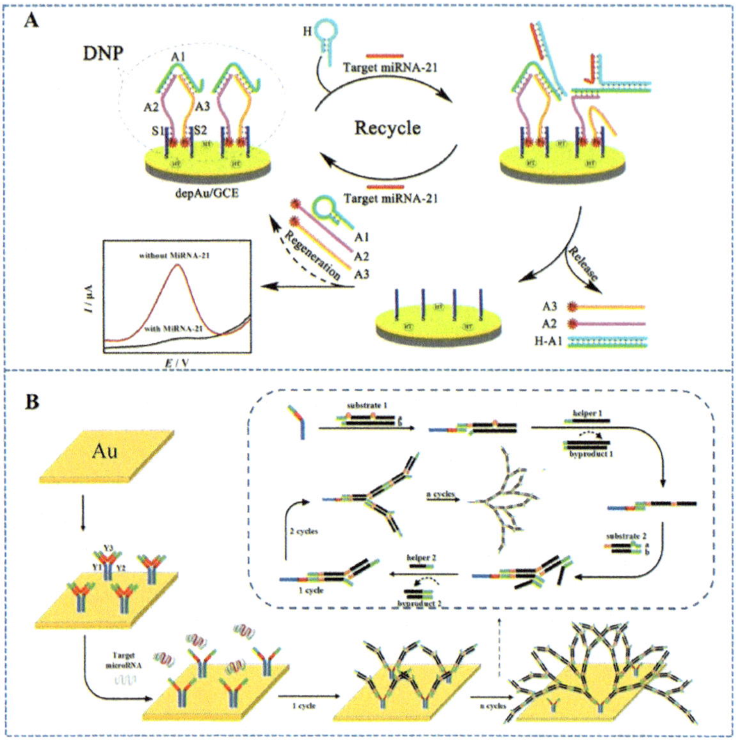

Fig. 8.7 Typical 2D DAN nanostructures-based electrochemical biosensor for miRNA detection. **A** A novel 2D DNA nanostructure combined with toehold-mediated displacement amplification for detection of miRNA. (Reprinted with the permission from Ref. [46] Copyright 2018 American Chemical Society) **B** Y-type DNA nanostructure combined with a nonlinear HCR for detection of miRNA (Reprinted with permission from Ref. [47] Copyright 2019 Elsevier)

Moreover, the sizes of TDN can be programmatically regulated based on Watson–Crick base pairing. Lin et al. constructed a typical electrochemical biosensor based on TDN with different sizes (Fig. 8.8A). Thus, the distance between neighbor probes appended at TDN can be precisely tuned by the size of TDN from 2.4 to 12.6 nm. The sensor performances, including the hybridization efficiency, the hybridization rate, and the limit of detection can be programmed by simply changing the size of TDN. The TDN-based electrochemical DNA sensors also possess high selectivity and stability when detecting target DNA in 50% FBS. However, the employed sandwich structure of TDN-based electrochemical biosensors requires multiple incubation and washing steps, which impedes their use in real-time analysis of target nucleic acids [51]. To solve this problem, Li et al. designed a TDN-based electrochemical sensor to achieve real-time detection of target nucleic acids in whole blood using direct detection mode. As shown in (Fig. 8.8B), the capture strand labeled with MB standing at the top vertex can recognize target DNA, and the assistant strand standing at the bottom can also bind to targets to bring the label redox in close proximity to the electrode surface, improving the signal [52]. After further optimization of the two functional strands in the TDNs, one-step detection of DNA in the picomolar range can be achieved in less than 10 min and directly in complex media. Moreover, due to the stability and the antifouling ability of TDN structure and the presence of three thiol vertexes for immobilization on the gold surface, the baseline drift of the TDN-based biosensor can be greatly decreased even after several hours in flowing whole blood compared to the traditional MCH-based electrochemical DNA sensors. Thus, TDN-based electrochemical biosensors in a single-step procedure hold great potential to be employed for real-time monitoring of target nucleic acids in live animals. However, in this sensor, one target DNA only produces one signal event, and this 1:1 binding model will limit the detection sensitivity, which can be improved by combined TDN and signal amplification methods.

Due to the rapid development of DNA technology, many framework nucleic acids that have well-defined frames and cavities have been designed and constructed [53]. Some other framework nucleic acids besides TDNs have also been used as probes for electrochemical biosensors. For example, Miao's group has immobilized regular and irregular triangular prism DNA nanostructures on the gold electrode surface for the detection of miRNA [54, 55], SARS-CoV-2 [56], and ctDNA [57]. The triangular

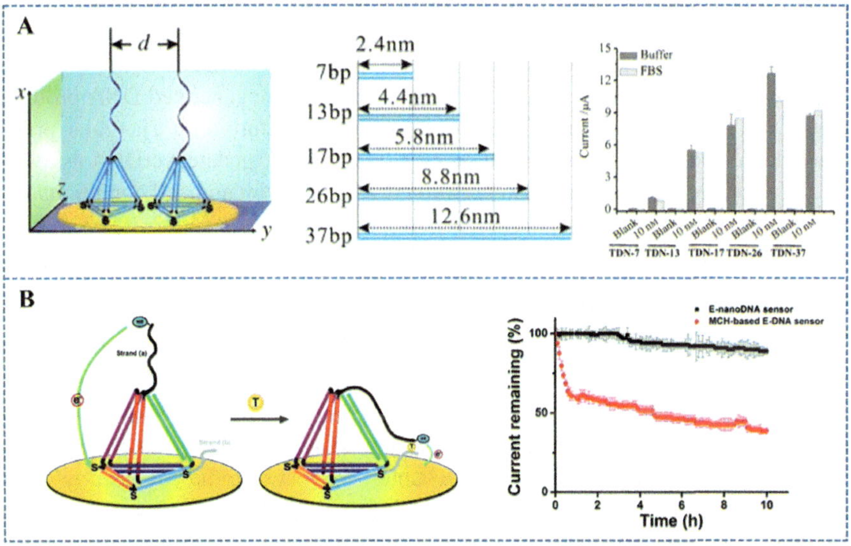

Fig. 8.8 Typical 3D probes electrochemical biosensor based on TDN nanostructure. **A** The distance between neighbor probes appending at the top of TDN can be precisely regulated by varying the size of TDN. TDN-based electrochemical biosensors can selectively recognize with target DNA in FBS, without an increase in background signal or a loss in signal gain. (Reprinted with the permission from Ref. [51] Copyright 2015 Wiley–VCH Verlag GmbH) **B** An electrochemical platform for real-time detection of target DNA with high baseline stability in whole blood was realized by rational design of TDN structure. (Reprinted with the permission from Ref. [52] Copyright 2018 The Royal Society of Chemistry)

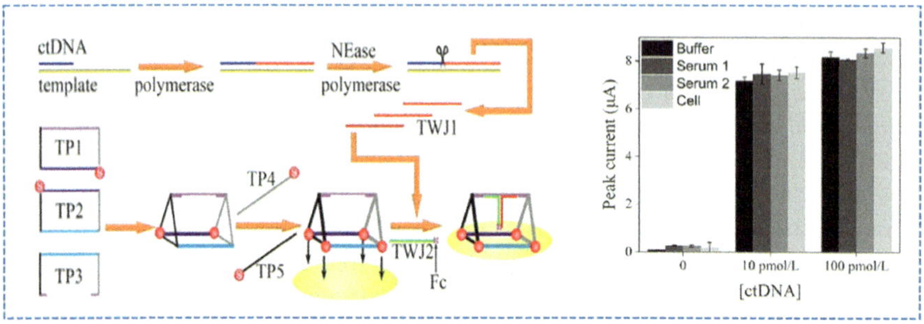

Fig. 8.9 Schematic illustration of a 3D DNA nanostructure-based electrochemical biosensor for ctDNA detection by combining triangular prism DNA nanostructures with strand displacement polymerization reaction (Reprinted with the permission from Ref.34 Copyright 2019 Elsevier)

prism DNA nanostructure is assembled from five ss DNA and can be anchored on the gold surface by the labeled thiol groups at the bottom of the structure. Due to the rigid structure and multiple thiol anchors, triangular prism DNA nanostructure can improve the sensing sensitivity and selectivity, especially in complex matrices. Moreover, it is easy to combine triangular prism DNA nanostructure-based electrochemical biosensors with signal amplification to further improve the sensitivity. As shown in (Fig. 8.9), they combined the merits of strand displacement amplification and framework nucleic acid for ultrasensitive detection of ctDNA. After the target initiated strand displacement polymerization reaction, the generated DNA products helped the formation of a three-way junction nanostructure on a triangular prism, which localized the electrochemical species. The sensor can detect ctDNA as low as 48 aM in buffer solution, and also performed well in serum and cell conditions.

8.4 Conclusion and Outlook

With the rapid development of high-throughput sequencing technology, more and more nucleic acid tumor markers of various types have been discovered. Electrochemical sensors as a powerful platform have been widely used for the detection of nucleic acid tumor markers. In this chapter, we focus on the engineering of the electrochemical biosensing interface with different DNA probes from 1 to 3D for the detection of nucleic acids in complex matrices, such as FBS and whole blood. Although the structures of 1D and 2D DNA probes are simple, the surface density and the lateral space between probes are difficult to control, which limits the sensitivity of the electrochemical biosensors. In addition, these sensors suffer baseline drift and poor robustness in complex matrices. The development of DNA technology offers unprecedented opportunities for developing DNA-based electrochemical biosensors. Specifically, the framework nucleic acids with a highly rigid and stable structure can be self-assembled on the electrode surface, allowing the appending recognized probes to keep an upright orientation with controllable lateral distance. And the nuclease resistance and protein resistance of framework nucleic acids, such as tetrahedral DNA nanostructure, allow the electrochemical biosensors to real-time detect target nucleic acids in complex matrices with minimal nonspecific adsorption. Therefore, 3D-based electrochemical sensors can detect nucleic acids with higher sensitivity and selectivity, even in complex matrices.

Considering the convenience of designing DNA nanostructures with different sizes and geometric shapes, we envision that DNA nanostructures will be a promising tool for electrochemical sensors. Programmable engineering of the electrochemical interface with highly ordered DNA nanostructures can improve nucleic acid hybridization in both thermodynamics and kinetics, which will support ultra-sensitive detection of nucleic acids for early diagnosis of cancer and rapid screening of infectious diseases. Thus, we envisage that the sensitivity of the electrochemical biosensors will in the future be comparable to that of the most sensitive PCR techniques. Moreover, with advances in materials, manufacturing, and nanotechnology, we believe that electrochemical biosensors can be further improved in terms of detection accuracy, instrument miniaturization and intelligence, and long-term monitoring in vivo. We hope that electrochemical biosensors can provide broad prospects for the diagnosis and treatment of cancer.

References

1. Djebali S, Davis CA, Merkel A, Dobin A, Lassmann T, Mortazavi A, Tanzer A, Lagarde J, Lin W, Schlesinger F, Xue C, Marinov GK, Khatun J, Williams BA, Zaleski C, Rozowsky J, Roder M, Kokocinski F, Abdelhamid RF, Alioto T, Antoshechkin I, Baer MT, Bar NS, Batut P, Bell K, Bell I, Chakrabortty S, Chen X, Chrast J, Curado J, Derrien T, Drenkow J, Dumais E, Dumais J, Duttagupta R, Falconnet E, Fastuca M, Fejes-Toth K, Ferreira P, Foissac S, Fullwood MJ, Gao H, Gonzalez D, Gordon A, Gunawardena H, Howald C, Jha S, Johnson R, Kapranov P, King B, Kingswood C, Luo OJ, Park E, Persaud K, Preall JB, Ribeca P, Risk B, Robyr D, Sammeth M, Schaffer L, See LH, Shahab A, Skancke J, Suzuki AM, Takahashi H, Tilgner H, Trout D, Walters N, Wang H, Wrobel J, Yu Y, Ruan X, Hayashizaki Y, Harrow J, Gerstein M, Hubbard T, Reymond A, Antonarakis SE, Hannon G, Giddings MC, Ruan Y, Wold B, Carninci P, Guigo R, Gingeras TR (2012) Landscape of transcription in human cells. Nature 489:101–108
2. Jayanthi V, Das AB, Saxena U (2017) Recent advances in biosensor development for the detection of cancer biomarkers. Biosens Bioelectron 91:15–23
3. Christie EL, Dawson SJ, Bowtell DD (2016) Blood worth bottling: circulating tumor DNA as a cancer biomarker. Cancer Res 76:5590–5591
4. Wu L, Qu X (2015) Cancer biomarker detection: recent achievements and challenges. Chem Soc Rev 44:2963–2997
5. Labib M, Sargent EH, Kelley SO (2016) Electrochemical methods for the analysis of clinically relevant biomolecules. Chem Rev 116:9001–9090
6. Jie J, Hu S, Liu W, Wei Q, Huang Y, Yuan X, Ren L, Tan M, Yu Y (2020) Portable and battery-powered PCR device for DNA amplification and fluorescence detection. Sensors (Basel) 20
7. Du Y, Dong S (2017) Nucleic acid biosensors: recent advances and perspectives. Anal Chem 89:189–215
8. Abi A, Mohammadpour Z, Zuo X, Safavi A (2018) Nucleic acid-based electrochemical nanobiosensors. Biosens Bioelectron 102:479–489

9. Han R, Li Y, Chen M, Li W, Ding C, Luo X (2022) Antifouling electrochemical biosensor based on the designed functional peptide and the electrodeposited conducting polymer for CTC analysis in human blood. Anal Chem 94:2204–2211
10. Wang DX, Wang J, Wang YX, Du YC, Huang Y, Tang AN, Cui YX, Kong DM (2021) DNA nanostructure-based nucleic acid probes: construction and biological applications. Chem Sci 12:7602–7622
11. Ma ZY, Xu F, Qin Y, Zhao WW, Xu JJ, Chen HY (2016) Invoking direct exciton-plasmon interactions by catalytic ag deposition on Au nanoparticles: photoelectrochemical bioanalysis with high efficiency. Anal Chem 88:4183–4187
12. Li S, Tian T, Zhang T, Cai X, Lin Y (2019) Advances in biological applications of self-assembled DNA tetrahedral nanostructures. Mater Today 24:57–68
13. Jing L, Xie C, Li Q, Yang M, Li S, Li H, Xia F (2022) Electrochemical biosensors for the analysis of breast cancer biomarkers: from design to application. Anal Chem 94:269–296
14. Hai X, Li Y, Zhu C, Song W, Cao J, Bi S (2020) DNA-based label-free electrochemical biosensors: from principles to applications. TrAC Trend Anal Chem 133
15. Li C, Wu D, Hu X, Xiang Y, Shu Y, Li G (2016) One-step modification of electrode surface for ultrasensitive and highly selective detection of nucleic acids with practical applications. Anal Chem 88:7583–7590
16. Pei H, Zuo XL, Zhu D, Huang Q, Fan CH (2013) Functional DNA nanostructures for theranostic applications. Acc Chem Res 47(2):550–559
17. Deng M, Li M, Li F, Mao X, Li Q, Shen J, Fan C, Zuo X (2019) Programming accessibility of DNA monolayers for degradation-free whole-blood biosensors. ACS Mater Lett 1:671–676
18. Li H, Dauphin-Ducharme P, Arroyo-Curras N, Tran CH, Vieira PA, Li S, Shin C, Somerson J, Kippin TE, Plaxco KW (2017) A biomimetic phosphatidylcholine-terminated monolayer greatly improves the in vivo performance of electrochemical aptamer-based sensors. Angew Chem Int Ed 56:7492–7495
19. Russo MJ, Han M, Desroches PE, Manasa CS, Dennaoui J, Quigley AF, Kapsa RMI, Moulton SE, Guijt RM, Greene GW, Silva SM (2021) Antifouling strategies for electrochemical biosensing: mechanisms and performance toward point of care based diagnostic applications. ACS Sens 6:1482–1507
20. Hui N, Sun X, Niu S, Luo X (2017) PEGylated polyaniline nanofibers: antifouling and conducting biomaterial for electrochemical DNA sensing. ACS Appl Mater Inter 9:2914–2923
21. Chen L, Lv S, Liu M, Chen C, Sheng J, Luo X (2018) Low-fouling magnetic nanoparticles and evaluation of their potential application as disease markers assay in whole serum. ACS Appl Nano Mater 1:2489–2495
22. Chen S, Li L, Zhao C, Zheng J (2010) Surface hydration: principles and applications toward low-fouling/nonfouling biomaterials. Polymer 51:5283–5293
23. Su S, Wu Y, Zhu D, Chao J, Liu X, Wan Y, Su Y, Zuo X, Fan C, Wang L (2016) On-electrode synthesis of shape-controlled hierarchical flower-like gold nanostructures for efficient interfacial DNA assembly and sensitive electrochemical sensing of MicroRNA. Small 12:3794–3801
24. Daggumati P, Appelt S, Matharu Z, Marco ML, Seker E (2016) Sequence-specific electrical purification of nucleic acids with nanoporous gold electrodes. J Am Chem Soc 138:7711–7717
25. Li L, Wang L, Xu Q, Xu L, Liang W, Li Y, Ding M, Aldalbahi A, Ge Z, Wang L, Yan J, Lu N, Li J, Wen Y, Liu G (2018) Bacterial analysis using an electrochemical DNA biosensor with poly-adenine-mediated DNA self-assembly. ACS Appl Mater Inter 10:6895–6903
26. Wang L, Wen Y, Yang X, Xu L, Liang W, Zhu Y, Wang L, Li Y, Li Y, Ding M, Ren S, Yang Z, Lv M, Zhang J, Ma K, Liu G (2019) Ultrasensitive electrochemical DNA biosensor based on a label-free assembling strategy using a triblock polyA DNA probe. Anal Chem 91:16002–16009
27. Wang Q, Wen Y, Li Y, Liang W, Li W, Li Y, Wu J, Zhu H, Zhao K, Zhang J, Jia N, Deng W, Liu G (2019) Ultrasensitive electrochemical biosensor of bacterial 16S rRNA gene based on polyA DNA probes. Anal Chem 91:9277–9283
28. Pei H, Li F, Wan Y, Wei M, Liu H, Su Y, Chen N, Huang Q, Fan C (2012) Designed diblock oligonucleotide for the synthesis of spatially isolated and highly hybridizable functionalization of DNA-gold nanoparticle nanoconjugates. J Am Chem Soc 134:11876–11879

29. Gu M, Yi X, Xiao Y, Zhang J, Lin M, Xia F (2023) Programming the dynamic range of nanobiosensors with engineering poly-adenine-mediated spherical nucleic acid. Talanta 256:124278
30. Mills DM, Martin CP, Armas SM, Calvo-Marzal P, Kolpashchikov DM, Chumbimuni-Torres KY (2018) A universal and label-free impedimetric biosensing platform for discrimination of single nucleotide substitutions in long nucleic acid strands. Biosens Bioelectron 109:35–42
31. Fan CH, Plaxco KW, Heeger AJ Electrochemical interrogation of conformational changes as a reagentless method for the sequence-specific detection of DNA. PNAS 100(16):9134–9137
32. Lin M, Yi X, Wan H, Zhang J, Huang F, Xia F (2020) Photoresponsive electrochemical DNA biosensors achieving various dynamic ranges by using only-one capture probe. Anal Chem 92:9963–9970
33. Yu S, Wang Y, Jiang LP, Bi S, Zhu JJ (2018) Cascade amplification-mediated in situ hot-spot assembly for MicroRNA detection and molecular logic gate operations. Anal Chem 90:4544–4551
34. Oishi M, Juji S (2021) Acceleration of DNA hybridization chain reactions on 3D nanointerfaces of magnetic particles and their direct application in the enzyme-free amplified detection of microRNA. ACS Appl Mater Inter 13:35533–35544
35. Zhao Y, Chen F, Li Q, Wang L, Fan C (2015) Isothermal amplification of nucleic acids. Chem Rev 115:12491–12545
36. Dong JB, Wen L, Yang HS, Zhao JY, He CJ, Hu ZK, Peng L, Hou CJ, Huo DQ (2022) Catalytic hairpin assembly-driven ratiometric dual-signalelectrochemical biosensor for ultra-sensitive detection of micrornabased on the ratios of Fe-MOFs and MB-GA-UiO-66-NH2. Anal Chem 94:5846–5855
37. Guo Q, Yu Y, Zhang H, Cai C, Shen Q (2020) Electrochemical sensing of exosomal MicroRNA based on hybridization chain reaction signal amplification with reduced false-positive signals. Anal Chem 92:5302–5310
38. Liu J, Xie G, Lv S, Xiong Q, Xu H (2023) Recent applications of rolling circle amplification in biosensors and DNA nanotechnology. TrAC Trends Anal Chem 160
39. Saadati A, Hassanpour S, Mdl G, Mosafer J, Hashemzaei M, Mokhtarzadeh A, Baradaran B (2019) Recent advances on application of peptide nucleic acids as a bioreceptor in biosensors development. TrAC Trend Anal Chem 114:56–68
40. Hu Q, Wang Q, Kong J, Li L, Zhang X (2018) Electrochemically mediated in situ growth of electroactive polymers for highly sensitive detection of double-stranded DNA without sequence-preference. Biosens Bioelectron 101:1–6
41. Das J, Ivanov I, Safaei TS, Sargent EH, Kelley SO (2018) Combinatorial probes for high-throughput electrochemical analysis of circulating nucleic acids in clinical samples. Angew Chem Int Ed 57:3711–3716
42. Das J, Cederquist KB, Zaragoza AA, Lee PE, Sargent EH, Kelley SO (2012) An ultrasensitive universal detector based on neutralizer displacement. Nat Chem 4:642–648
43. Das J, Ivanov I, Montermini L, Rak J, Sargent EH, Kelley SO (2015) An electrochemical clamp assay for direct, rapid analysis of circulating nucleic acids in serum. Nat Chem 7:569–575
44. Das J, Ivanov I, Sargent EH, Kelley SO (2016) DNA clutch probes for circulating tumor DNA analysis. J Am Chem Soc 138:11009–11016
45. Zhang B, Chen M, Cao J, Liang Y, Tu T, Hu J, Li T, Cai Y, Li S, Liu B, Xu J, Liang B, Ye X, Cai X (2021) An integrated electrochemical POCT platform for ultrasensitive circRNA detection towards hepatocellular carcinoma diagnosis. Biosens Bioelectron 192:113500
46. Zhang X, Yang Z, Chang Y, Qing M, Yuan R, Chai Y (2018) Novel 2D-DNA-nanoprobe-mediated enzyme-free-target-recycling amplification for the ultrasensitive electrochemical detection of MicroRNA. Anal Chem 90:9538–9544
47. Zhou L, Wang Y, Yang C, Xu H, Luo J, Zhang W, Tang X, Yang S, Fu W, Chang K, Chen M (2019) A label-free electrochemical biosensor for microRNAs detection based on DNA nanomaterial by coupling with Y-shaped DNA structure and non-linear hybridization chain reaction. Biosens Bioelectron 126:657–663

48. Athanasiadou D, Carneiro KMM (2021) DNA nanostructures as templates for biomineralization. Nat Rev Chem 5:93–108
49. Lin M, Song P, Zhou G, Zuo X, Aldalbahi A, Lou X, Shi J, Fan C (2016) Electrochemical detection of nucleic acids, proteins, small molecules and cells using a DNA-nanostructure-based universal biosensing platform. Nat Protoc 11:1244–1263
50. Pei H, Lu N, Wen Y, Song S, Liu Y, Yan H, Fan C (2010) A DNA nanostructure-based biomolecular probe carrier platform for electrochemical biosensing. Adv Mater 22:4754–4758
51. Lin M, Wang J, Zhou G, Wang J, Wu N, Lu J, Gao J, Chen X, Shi J, Zuo X, Fan C (2015) Programmable engineering of a biosensing interface with tetrahedral DNA nanostructures for ultrasensitive DNA detection. Angew Chem Int Ed 54:2151–2155
52. Li C, Hu X, Lu J, Mao X, Xiang Y, Shu Y, Li G (2018) Design of DNA nanostructure-based interfacial probes for the electrochemical detection of nucleic acids directly in whole blood. Chem Sci 9:979–984
53. Guo Z, Zhang L, Yang Q, Peng R, Yuan X, Xu L, Wang Z, Chen F, Huang H, Liu Q, Tan W (2022) Manipulation of multiple cell-cell interactions by tunable DNA scaffold networks. Angew Chem Int Ed 61:e202111151
54. Chai H, Chen X, Shi R, Miao P (2023) Irregular DNA triangular prism/triplex assembly for duplicate MiRNA analysis with nicking endonuclease-mediated amplification. Anal Chem 95:4564–4569
55. Chai H, Zhu J, Guo Z, Tang Y, Miao P (2023) Ultrasensitive miRNA biosensor amplified by ladder hybridization chain reaction on triangular prism structured DNA. Biosens Bioelectron 220:114900
56. Jiang Y, Chen X, Feng N, Miao P (2022) Electrochemical aptasensing of SARS-CoV-2 based on triangular prism DNA nanostructures and dumbbell hybridization chain reaction. Anal Chem 94:14755–14760
57. Chai H, Miao P (2021) Ultrasensitive assay of ctDNA based on DNA triangular prism and three-way junction nanostructures. Chin Chem Lett 32:783–786

Chapter 9
Electrochemical Biosensors for Hepatic and Cardiac Biomarkers Detection

Jing Pan, Qitao Zhou, Hui Li, Shaoguang Li, and Fan Xia

Abstract Hepatic and cardiac diseases are two of the main serious diseases harming people's health and even life. Electrochemical biosensors, which have sufficient sensitivity, high specificity, simple operation process, and fast response, are promising in the diagnosis of hepatic and cardiac diseases through the detection of hepatic and cardiac biomarkers. This chapter outlines the recent development of electrochemical sensors for the detection of different hepatic and cardiac biomarkers. Firstly, the biomarkers of different hepatic and cardiac diseases have been listed. Then, the recent development of electrochemical sensors for the detection of some typical hepatic and cardiac biomarkers has been introduced. The typical biomarkers include cardiac troponins (troponin T, troponin I), creatine kinase, C-reactive protein, and typical hepatitis biomarkers (hepatitis B surface antigen, hepatitis B virus DNA, hepatitis C virus core antigen). In addition, a discussion and outlook of related electrochemical sensors on future research direction are also included.

Keywords Electrochemical biosensors · Hepatic biomarkers · Cardiac biomarkers · Troponins · Creatine kinase · C-reactive protein

9.1 Introduction

The liver and heart are two of the most important organs for humans. Specifically, the liver as an endocrine organ can produce various systemic factors to coordinate physiology and metabolism. Heart is the power of the body's circulatory system. Hepatic and cardiac diseases can lead to dysfunctions of the human body or even death. Actually, hepatic and cardiac diseases are two of the leading causes of human mortalities throughout the world. For example, liver disease and acute myocardial

J. Pan · Q. Zhou (✉) · H. Li · S. Li · F. Xia
State Key Laboratory of Biogeology and Environmental Geology, Engineering Research Center of Nano-Geomaterials of the Ministry of Education, Faculty of Materials Science and Chemistry, China University of Geosciences, Wuhan 430074, China
e-mail: zhouqitao@cug.edu.cn

F. Xia et al. (eds.), *Electrochemical Biosensors for Whole Blood Analysis*,
https://doi.org/10.1007/978-981-99-5644-9_9

infarction account for approximately 2 million and 8.9 million deaths per year worldwide, respectively [1, 2]. Besides, a potential cross-talk between the liver and heart has been reported very recently [3]. It indicates an accelerated effect of abnormal liver function on cardiac aging. Thus, early diagnosis and detection of related risk factors are crucial to the hepatic and cardiac diseases' prevention as well as morbidity reduction.

Biomarkers in blood samples could offer a clear guideline for the risk assessment, diagnosis, disease evolution analysis, prognosis, and therapy effectiveness evaluation of hepatic and cardiac diseases [4]. At present, the detection of these biomarkers can be accomplished by enzyme-linked immunosorbent assay, surface plasmon resonance, fluorescence, colorimetric assay, and chemiluminescence. However, most of them exhibit a variety of limitations in the detection of hepatic and cardiac biomarkers, such as limited sensitivity, low specificity, complicated operation, time-consuming process, high cost, or requirement for sophisticated equipment. Recently, electrochemical sensors have attracted more and more interest ascribing to their simple operation process, fast response time, low cost, and easy to be scalable [5]. For the detection of hepatic and cardiac biomarkers, they can also exhibit both high sensitivity and specificity. Therefore, this chapter introduces the biomarkers of different hepatic and cardiac diseases as well as the recent development of electrochemical sensors for the detection of typical biomarkers.

9.1.1 Biomarkers of Myocardial Necrosis/Myocardial Damage

Myocardial necrosis, for example, acute myocardial infarction, is one of the main reasons for the global morbidity and mortality. Generally speaking, myocardial necrosis and myocardial damage are derived from myocardial ischemia and cell death, which are often caused by atherosclerotic plaques [6]. Currently, the most commonly used biomarkers in the diagnosis of myocardial necrosis/myocardial damage in clinical practice (Fig. 9.1) include cardiac troponins, creatine kinase (CK), myoglobin (Myo), heart-type fatty acid-binding protein and ischaemia modified albumin (BSA). All of them can be detected with high sensitivity and specificity by employing electrochemical sensors [7].

It is evidenced that the management of myocardial Necrosis is extremely time critical and the US National Heart Attack Alert Program has even published a "60 min to treatment Working Group report" [8]. Consequently, the high sensitivity of the sensors, which is related to the early diagnosis of related diseases, is highly required to the effective treatment of myocardial necrosis/myocardial damage.

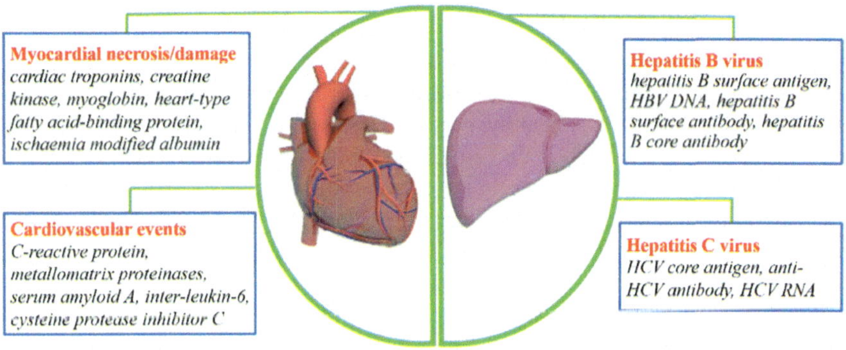

Fig. 9.1 The hepatic and cardiac diseases discussed in this chapter and the corresponding biomarkers

9.1.2 Predictive Biomarkers of Cardiovascular Events

The Cardiovascular diseases (CVD) have complex pathology and implicate multiple biological pathways, which include, inflammation, oxidative stress, and so on. As a result, inflammation and oxidative stress biomarkers could help to predict the risk of cardiovascular diseases [4]. Until now, a variety of predictive biomarkers for cardiovascular events have been found, including C-reactive protein (CRP), metallomatrix proteinases, serum amyloid A, interleukin-6, cysteine protease inhibitor C, and so on (Fig. 9.1).

9.1.3 Biomarkers of Hepatitis

Liver disease has a high fatality and infection rate. About 25% of the global deaths are derived from the end-stage liver disease. Among all the liver diseases, chronic infections with hepatitis B virus (HBV) and hepatitis C virus (HCV) are the highest risk factors for liver cancer and cirrhosis [9]. Specifically, 57% of global cirrhosis was caused by HBV (30%) and HCV (27%). In addition, 78% of hepatocellular carcinoma was the result of HBV (53%) and HCV (25%) [9]. Besides, the worldwide prevalence of HCV is approximately 2%-3% and HCV infection exhibits a low cure rate [10]. Thus, the accurate and early diagnosis of hepatitis is of great importance. The measured concentration of hepatitis B surface antigen (HBsAg), HBV DNA, hepatitis B surface antibody, and hepatitis B core antibody in serum are identified as effective biomarkers of HBV (Fig. 9.1) [11]. On the other hand, HCV core antigen (HCVcAg), anti-HCV antibody, and HCV RNA are mainly employed as the biomarker of HCV (Fig. 9.1) [10].

9.1.4 Cardiac Troponins

Cardiac troponin is a kind of protein specific to heart muscle cells and almost undetectable in the blood of a healthy person. However, when myocardial Necrosis/myocardial damage occurs, abundant cardiac troponin would be released into the blood. Thus, the obvious increase in the concentration of cardiac troponin would be a signal of myocardial necrosis/damage [12]. Among the three regulatory proteins of cardiac troponin, clinical tests have proved the concentration of troponin I (cTnI) and troponin T (cTnT) as two of the most specific indicators for myocardial necrosis/damage. Specifically, the normal concentration of cTnI in the blood is less than 0.4 ng/mL. If the detected concentration for cTnI is higher than 2.0 ng/mL, the risk for myocardial damage is indicated [7].

Electrochemical sensors refer to the sensors using electrochemical principles, which include electrochemical impedance spectroscopy (EIS), amperometric (i-t) method, differential pulse voltammetry (DPV), square wave voltammetry (SWV), cyclic voltammetry (CV), and so on. The employment of proper recognition parts could simultaneously ensure high specificity and sensitivity of the sensor. For the detection of cardiac troponin, antibodies, and aptamers are commonly employed as the recognition part of the corresponding sensors. For example, a single-stranded DNA aptamer against cTnI is developed with a much lower dissociation constant value (270 pM) than that of a cTnI antibody (20.8 nM) [5]. When applied to an electrochemical cTnI biosensor, a low limit of detection (LOD) of 1.0 pM (24 pg/mL) and a wide linear range of 1–10,000 pM are demonstrated in a buffer solution (Fig. 9.2A). The specificity test of this kind of biosensor also indicates the high specificity for cTnI compared to other interfering proteins. Furthermore, the biosensor can effectively measure the concentrations of cTnI in the solution with the addition of human serum.

Although most of the related reports have published sensitive and specific detection of one regulatory protein of cardiac troponin, simultaneous detection of the two regulatory proteins of cardiac troponin is more recommended for reliable and accurate clinical diagnosis. Furthermore, it can reduce blood consumption (sample volume) and cost. Screen-printed electrodes, which have easily modified surfaces, offer a promising pathway to achieve simultaneous detection of cTnI and cTnT [14]. For example, employing screen-printed electrodes (Fig. 9.2B), simultaneous detection of cTnI and cTnT in human serum can be achieved with a very low LOD of 1 pg mL^{-1} [13] (Table 9.1).

9.1.5 CK

CK (adenosine-5-triphosphate-creatine phosphotransferase) is a significant enzyme produced by muscle and brain. It contains three isoenzymes (CK-BB, CK-MB, and CK-MM) [27]. The three isoenzymes are produced by the muscles at different places.

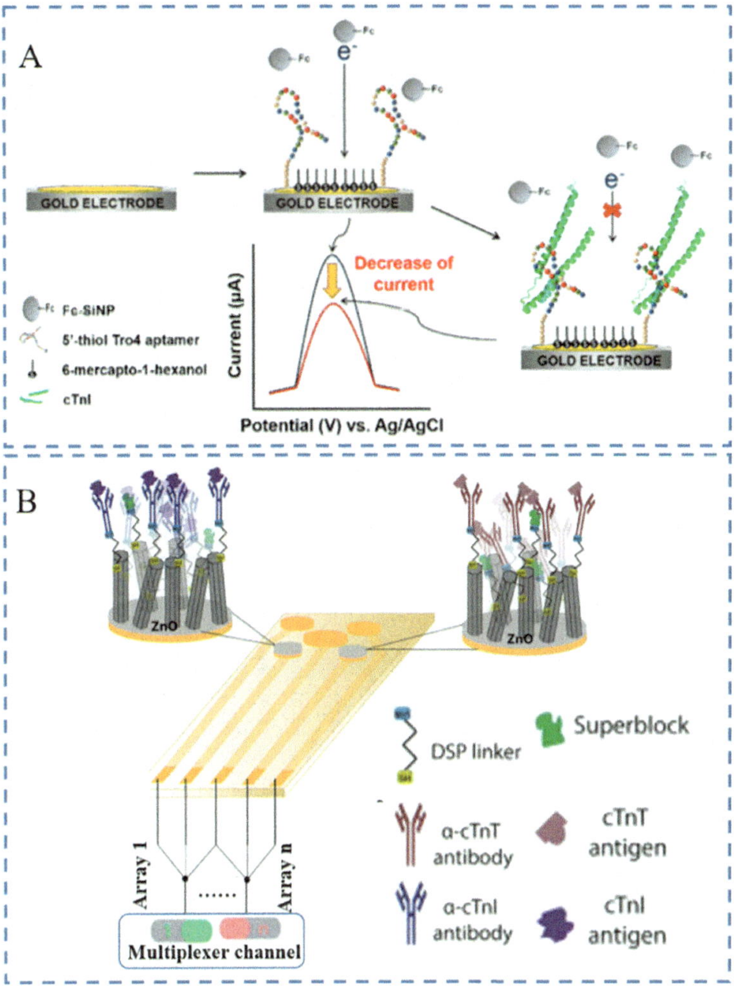

Fig. 9.2 **A** The DNA aptamer-based electrochemical cTnI sensor [5]. **B** The screen-printed electrodes-based electrochemical sensor for simultaneous detections of cTnI and cTnT [13]

Among them, only CK-MB is primarily produced by the heart muscle. In general, the concentration of normal blood CK for a healthy person is in the range of 0.038–0.174 U/mL for males and 0.026–0.14 U/mL for females [28]. However, when the related muscle or brain is injured, the concentration of blood CK would be much increased. Therefore, the concentration of CK-MB is recognized as an effective and specific biomarker for the diagnosis of myocardial necrosis and damage [29].

On account of that CK is an enzyme [28], which has many specific properties, the method for the detection of CK-MB can be different from the traditional electrochemical biosensors. For example, for traditional electrochemical biosensors, the

Table 9.1 The performance and employed methods of the existing electrochemical biosensors for cardiac troponins detecting

Biomarker	Electrochemical method	Nanostructures	Recognition method	LOD	Detection range	References
cTnI	SWV	SiO_2NPs	DNA aptamer	1.0 pM (24pg/mL)	1–1000 pM	[5]
	EIS	Ab-NH_2-MIL-88B(Fe_2Co)-MOF/NF	Antibody	13 fg/mL	10 ng–0.1 fg/mL	[15]
	EIS	Mn_3O_4-RGO	Antibody	8.0 pg/mL	0.008–20 ng/mL	[16]
	i-t	Mesoporous Fe_3O_4	Antibody	0.39 pg/mL	1.0 pg/mL–100 ng/mL	[17]
	DPV	–	Antibody	0.01 ng/mL	0.01–100 ng/mL	[18]
	DPV	PtCoCuPd HBTPs	Antibody	0.2 pg/mL	0.001–100.0 ng/mL	[19]
	EIS	MoS_2 NSs	DNA aptamer	0.95 pM	10 pM–10.0 μM	[20]
	DPV	COFs	Aptamer	2.6 fg/mL	10 fg/mL–10 ng/mL	[21]
	DPV	N-rGO	DNA aptamer	1 pg/mL	–	[22]
	DPV	Au NDs	DNA aptamer	8.0 pg/mL	0.05–500 ng/mL	[23]
cTnT	CV	Streptavidin-microsphere	Antibody	0.2 ng/mL	0.1–10 ng/mL	[24]
	SWV	Nanoclay	Antibody	0.35 pg/mL	1.0–10 pg/ mL	[25]
	DPV	MWCNTs	molecularly imprinted polymer	0.04 pg/mL	0.1–8.0 pg/mL	[26]

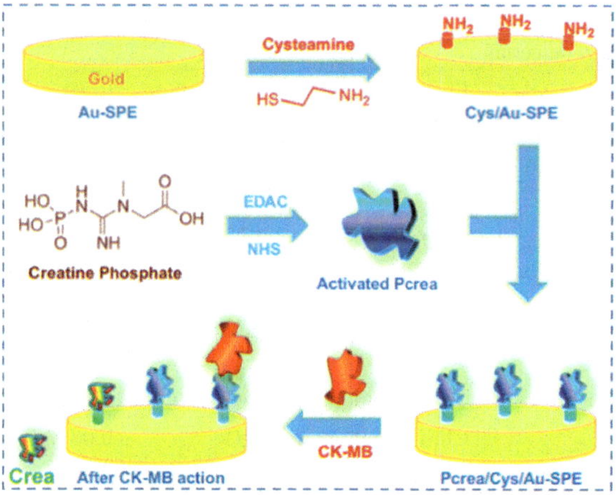

Fig. 9.3 The preparation and detection process of a Cys-aminated Au/-based electrochemical CK-MB biosensor [31]

detection of the biomarkers is realized by the binding of the recognition part of the sensor and the biomarkers in the electrolyte. However, an enzyme can directly be an alternative to the recognition part of biosensors and form an enzymatic-based biosensor [30]. When coating the proper enzyme on the surface of the electrode, the analyte in the electrolyte can diffuse to the enzyme membrane and contact the enzyme to accomplish enzymatic catalytic reaction or produce detectable ions. In this way, the signal intensity of the sensor could be changed and the specific detection can be realized. Based on this mechanism, if the substrate is immobilized on the surface of the electrode, the corresponding enzyme can be specifically detected. For instance, a cysteamine (Cys)-aminated Au/SPE is coated by phosphorylated form of creatine (Pcrea) and fabricate a CK-MB biosensor (Fig. 9.3) [31]. Considering the electroactive nature of Pcrea at low potential, the consumption of Pcrea could adjust the electrical response of the corresponding sensor. Based on the SWV method, the sensor exhibits a LOD of 0.11 μg/mL. Besides, the specificity of the sensor is illustrated under the interference of cTnT, bovine serum, BSA, and Myo.

The performance of other CK-MB electrochemical biosensors can be referred to Table 9.2. It needs to be mentioned that enzyme is unstable in certain physical or chemical conditions (high temperature, strong acid and alkali, etc.) [32]. Thus, the operating conditions of CK biosensors should be paid attention.

9.1.6 CRP

CRP is an acute-phase protein generated within the liver and an important biomarker of CVD. Actually, the American Heart Association as well as the United States Center for Disease Control have proposed that the risk of CVD can be leveled by the concentration of CRP in blood [39]. When the CRP concentration is lower than 1.0 mg/L, the risk of evolving CVD can be regarded as very low. When a person's

Table 9.2 The performance and employed methods of the existing electrochemical biosensors for CK detecting

Biomarker	Electrochemical method	Recognition method	LOD	Detection range	References
CK-MB	EIS	Antibody	3700 pg/mL	0.01–0.5 μg/mL	[29]
	SWV	Enzymic catalytic reaction	0.11 μg/mL	–	[31]
	i-t	Antibody	0.88 pg/mL	0.001–2000 ng mL	[33]
	Chronocoulogram	IgG	10 pg/mL	0.00001–1 μg/mL	[34]
	Chronocoulogram	IgG	0.08 pg/mL	–	[35]
	SWV	Antibody	0.5 pg/mL	0.001–50 ng/mL	[36]
	DPV	Antibody	0.62 pg/mL	0.000001–2.5 μg/mL	[37]
	Conductometric	Antibody	0.15 pg/mL	–	[38]
	i-t	Glucose oxidase and hexokinase	–	0.3268–5.229 μg/mL	[39]

CRP concentration is higher than 3.0 mg/L within a long period, he has a very high risk of developing CVD. Besides, the CRP concentration can be largely different for different persons or a person in different conditions [41]. The serum CRP concentration of a normal person is lower than 1 mg/L or even 0.01 mg/L [41]. If an acute-phase event is triggered, the CRP concentration can be obviously increased to 300–500 mg/L within 48 h [40]. The value can be even as high as 2500 mg/L for patients with tumor [41]. Thus, sufficient detection range is the key to the CRP detection. Recently, nanostructures have been proven to widen the detection range of electrochemical sensors. For instance, metal–organic framework-derived ZnO/porous carbon composite is applied to a CRP biosensor (Fig. 9.4A) [42]. The specific structure greatly enhanced the active surface area of the electrode and the immobilization of biomolecules. In consequence, electrochemical signals have been effectively amplified and the detection range has been widened. The linear range can be as wide as 0.01–1000 ng mL^{-1} and the detection limit can be as low as 5.0 pg mL^{-1}.

To detect the CRP in blood, the blood consumption should be considered. That's because the less blood is needed, the less pain the patient will suffer. Thus, the biosenors for the markers in blood should have low sample consumption. Taking this into consideration, microfluidic device, which has the advantages of portability,

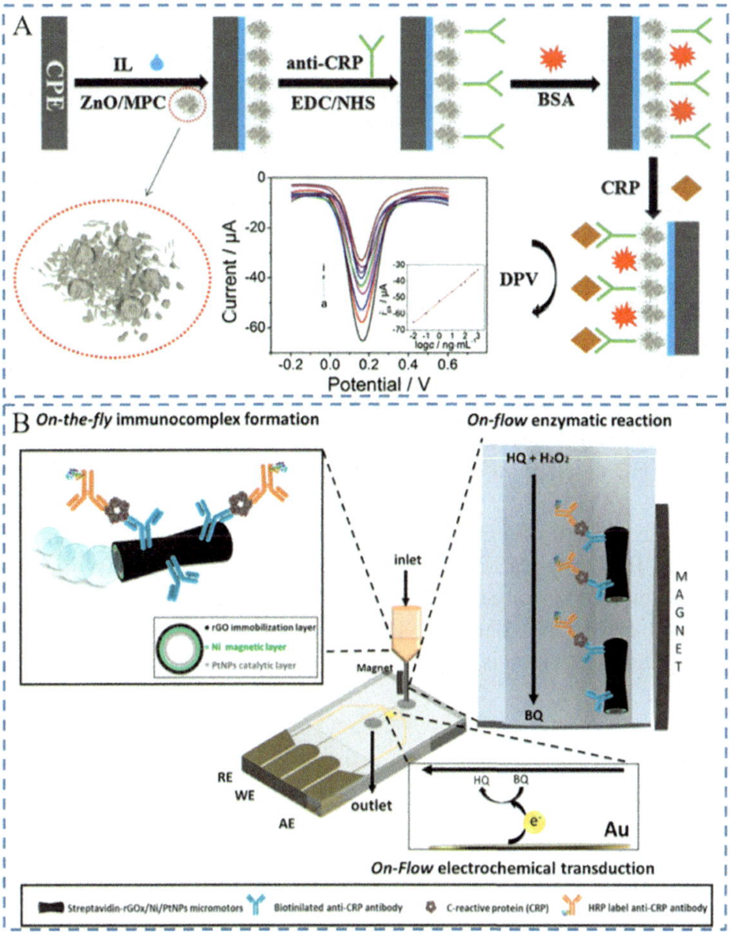

Fig. 9.4 **A** Schematic diagram of the preparation procedures for electrochemical CRP sensor based on ZnO/porous carbon composite [42]. **B** the illustration of a microfluidic device-based CRP biosensor [43]

massive parallelization, and low reagents consumption, could bring new opportunities for cardiac troponin biosensors. For instance, Escarpa et al. have fabricated a CRP biosensor based on a microfluidic device [43]. An outstanding sensitivity (LOD = 0.54 μg/mL) can be achieved with very low volume clinical samples (<10 μL) and very fast assay process (8 min).

9.1.7 Hepatitis Biomarkers

Compared with other detection methods, the electrochemical sensor for hepatitis diagnosis has more potential for point-of-care (POC) applications. The rapid, portable, and easy diagnostic process of POC testing allows rapid clinical decisions and more possibility for the patients to return the hepatitis test results on time [44]. Thus, the POC electrochemical sensor becomes the current development direction for hepatitis detection.

Very currently, a portable electrochemical biosensors for HCV detection has been reported [45]. Employing the i-t method, a potentiostat circuit, a Bluetooth module, and a controller unit, the biosensor can not only be portable, but also realize wireless data transmission. Additionally, the functionalization with cystamine and glutaraldehyde brought the sensor a covalent binding ability with recombinant core-HCV protein and high sensitivity. A similar smartphone-based portable electrochemical biosensor for HBV detection has also been developed for real-time detection very recently (Fig. 9.5A) [46].

In 2007, the concept of a microfluidic paper-based analytical device (PAD) was introduced to POC application [47]. On account of its simplicity, low cost, and portability form, its emerging attracted a lot of attention. Thereafter, the PAD started to be used for HVB DNA detection [49]. For instance, a pop-up PAD device for HVB DNA detection, which is disposable, affordable, and simple, has been reported

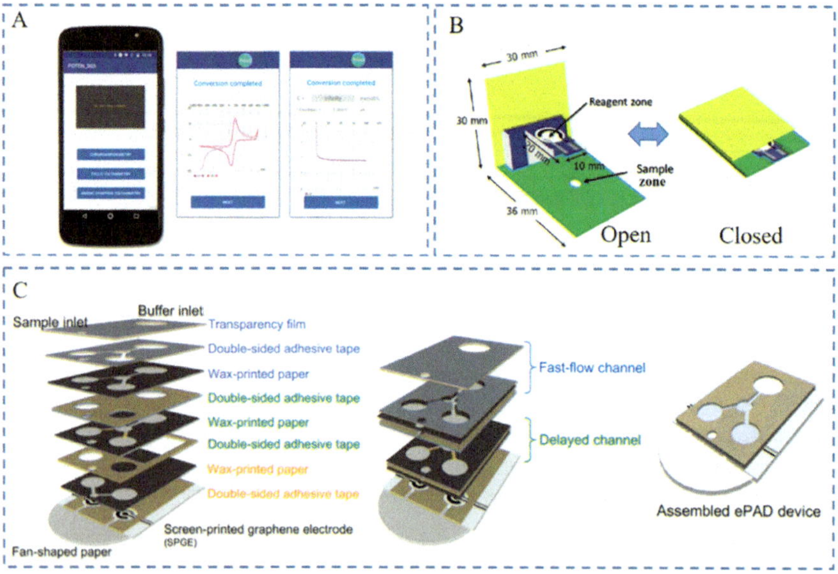

Fig. 9.5 The illustration of **A** a portable electrochemical biosensor for HBV detection has been reported [46], **B** a pop-up PAD device for HVB DNA detection [47], and **C** a PAD platform for simultaneous sensing of HBsAg and HCVcAg [48]

very recently (Fig. 9.5B) [47]. Moreover, the use of pyrrolidinyl peptide nucleic acid as the probe endows the sensor with high binding affinity and specificity. The LOD and linear range of the sensor are 1.45 pM and 50 pM–100 nm, respectively. What's more, it has realized the sensing of HBV DNA, which is extracted from plasmid constructs.

However, the PADs for hepatitis diagnosis still need multistep reagent manipulation, which limits their use for end users. Thus, a series of methods have been developed to further simplify the manipulation or realize the simultaneous detection of multiple targets. For instance, a PAD platform for simultaneous sensing of hepatitis B surface antigen (HBsAg) and HCV core antigen (HCVcAg) has been reported (Fig. 9.5C) [48]. Furthermore, the multi-step reagent manipulation has been simplified by this sensor and only a single buffer loading is needed. This simplification is achieved by the cooperation of a fast-flow channel and a delayed channel. The fast-flow channel aims to automatically wash the unbound antigens. The delayed channel is applied to offer redox regent for further analysis. In addition, the sensor exhibits high sensitivity for HBsAg and HCVcAg detection. The LODs for HBsAg and HCVcAg detection are as low as 18.2 pg mL^{-1} and 1.19 pg mL^{-1}, respectively. The performance of other existing electrochemical biosensors for hepatitis biomarker detection can refer to Table 9.3.

9.1.8 Conclusion and Outlook

Electrochemical biosensors with high specificity and sensitivity for hepatic and cardiac biomarkers detection are of great importance for clinical precaution, diagnosis, and prognosis of related diseases. The typical hepatic and cardiac diseases include myocardial necrosis/myocardial damage, myocardial ischemia, cardiovascular diseases, and hepatitis (HBV and HCV infections). Different diseases have different biomarkers. The representative electrochemical biomarkers for cardiac diseases are cardiac troponins and CK. Besides, the typical predictive biomarkers for cardiovascular events include CRP. In addition, the representative electrochemical biomarkers for hepatic diseases are HBsAg, HBV DNA, and HCVcAg. On account of the different properties of different markers, the corresponding biochemical sensors have different research priorities. For cardiac troponins detection, the specificity and sensitivity can be enhanced by using proper recognition parts and employing nanostructures, respectively. While, for the detection of CK, which is an enzyme and has many specific properties, an enzymatic catalytic reaction can be utilized. The CRP detection is usually accomplished in serum or plasma samples. The blood consumption can be reduced by microfluidic devices. On account of the infectivity and chronic property of viral hepatitis for most cases, the POC devices with a rapid, portable, and easy diagnostic process would benefit the compliance of patients and result feedback. Thus, POC testing is one of the main development directions for HBV and HCV detection.

Table 9.3 The performance and employed methods of the existing electrochemical biosensors for hepatitis biomarkers detection

Biomarker	Electrochemical method	Sensor form	Recognition method	LOD	Detection range	References
HBsAg	i-t	Smartphone-controlled and portable device	Antibody	0.17 μg/mL	10–200 μg/mL	[46]
	Chronoamperometric charges	PADs	Antibody	18.2 pg /mL	0.1–250 ng/mL	[48]
	DPV	Sandwich-type	Antibody	0.16 pg/mL	0.5 pg/mL –0.25 ng/mL	[50]
HBV DNA	DPV	PADs	acpcPNA	1.45 pM	50 pM–100 nM	[47]
	CV	Recombinase polymerase amplification method	ssDNA	0.1 fM	–	[51]
	CV	Traditional device	ssDNA	10 ng/mL	0.05–1 μg/mL	[52]
	DPV	Traditional device	DNA molecular beacon	3.00×10^{-13} M	4.0×10^{-13}–4.0×10^{-9} M	[53]
HCVcAg	Chronoamperometric charges	PADs	Antibody	1.19 pg/ mL	0.001–250 ng/ mL	[48]
	DPV	Traditional device	Antibody	3 fg/mL	0.05 pg mL–60 ng mL	[54]
	SWV	Traditional device	Antibody	0.01 pg/mL	0.25–300 pg/mL	[55]
	DPV	Sandwich-type	Antibody	32 fg/mL	0.1–312.5 pg/mL	[56]
	DPV	Traditional device	Antibody	0.015 pg/ mL (0.71 fmol/L)	0.05–1000 pg/mL	[57]

Although great progress has been achieved these years, some challenges still remain. For example, most sensors can only realize the detection of a single cardiac biomarker, which can only reflect the physiological changes partially. Considering the complex pathological and physiological mechanisms, the simultaneous detection of multiple biomarkers could be more accurate for the diagnosis of the hepatic and cardiac diseases. Additionally, the practical applications require high stability for the long-term usage of the biosensors. However, the electrochemical biosensors are usually environmentally sensitive. The performance would be influenced by the temperature, pH, ionic strength, and so on. Thus, the stability of the related sensors should be investigated and improved. It is believed that the intensive efforts on the electrochemical biosensors for hepatic and cardiac biomarkers detection would result in continuous innovations and provide more clinical application potential.

References

1. Asrani SK, Devarbhavi H, Eaton J, Kamath PS (2019) Burden of liver diseases in the world. J Hepatol 70(1):151–171
2. Du X, Su X, Zhang W, Yi S, Zhang G, Jiang S, Li H, Li S, Xia F (2021) Progress, opportunities, and challenges of troponin analysis in the early diagnosis of cardiovascular diseases. Anal Chem 94(1):442–463
3. Huang K, Miao T, Chang K, Kim J, Kang P, Jiang Q, Simmonds AJ, Di Cara F, Bai H (2020) Impaired peroxisomal import in Drosophila oenocytes causes cardiac dysfunction by inducing upd3 as a peroxikine. Nat Commun 11(1):1–15
4. Stoner L, Lucero AA, Palmer BR, Jones LM, Young JM, Faulkner J (2013) Inflammatory biomarkers for predicting cardiovascular disease. Clin Biochem 46(15):1353–1371
5. Jo H, Gu H, Jeon W, Youn H, Her J, Kim S-K, Lee J, Shin JH, Ban C (2015) Electrochemical aptasensor of cardiac troponin I for the early diagnosis of acute myocardial infarction. Anal Chem 87(19):9869–9875
6. Hasanzadeh M, Shadjou N, Eskandani M, de la Guardia M, Omidinia E (2013) Electrochemical nano-immunosensing of effective cardiac biomarkers for acute myocardial infarction. Trends Analyt Chem 49:20–30
7. Pourali A, Rashidi MR, Barar J, Pavon-Djavid G, Omidi Y (2021) Voltammetric biosensors for analytical detection of cardiac troponin biomarkers in acute myocardial infarction. Trends Analyt Chem 134:116123
8. Friess U, Stark M (2009) Cardiac markers: a clear cause for point-of-care testing. Anal Bioanal Electrochem 393(5):1453–1462
9. Perz JF, Armstrong GL, Farrington LA, Hutin YJ, Bell BP (2006) The contributions of hepatitis B virus and hepatitis C virus infections to cirrhosis and primary liver cancer worldwide. J Hepatol 45(4):529–538
10. Uliana CV, Riccardi CS, Yamanaka H (2014) Diagnostic tests for hepatitis C: recent trends in electrochemical immunosensor and genosensor analysis. World J Gastroenterol 20(42):15476–15491
11. Coffin CS, Zhou K, Terrault NA (2019) New nd old biomarkers for diagnosis and management of chronic hepatitis B virus infection. Gastroenterology 156(2):355–368. e3
12. Zhang GH, Zhang L, Yu Y, Lin BX, Wang YM, Guo ML, Cao YJ (2020) Dual-mode of electrochemical-colorimetric imprinted sensing strategy based on self-sacrifice beacon for diversified determination of cardiac troponin I in serum. Biosens Bioelectron 167:112502

13. Shanmugam NR, Muthukumar S, Chaudhry S, Anguiano J, Prasad S (2017) Ultrasensitive nanostructure sensor arrays on flexible substrates for multiplexed and simultaneous electrochemical detection of a panel of cardiac biomarkers. Biosens Bioelectron 89:764–772
14. Taleat Z, Khoshroo A, Mazloum-Ardakani M (2014) Screen-printed electrodes for biosensing: a review (2008–2013). Microchim Acta 181:865–891
15. Palanisamy S, Raja DS, Subramani B, Wu TH, Wang YM (2020) Label-free bimetallic in situ-grown 3D nickel-foam-supported NH2-MIL-88B(Fe2Co)-MOF-based Impedimetric immunosensor for the detection of cardiac troponin I. ACS Appl Mater Interfaces 12(29):32468–32476
16. Singh N, Ali MA, Ra P, Sharma A, Malhotra BD, John R (2017) Microporous nanocomposite enabled microfluidic biochip for cardiac biomarker detection. ACS Appl Mater Interfaces 9(39):33576–33588
17. Ma N, Zhang T, Yan T, Kuang X, Wang H, Wu D, Wei Q (2019) Novel electrochemical immunosensor for sensitive monitoring of cardiac troponin I using antigen response cargo released from mesoporous Fe3O4. Biosens Bioelectron 143:111608
18. Gholami MD, O'mullane AP, Sonar P, Ayoko GA, Izake EL (2021) Antibody coated conductive polymer for the electrochemical immunosensing of Human Cardiac Troponin I in blood plasma. Anal Chim Acta 1185:339082
19. Chen Y, Mei L-P, Feng J-J, Yuan P-X, Luo X, Wang A-J (2019) Simple one-pot aqueous synthesis of 3D superstructured PtCoCuPd alloyed tripods with hierarchical branches for ultrasensitive immunoassay of cardiac troponin I. Biosens Bioelectron 145:111638
20. Qiao X, Li K, Xu J, Cheng N, Sheng Q, Cao W, Yue T, Zheng J (2018) Novel electrochemical sensing platform for ultrasensitive detection of cardiac troponin I based on aptamer-MoS2 nanoconjugates. Biosens Bioelectron 113:142–147
21. Song Z, Song J, Gao F, Chen X, Wang Q, Zhao Y, Huang X, Yang C, Wang Q (2022) Novel electroactive ferrocene-based covalent organic frameworks towards electrochemical label-free aptasensors for the detection of Cardiac Troponin I. Sens Actuators B Chem 368:132205
22. Chekin F, Vasilescu A, Jijie R, Singh SK, Kurungot S, Iancu M, Badea G, Boukherroub R, Szunerits S (2018) Sensitive electrochemical detection of cardiac troponin I in serum and saliva by nitrogen-doped porous reduced graphene oxide electrode. Sens Actuators B Chem 262:180–187
23. Negahdary M, Behjati-Ardakani M, Sattarahmady N, Yadegari H, Heli H (2017) Electrochemical aptasensing of human cardiac troponin I based on an array of gold nanodumbbells-applied to early detection of myocardial infarction. Sens Actuators B Chem 252:62–71
24. Silva BVM, Cavalcanti IT, Mattos AB, Moura P, Sotomayor MDT, Dutra RF (2010) Disposable immunosensor for human cardiac troponin T based on streptavidin-microsphere modified screen-printed electrode. Biosens Bioelectron 26(3):1062–1067
25. Landim VP, Foguel MV, Prado CM, Sotomayor MP, Vieira IC, Silva BV, Dutra RF (2022) A polypyrrole/nanoclay hybrid film for ultra-sensitive cardiac troponin T electrochemical immunosensor. Biosensors 12(7):545
26. Phonklam K, Wannapob R, Sriwimol W, Thavarungkul P, Phairatana T (2020) A novel molecularly imprinted polymer PMB/MWCNTs sensor for highly-sensitive cardiac troponin T detection. Sens Actuators B Chem 308:127630
27. Rahbani JF, Roesler A, Hussain MF, Samborska B, Dykstra CB, Tsai L, Jedrychowski MP, Vergnes L, Reue K, Spiegelman BM (2021) Creatine kinase B controls futile creatine cycling in thermogenic fat. Nature 590(7846):480–485
28. Liu Y, Luo X, Zhang Y, Li Z, Xiao H, Cui Y (2022) Electrochemical protein-based biosensors for creatine kinase: a review. IEEE Sens J 22:10280–10291
29. Geng T, Song Z, Zhang J, Xu Z (2017) Creatine kinase determination based on an electrochemical impedance immunosensor. Int J Electrochem Sci 12(9):8552–8563
30. Hassan MH, Vyas C, Grieve B, Bartolo P (2021) Recent advances in enzymatic and non-enzymatic electrochemical glucose sensing. Sensors 21(14):4672
31. Moreira FT, Dutra RA, Noronha JP, Sales MGF (2014) Novel sensory surface for creatine kinase electrochemical detection. Biosens Bioelectron 56:217–222

32. Hanefeld U, Gardossi L, Magner E (2009) Understanding enzyme immobilisation. Chem Soc Rev 38(2):453–468
33. Cen S-Y, Feng Y-G, Zhu J-H, Wang X-Y, Wang A-J, Luo X, Feng J-J (2021) Eco-friendly one-pot aqueous synthesis of ultra-thin AuPdCu alloyed nanowire-like networks for highly sensitive immunoassay of creatine kinase-MB. Sens Actuators B Chem 333:129573
34. Haque A-MJ, Kim J, Dutta G, Kim S, Yang H (2015) Redox cycling-amplified enzymatic Ag deposition and its application in the highly sensitive detection of creatine kinase-MB. ChemComm 51(77):14493–14496
35. Seo J, Ha H, Park S, Haque A-MJ, Kim S, Joo JM, Yang H (2017) Immunosensor employing stable, solid 1-amino-2-naphthyl phosphate and ammonia-borane toward ultrasensitive and simple point-of-care testing. ACS Sens 2(8):1240–1246
36. Dou YH, Haswell SJ, Greenman J, Wadhawan J (2012) Voltammetric immunoassay for the detection of protein biomarkers. Electroanalysis 24(2):264–272
37. Wang X-Y, Chen Y, Mei L-P, Wang A-J, Yuan P-X, Feng J-J (2020) Confining signal probe in porous PdPtCoNi@ Pt-skin nanopolyhedra to construct a sandwich-type electrochemical immmunosensor for ultrasensitive detection of creatine kinase-MB. Sens Actuators B Chem 315:128088
38. Lee I, Luo X, Huang J, Cui XT, Yun M (2012) Detection of cardiac biomarkers using single polyaniline nanowire-based conductometric biosensors. Biosensors 2(2):205–220
39. Kucherenko IS, Soldatkin OO, Lagarde F, Jaffrezic-Renault N, Dzyadevych S, Soldatkin A (2015) Determination of total creatine kinase activity in blood serum using an amperometric biosensor based on glucose oxidase and hexokinase. Talanta 144:604–611
40. Dhara K, Mahapatra DR (2020) Review on electrochemical sensing strategies for C-reactive protein and cardiac troponin I detection. Microchem J 156:104857
41. Pan J, Xu W, Li W, Chen S, Dai Y, Yu S, Zhou Q, Xia F (2023) Electrochemical aptamer-based sensors with tunable detection range. Anal Chem 95(1):420–432
42. Dong S, Zhang D, Cui H, Huang T (2019) ZnO/porous carbon composite from a mixed-ligand MOF for ultrasensitive electrochemical immunosensing of C-reactive protein. Sens Actuators B Chem 284:354–361
43. Molinero-Fernandez A, Lopez MA, Escarpa A (2020) Electrochemical microfluidic micromotors-based immunoassay for C-reactive protein determination in preterm neonatal samples with sepsis suspicion. Anal Chem 92(7):5048–5054
44. Fisher DG, Hess KL, Erlyana E, Reynolds GL, Cummins CA, Alonzo TA (2015) Comparison of rapid point-of-care tests for detection of antibodies to hepatitis C virus. Open Forum Infect Di 2(3):101
45. da Costa JPdC, Bastos WB, Da Costa PI, Zaghete MA, Longo E, Carmo JP (2019) Portable laboratory platform with electrochemical biosensors for immunodiagnostic of hepatitis C virus. IEEE Sens J 19(22):10701–10709
46. Teengam P, Siangproh W, Tontisirin S, Jiraseree-amornkun A, Chuaypen N, Tangkijvanich P, Henry CS, Ngamrojanavanich N, Chailapakul O (2021) NFC-enabling smartphone-based portable amperometric immunosensor for hepatitis B virus detection. Sens Actuators B Chem 326:128825
47. Srisomwat C, Teengam P, Chuaypen N, Tangkijvanich P, Vilaivan T, Chailapakul O (2020) Pop-up paper electrochemical device for label-free hepatitis B virus DNA detection. Sens Actuators B Chem 316:128077
48. Boonkaew S, Yakoh A, Chuaypen N, Tangkijvanich P, Rengpipat S, Siangproh W, Chailapakul O (2021) An automated fast-flow/delayed paper-based platform for the simultaneous electrochemical detection of hepatitis B virus and hepatitis C virus core antigen. Biosens Bioelectron 193:113543
49. Li X, Scida K, Crooks RM (2015) Detection of hepatitis B virus DNA with a paper electrochemical sensor. Anal Chem 87(17):9009–9015
50. Li X, Liu T, Zhang Y, Ni X, Hossain MN, Chen X, Huang H, Kraatz H-B (2021) A novel electrochemical immunosensor for hepatitis B surface antigen based on Fe3O4 nanoflowers and heterogeneous chain reaction signal amplification strategy. Talanta 221:121459

51. Ichzan AM, Hwang S-H, Cho H, San Fang C, Park S, Kim G, Kim J, Nandhakumar P, Yu B, Jon S (2021) Solid-phase recombinase polymerase amplification using an extremely low concentration of a solution primer for sensitive electrochemical detection of hepatitis B viral DNA. Biosens Bioelectron 179:113065
52. Wu Y, Li N, Xu S, Chong X, Zhang C (2022) Detection of hepatitis B virus DNA in human serum based on highly sensitive electrochemical sensor. Int J Electrochem Sci 17:220919
53. Zheng J, Chen C, Wang X, Zhang F, He P (2014) A sequence-specific DNA sensor for Hepatitis B virus diagnostics based on the host–guest recognition. Sens Actuators B Chem 199:168–174
54. Valipour A, Roushani M (2017) Using silver nanoparticle and thiol graphene quantum dots nanocomposite as a substratum to load antibody for detection of hepatitis C virus core antigen: electrochemical oxidation of riboflavin was used as redox probe. Biosens Bioelectron 89:946–951
55. Ma C, Liang M, Wang L, Xiang H, Jiang Y, Li Y, Xie G (2013) MultisHRP-DNA-coated CMWNTs as signal labels for an ultrasensitive hepatitis C virus core antigen electrochemical immunosensor. Biosens Bioelectron 47:467–474
56. Wang Y, Li L, Dong Z, Yu Y, Zhou A, Zhao X, Zhang J (2021) Ultrasensitive electrochemical detection of hepatitis C virus core antigen using terminal deoxynucleotidyl transferase amplification coupled with DNA nanowires. Microchim Acta 188(8):1–8
57. Pusomjit P, Teengam P, Chuaypen N, Tangkijvanich P, Thepsuparungsikul N, Chailapakul O (2022) Electrochemical immunoassay for detection of hepatitis C virus core antigen using electrode modified with Pt-decorated single-walled carbon nanotubes. Microchim Acta 189(9):1–10

Chapter 10
Electrochemical Biosensors for Inflammatory Biomarkers Detection

Zhijuan Duan, Danlong Chen, Hong Liu, Huiying Xue, Fujian Huang, and Fan Xia

Abstract Detection of inflammatory biomarkers by electrochemical biosensors is vital for the early detection of disease, gene mutations, and biological targets. In this chapter, we summarized the detection of inflammatory biomarkers by electrochemical biosensors, especially the inflammatory biomarkers in whole blood. First, different inflammatory biomarkers related to different diseases were introduced, especially the inflammatory biomarkers in the whole blood. Then electrochemical biosensors were introduced in detail. Next, inflammatory biomarkers in whole blood were introduced in detail. And the application of electrochemical sensors in life analysis and clinical application was summarized. Finally, we discussed the prospects and challenges for technologies based on electrochemical biosensors for inflammatory biomarkers detection.

Keywords Inflammatory biomarkers · Electrochemical biosensor · Whole blood

10.1 Brief Introduction of Inflammatory Biomarkers

10.1.1 Inflammatory Biomarkers

The process of inflammation is the defense process of the body caused by tissue damage. Inflammation can be caused by various factors, such as physical factors (trauma or high temperature or cold), chemical factors (endogenous or exogenous acid–base balance), or infectious factors (bacteria, viruses or fungi) [1]. The course

Z. Duan · D. Chen · H. Liu · H. Xue
Faculty of Materials Science and Chemistry, China University of Geosciences, Wuhan 430074, China

F. Huang (✉) · F. Xia
State Key Laboratory of Biogeology and Environmental Geology, Engineering Research Center of Nano-Geomaterials of Ministry of Education, Faculty of Materials Science and Chemistry, China University of Geosciences, Wuhan 430074, China
e-mail: huangfj@cug.edu.cn

F. Xia et al. (eds.), *Electrochemical Biosensors for Whole Blood Analysis*,
https://doi.org/10.1007/978-981-99-5644-9_10

of acute inflammation has a sequence of symptoms such as fever, pain, or swelling. Chronic inflammation has none of these symptoms and may involve fibrosis, tissue destruction, and eventually necrosis [2]. When organisms are stimulated by the outside world, a series of complex cell biological effects are initiated in vivo, including the activation and conduction of stress signals, the imbalance of inflammatory response, and the release of various harmful factors. Currently, it is believed that various inflammatory mediators and cellular mediators are released during external stimulation and interact with each other to form a complex cytokine network, which plays an important role in inflammatory response and immune disorders. Biological immune disorders are likely to cause abnormal activation of various biomarkers, which in turn lead to a cascade of immune responses through the release of a large number of inflammatory cytokines [3]. If the production of these inflammatory factors is blocked in time, the inflammatory response can be reduced and the vicious cycle can be blocked, which is expected to provide new ideas and new methods for the treatment of various inflammation-induced diseases [4].

Inflammatory biomarkers have some common characteristics: (1) biomarkers are mostly secreted proteins with low molecular weight (<30 kd), and are generally glycoproteins. (2) Biomarkers are usually involved in immune and inflammatory responses. (3) The production of cytokines is transient and regional, and the main form of secretion is paracrine or autocrine. (4) Biomarkers have high biological activity. (5) Biomarkers must bind to highly affinity-specific receptors on the surface of their target cells. (6) The effects of different factors often overlap each other.

The levels of these biomarkers can be measured in the blood and used to diagnose and monitor various inflammatory diseases, such as rheumatoid arthritis, inflammatory bowel disease, and others. Inflammatory biomarkers can also be used to monitor the effectiveness of treatment and to predict the risk of future health problems [5].

10.1.2 Inflammatory Biomarkers in Whole Blood

Inflammatory biomarkers can be measured in whole blood, which is a sample of blood that has not been separated into its component parts. Blood is made up of plasma and blood cells. In the whole blood, 70% is plasma and about 30% is blood cells. In whole blood, inflammatory biomarkers are typically found in the plasma, which contains a variety of electrolytes, fibrinogen, coagulation factors, and so on. We will briefly introduce the inflammatory factors in whole blood, serum, plasma, and platelets. At present, serum is the main sample for detecting inflammation, which involves many inflammatory biomarkers, and has a great relationship with cancer, nervous system, and COVID-19 diagnostic biomarker diseases. Some of the most commonly measured inflammatory biomarkers in whole blood include:

C-reactive protein (CRP) is an acute-phase protein synthesized by the liver and released into the blood when tissue is damaged by trauma or inflammation. Serum CRP concentrations can be hundreds or thousands of times higher during acute bacterial infection or tissue trauma. Therefore, CRP is an important biomarker for the identification of infection [6].

Interleukin-6 (IL-6) is a cytokine, a type of signaling molecule that is involved in the immune response. Elevated levels of IL-6 in whole blood can indicate the presence of systemic inflammation [7].

Tumor necrosis factor-α (TNF-α) is another cytokine involved in the immune response. Elevated levels of TNF-α in the whole blood indicate the presence of inflammation [8].

Erythrocyte Sedimentation Rate (ESR) is a measure of the rate at which red blood cells settle to the bottom of a test tube. Elevated ESR can indicate the presence of inflammation [9].

In addition to these inflammatory factors, there are many other inflammatory factors in the whole blood, such as inflammatory cytokines PSA [10] and P53 protein (AβO) [11] in blood plasma have been linked to cancer and neurological diseases. ORF1ab RNA in whole blood is associated with disease in diagnostic biomarkers of COVID-19 [12], while inflammatory cytokines IL-6 [13], IL-8 [14], and IL-1β [15] in platelets are associated with atherosclerosis. Examples of the detection of inflammatory biomarkers in different samples and their analytical parameters and methods will be detailedly discussed in the following context.

10.2 Principles for Inflammatory Biomarkers Detection by Electrochemical Biosensors

A biosensor is an apparatus that transforms bioinformation, for instance, the concentration of sample components into viable signals. As shown in Fig. 10.1, electrochemical biosensor mainly includes recognition elements and signal transducers. Here, we review biosensors for inflammatory biomarkers detection including transducer elements, biorecognition elements, and detection methods.

10.2.1 Transducer Elements-Electrode

The different materials and structures greatly affect the properties of the electrode, and ultimately affect the sensibility, selectivity of the text of the electrochemical biosensor. The differences in electrode materials have been introduced in detail in different literature, and will not be repeated here. Electrode structure, including planar structure, linear structure, nanostructure, and array structure, can also affect electrode performance. Here we focus on nanoelectrodes.

The surface morphology of the electrode with micron and nanostructure affects the detection of inflammatory biological targets. The nanomodification of the electrode greatly increased the specific surface area but did not increase the volume of the

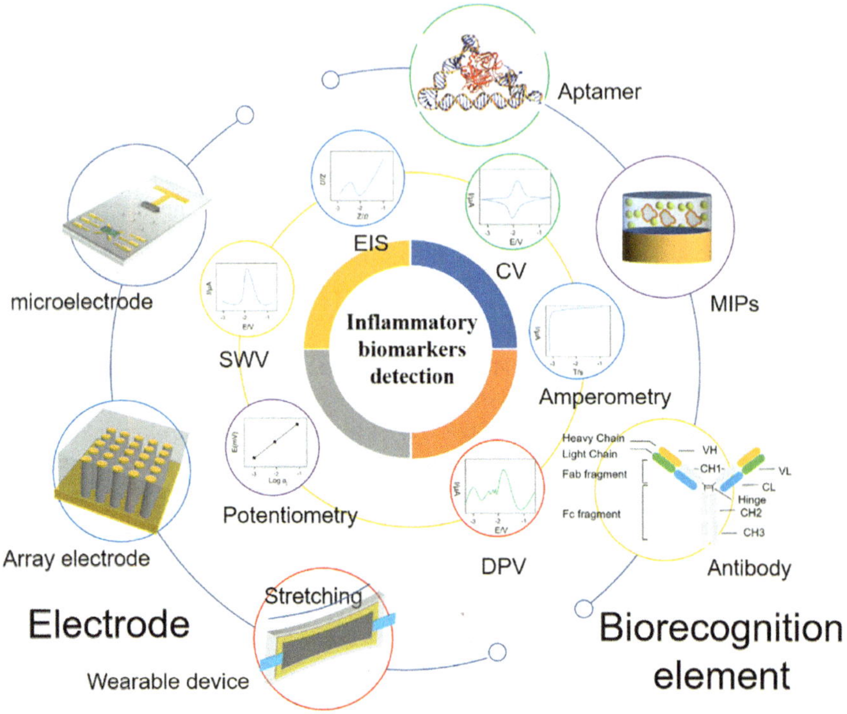

Fig. 10.1 The principle of electrochemical biosensors detecting inflammatory biomarkers. CV: Cyclic voltammetry; SWV: Square wave voltammetry; DPV: Differential pulse voltammetry; EIS: Electrochemical impedance spectroscopy.

electrode, thus increasing the reaction area of the fluid and surface. The preparation of electrode nanostructures is mainly achieved by bottom-up wet chemistry and electrochemical methods. In the process of wet chemistry approaches the preparation of electrode nanostructures, nanoparticles are usually deposited or coupled to the surface of the planar electrode [16]. For example, gold nanoparticles are usually deposited on the surface of a planar electrode and fixed to a biorecognition element. In these studies, it is usually done by physical adsorption [17] or chemical methods [18]. Compared with planar electrodes, nanostructured electrodes with high curvature can reduce the aggregation of identifying elements in experimental studies and molecular dynamics simulations of DNA sensing [19, 20].

In addition to coupling treated nanomaterials to the electrode surface, nanoelectrodes are often prepared by bottom-up electrochemical methods. Although nanomodification of the electrode surface by physical or chemical methods is effective, the introduction of porosity, such as nanopores, into the electrode is also the way to prepare nanostructured electrodes. Carbon nanotubes (CNT) are ideal candidates for label-free sensing. The sensing mechanism is the gating effect of the target proteins as they bind to the antibodies, bringing them near the CNT channels within the Debye length. When using antibodies as recognition elements, the large size of the antibody can enable the target protein to interact outside the space charge layer or the Debye length of the sensor.

Electrode nanostructures can improve electrode properties, such as sensitivity and detection limits, but fixation of biometric elements and target binding by nanostructures should also be considered. Differences between batches of nanostructured electrodes also need to be considered. It is not clear how the surface of nanostructures, such as morphology, crystal structure, and material properties, such as electrical conductivity, change between mass-produced electrodes.

10.2.2 Biorecognition Elements

The previous section discussed transduction components, the most important part of electrochemical detection. Complete biosensors include sensors and biometric elements. This section will discuss biometric elements that selectively bind inflammatory biomolecules.

The mechanisms of biometric components include biocatalytic and biocomplexing. In the biocatalytic mode, the biosensor responds to chemical reactions based on the catalytic target. Enzymes, cells, and tissues are commonly used as biocatalytic elements. Enzymes often label inflammatory molecules by a secondary binding step as biosensor recognition elements. In the biocomplexing model, the response of the sensor is based on the interaction between the target and macromolecules. Antibodies, peptides, phages, and imprinted polymers are broadly used biorecognition elements.

10.2.2.1 Biorecognition Elements

Antibodies and antibody fragments are the most widely used recognition elements. Antibodies have strong selectivity and high affinity and can bind specifically to epitopes of antigens [21]. Antibodies can be labeled with fluorescence or enzymes, which may lead to additional reagent and treatment-step-related measurement limitations, and may also alter the binding affinity to the antigen, affecting the selectivity of the biosensor.

Enzymes are another most commonly used biometric element, and their specificity comes from non-covalent recognition patterns in their three-dimensional structure [22]. When the enzyme is integrated into a biosensing device, it converts the target into an electrochemically measurable product through a biocatalytic effect. Considering the stability of the recognition element, aptamers have attracted much attention. It is a short-chain oligonucleotide composed of synthetic bases [23].

10.2.2.2 Non-biorecognition Elements

Aptamers can be customized to the targets and have similar chemical, thermal, and mechanical stability to molecularly imprinted polymers (MIPs). They can be reused and are relatively inexpensive [24]. MIPs are synthetic recognition elements that polymerize around the target, effectively freezing inflammatory molecules. There are many different ways to synthesize MIPs, such as thermal polymerization [25], UV-polymerization [26], and electropolymerization [27]. The number of studies on MIPs has increased significantly in recent years, and their performance has improved, but they still have the disadvantages of long production time, template leakage, complex manufacturing processes, and relatively poor synergies with electrochemical platforms. In addition to the recognition elements mentioned above, phages [28] and affibodies [29] are also included. The selection of biometric components is one of the most important steps in developing a biosensor, which depends on the properties of inflammatory molecules and the desired properties of the biosensor, such as cost, lifetime, and sensitivity. The ideal recognition element should have high stability and high affinity for the target molecule [30]. Table 10.1 compares the advantages and limitations of the various recognition elements.

10.2.3 Electrochemical Methods

Inflammatory molecules can be detected using a variety of electrochemical methods [32]. These methods differ in electrode configuration, applied signal, measured signal, mass transmission mechanism, binding information provided, and target size selectivity [33]. Electrochemical methods can be classified as potentiometric, voltammetry, electrochemical impedance spectroscopy, or conductometry, and these methods usually represent the measured signal. The signals applied on the electrode surface include both stabilization signals and change signals. The conclusion may require analysis of the output of transient signals, steady-state signals, and sometimes a combination of them. This section provides a brief introduction to the most commonly used electrochemical methods for the detection of inflammatory molecules.

10.2.3.1 Voltammetry

Voltammetry, also known as control potentiometry, is a method of measuring current by applying a rated voltage [32]. The current represents the charge transfer between the sample and the electrode and thus indicates the sample concentration. In chronoamperometry, the potential on the electrode is applied in steps, so that the current is measured over time. The potential in voltammetry can remain constant or change over time.

Table 10.1 Advantages and limitations of recognition elements in chemical sensors and biosensors

Recognition elements		Sensor designation	Advantages	Limitations
Classical	Enzymes	Enzymatic biosensor	–Specificity –Simple apparatus and procedures	–Purification is costly and time consuming –Poor stability –Efficient only at optimum pH and temperature
	Antibodies	Immunosensor	High affinity –Specificity	–Limited target (protein) –Laborious production –Production requires use of animals –Lack of stability
	Nucleic acids	Genosensor	–Stability	–Limited target (complementary nucleic acid)
	Whole cells	Whole-cell biosensor	–Low-cost preparation –Reduced purification requirements	
Recent	Phages	Phage biosensor	–Specificity and sensitivity –Stability	
	Aptamers (DNA, RNA, or peptides)	Aptasensor (particularly, DNA sensor or RNA sensor) Peptide sensor	–Easy to modify –Possibility to design structure –Possibility to denaturalize and to rehybridize –Possibility to distinguish targets with different functional groups –Thermally stable –In-vitro synthesis	

(continued)

Table 10.1 (continued)

Recognition elements		Sensor designation	Advantages	Limitations
	Molecularly imprinted polymers (MIPs)	MIP sensor	–High thermal, chemical, and mechanical tolerance tolerance –Reusability –Low cost	–Complex fabrication methodology –Time-consuming process –Incompatibility with aqueous media –Leakage of template molecules
	Affibodies	Affibody sensor	–Lack of disulfide bonds that enable intracellular applications –Long shelf-life	

Adapted and reproduced from ref. [31]

Although voltammetry is compatible with a variety of biosensors, it is most commonly used in conjunction with field effect transistor (FET) -based biosensors. FET is a semiconductor device that uses the electric field effect in the input loop to control the current in the output loop, conducting electricity only by the majority of charge carriers in the semiconductor. FET biosensors detect Inflammatory molecules via measured changes in source-drain channel conductivity that arise from the electric field of the sample environment [34].

The linear sweep voltammetry (LSV) has a linear change of applied voltage with time and measures the response current. Manickam et al. detected the amount of NO in blood by LSV, and the oxidation potential of the electrode changed with the concentration of NO. The anode current increases linearly and proportionally with the increase of NO concentration [35].

Pulsed voltammetry applies a voltage to the electrode surface in the form of a pulse. Compared with traditional voltammetry, pulse voltammetry is faster and more sensitive [36]. In step voltammetry, the potential changes step by step, which reduces the influence of the charging capacitor and a non-equilibrium electrode on the current signal.

10.2.3.2 Electrochemical Impedance Spectroscopy (EIS)

The previously mentioned methods, including methods of applying signals with continuous or stepped variation, change the electrode balance. A frequency response method, often referred to as EIS, is the response to a current or potential waveform applied periodically at a fixed frequency or within a certain frequency range [32]. A small amplitude sinusoidal potential signal is applied to the electrode so that opposite chemical reactions (i.e., oxidation and reduction) alternate on the electrode.

Therefore, even if the electrode works for a long time, the surface properties of the electrode will not change greatly. So, the EIS method is called the equilibrium method. In addition, the EIS method can be considered that the electrode is always in equilibrium, thus greatly simplifying data processing. EIS can be measured over a wide range of frequencies, allowing more analyte information and electrode structure information to be obtained.

10.3 Electrochemical Biosensors for Inflammatory Biomarkers Detection in Whole Blood

Blood is a significant source of clinical information for monitoring human disease and health [37]. Measuring biomarker levels within the blood, for instance, is an advantageous way of detecting illnesses such as cancer early, allowing for better treatment choices and improved rates of survival for patients [38]. Moreover, the curative effect has been proven by tracking the levels of selected markers in whole blood before and following treatments, which assists doctors in choosing the best options for therapy [39]. Monitoring relevant biomarkers on a regular basis is useful since it offers timely and quick feedback. The method necessitates the use of a user-friendly sensory platform capable of monitoring biomarker levels in a cheap, noninvasive, and time-efficient way with minimum pain for patients. However, few existing strategies meet all of the mentioned requirements. As a result, more research is required to realize such a strategy.

Biosensors have the possibility to satisfy the required parameters since they provide cheap, label-free, quick, and electrical detection. Using a capture agent on the sensor surface, these devices connect the targets with specificity and selectivity. Current biomedical research fields include drug discovery and clinical diagnostics. Protein, glucose, inorganic salt, and hormones are among the many different constituents that make up whole blood, making it exceptionally complicated. As a consequence, studying the complete blood can reveal a lot of information that is very helpful for determining the patient's health status. Utilizing a capture-release microfluidic system, significant advancements have recently been realized in biosensing from desalted serum or whole blood [40].

10.3.1 Potentiometry

Potentiometric biosensors that rely strongly on ion-selective electrodes, which are membrane electrodes whose potential reacts specifically to the concentration of a particular ion [41], monitor the buildup of charges at an electrode under the circumstances of little current flow. Ion-selective electrodes for ions like Na^+, K^+, and Cl^- arc typical examples, as well as the glass pH electrode. The sensors assess the voltage across a membrane that only responds with the charged ion of interest using an electrochemical cell with two electrodes as references.

In recent years, advances in the field of electrochemical biosensing consisting of different nanostructures have piqued the interest of biosensor researchers due to their potential properties such as biological safety, biological compatibility, low cost, improved analytical sensitivity, high electron transfer communication, and ease of fabrication [42]. For the first time, Ibupoto et al. [43] established the first potentiometric biosensor for CRP detection by applying ZnO nanotubes coated with antibodies. This scheme uses physisorption techniques to anchor monoclonal anti-C-reactive protein (CRP-8) modified with glutaraldehyde onto nanotubes. (Fig. 10.2).

Pro-inflammatory cytokines such as IL-1 and IL-6 have been shown to be activated in depressive patients during the acute phase [44, 45] Free radicals are extremely active substances, produced during inflammation and mitochondrial metabolism. According to experimental and clinical studies, these types of inflammatory reactions are associated with the nitrosative stress pathways and activation of oxidative [46]. The ability of a compound to neutralize damaging free radicals or to inhibit the interaction of a substrate with nitrogen species or reactive oxygen is referred to as antioxidant activity. Research on total antioxidant activity (TAA) might provide insight into the molecular mechanism that distinguishes severe depressive illness from healthy individuals [47].

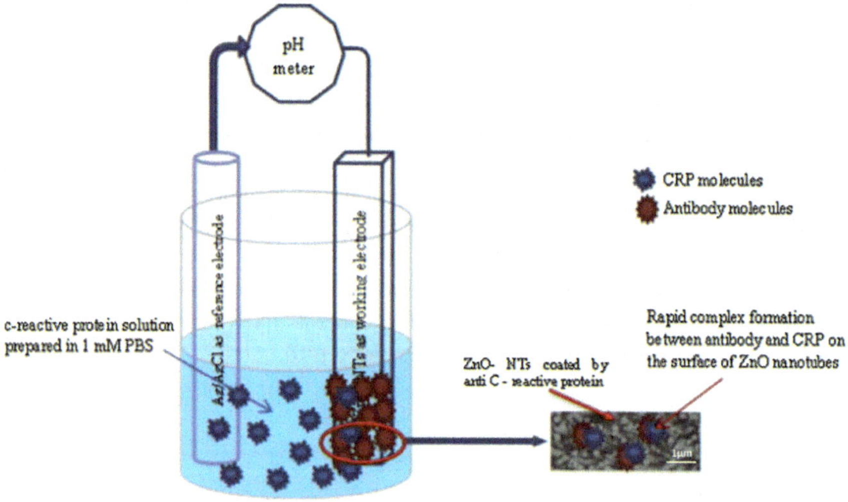

Fig. 10.2 The schematic illustration of the biosensor for CRP detection based on a potentiometric antibody attached to ZnO nanotubes [43]

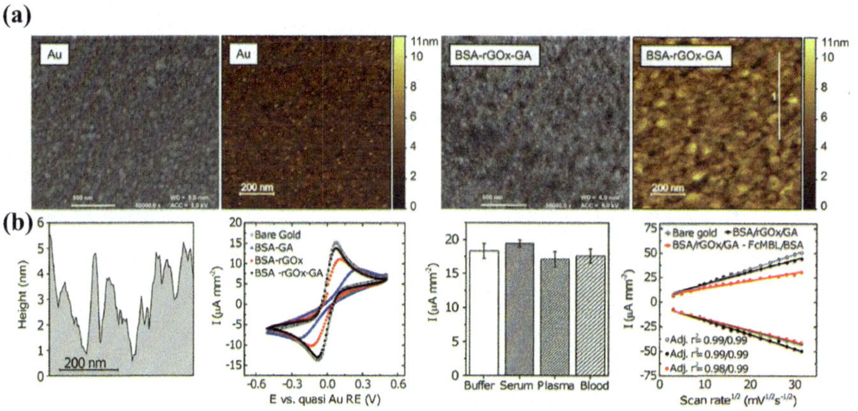

Fig. 10.3 Characterization of the nanocomposite coating [50]

10.3.2 Voltammetry

Voltammetric sensors, in general, work by triggering the generation or interference of the redox current output being measured. In its most basic form, this method relies on the direct oxidation or decrease of the target oxidation-active residues on a sufficiently biased electrode surface [48].

Both the antibody and the aptamer were applied to capture the target protein from whole blood, as previously stated. The aptamer is capable of detecting bioanalytes as well as remaining stable and having good biocompatibility [49]. As a result, by combining the function of the plasma separation membrane and the biometric probe, an assembled paper microfluidic chip is capable of plasma separation as well as simultaneously identifying the presence of PAB and CRP, making the device ideal for whole blood testing. An electrochemical sensor platform was described by Uroš et al. [50] By integrating a nanocomposite coating made of cross-linked bovine serum albumin with a network of reduced graphene oxide nanoparticles that avoid biological contamination while retaining electrical conductivity, the coating is able to detect various sepsis biomarkers concurrently. Cyclic voltammetry (CV) in ferri-/ferrocyanide solution was used to analyze the electrical characteristics of the nanocomposite (Fig. 10.3).

10.3.3 Electrochemical Impedance Spectroscopy (EIS)

EIS becomes an extremely potent tool as electrode surfaces are controllably changed with biological receptors, as previously stated, since it has an inherent ability to detect changes in the conductivity/resistance or capacity for charging of an electrochemical surface. EIS-based platforms hold great promise for the development of label-free

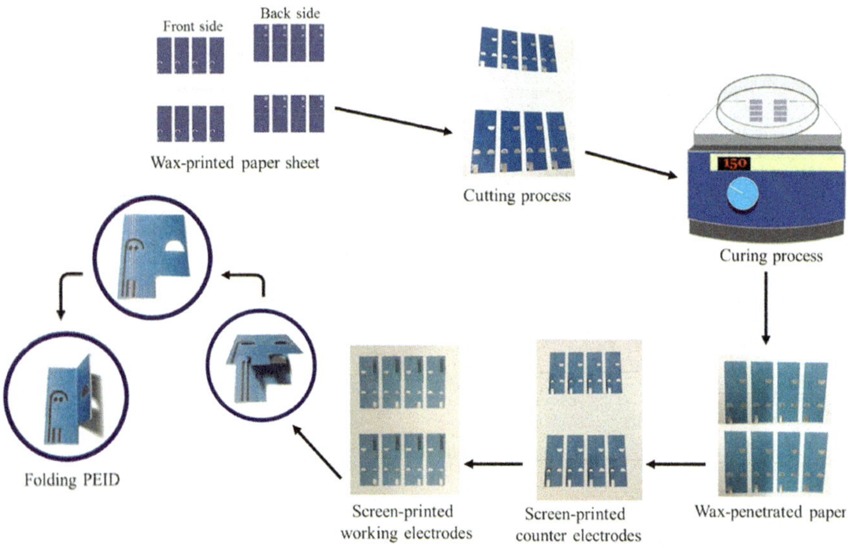

Fig. 10.4 Schematic diagram of a folded affinity-based PEID electrode sensor [58]

ultrasensitive diagnostic sensors for identifying protein biomarkers including diabetic markers [51], inflammatory markers [52], cancer markers [53], and cardiac markers [54]. The amplitude and phase of the response reflect the chemical and physical structure, hence EIS offers an excellent understanding of the signal transmission mechanisms that occur at the electrode contact [55].

There have been several reports of EIS assays for detecting inflammatory biomarkers in whole blood. Kanyong et al. [56] used a simple strategy for surface-initiated reversible addition-fragmentation chain transfer polymerization of HEMA using 4-cyano-4-(phenylcarbonothioylthio) pentanoic acid as a chain transfer agent, which was grafted directly to the surface of gold substrates via its dithio-group. Thomas et al. [57] created a label-free, sensitive, and recyclable electrochemical biosensor for detecting CRP in whole blood. Meanwhile, yuwadee et al. [58] developed a straightforward folded paper-based electrochemical impedance device with a double screen-printed electrode for electrochemical impedance measurement without labels (Fig. 10.4).

10.3.4 Conductometry

Conductometric sensors have long been used for measuring current flow (or conductance) via an analytical solution or medium connecting two electrodes [59].

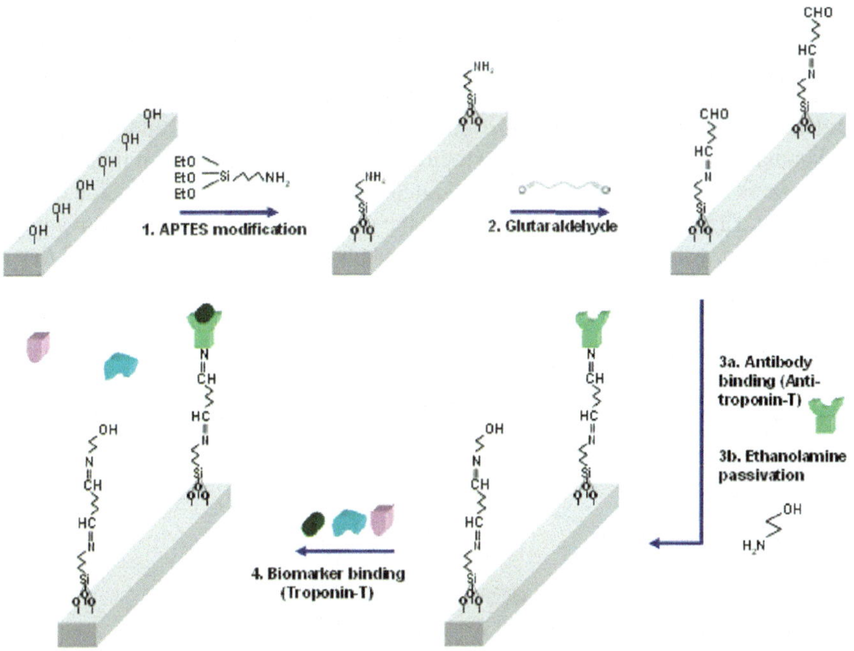

Fig. 10.5 Diagrammatic representation of the biochemical method for silicon nanowire devices' surface functionalization [60]

Traditionally, proteins, antibodies, disease biomarkers or different biomolecules have been detected using conductometric electrochemical biosensors with electrochemical sensor forms. Silicon nanowires can be created using either top-down lithography methods or catalyzed bottom-up growth techniques. Jay et al. [60] demonstrated a complementary metal-oxide semiconductor-compatible silicon nanowires array platform for ultrasensitive, label-free, real-time detection of cTnT in undiluted serum samples (Fig. 10.5).

Electrochemistry is a surface-based method that has a few advantages in the detection of biosensors. The measurement can be performed with very small sample volumes and is not significantly affected by the reaction volume [61]. Due to the lack of a separation process to separate the antigen–antibody complex from the unbound assay elements, chromophores, fluorophores, and other sample components that frequently obstruct spectrophotometric detection are unaffected by electrochemical detection in homogeneous immunoassays. As a result, hemoglobin, fat globules, and red blood cells have no effect on electrochemical measurements of colorful or turbid materials such as whole blood [62]. Electrochemistry, in our opinion, will be the most promising method for detecting inflammatory markers in whole blood.

10.4 Bioanalytical and Clinical Applications

The relationship between inflammatory biomarkers and diseases has received more and more attention with the improvement of scientific research and the study of inflammatory factors [63]. Early and rapid identification of the type of inflammatory biomarkers and the occurrence of associated diseases is important to guide clinical treatment. Electrochemical apparatus and the developed biosensor with advantages like rapidity, cost-effectiveness, stability, repeatability, good sensitivity, lower detection limit, and wide linear detection range [64]. Therefore, the development of reliable, convenient, and rapid electrochemical methods for the detection of inflammatory markers in clinical laboratories is particularly critical for the auxiliary diagnosis, differential diagnosis, guidance of treatment and prognostic evaluation of related diseases and avoidance of misdiagnosis (Fig. 10.6).

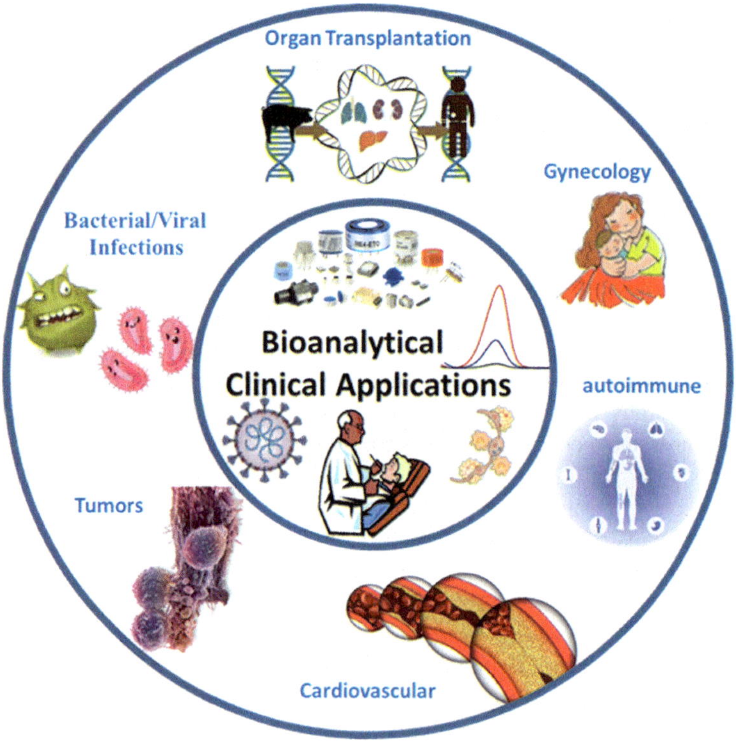

Fig. 10.6 Bioanalytical and clinical applications of electrochemical sensors for inflammatory biomarkers detection

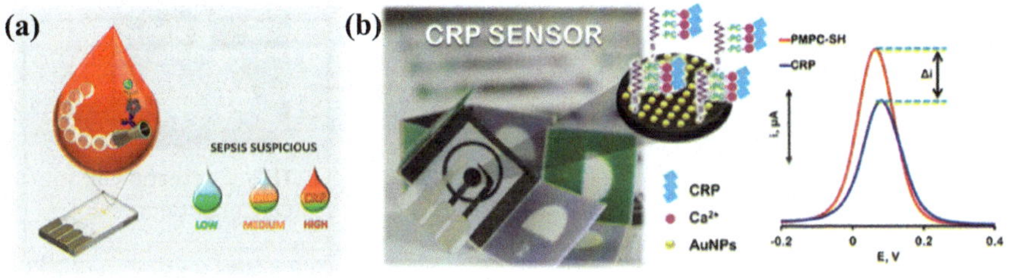

Fig. 10.7 **a** Schematic diagram of CRP analysis in sepsis samples; **b** Immunoassay for CRP [31]

10.4.1 Analysis and Application in the Diagnosis of Bacterial or Viral Infections

Biomarkers such as CRP, procalcitonin (PCT), and SAA are helpful in the early detection of sepsis. Once sepsis occurs in the body, the levels of biomarkers are elevated. By utilizing micromotors and microfluidics, Molinero et al. achieved reliable analysis of CRP in trace blood samples from patients with sepsis (Fig. 10.7) [31]. Balayan et al. showed the development of electrochemical biosensors based on nanomaterials integrated with molecularly imprinted polymer technology [65]. The developed sensor has an operating range of 0.01–10^6 pM as well as an LOD of 0.01 pM. Medetalibeyoglu et al. constructed an extremely sensitive sandwich-type electrochemical immunosensor that was used to assay the PCT [66]. The fabricated sensor exhibited excellent repeatability, reproducibility, stability and reusability and a wide linear detection range.

10.4.2 Analysis and Application in Malignant Tumors

Serum concentrations of TNF- and hs-CRP may be associated with pathophysiology and other functions in cancer patients [67]. The threat of radiation pneumonia is not validly diagnosed owing to early, non-specific clinical signs. Serum biomarkers interleukin-6, C-reactive protein, and procalcitonin are potentially an effective approach both for immediate monitoring and for managing RP risk in clinical practice [68]. Valverde et al. constructed an immunosensor using magnetic microbeads coupled to a carbon electrode in order to identify tumor necrosis factor-alpha (TNF-α) associated with breast cancer (BC) and metastasis [69]. The developed sensor shows the advantages of rapidity, cost-effectiveness, and high sensitivity (3.0 pg mL^{-1}) and has the potential to improve the reliability of BC diagnosis. Sri et al. fabricated a MoS_2 nanoflower-based biosensor by electrophoretically depositing it onto an indium tin oxide coated glass [70]. It serves to assay TNF-α in cancer samples (Fig. 10.8). The produced biosensor exhibited high sensitivity, high detectivity along with an extensive detection range. Altintas et al. developed and optimized a novel

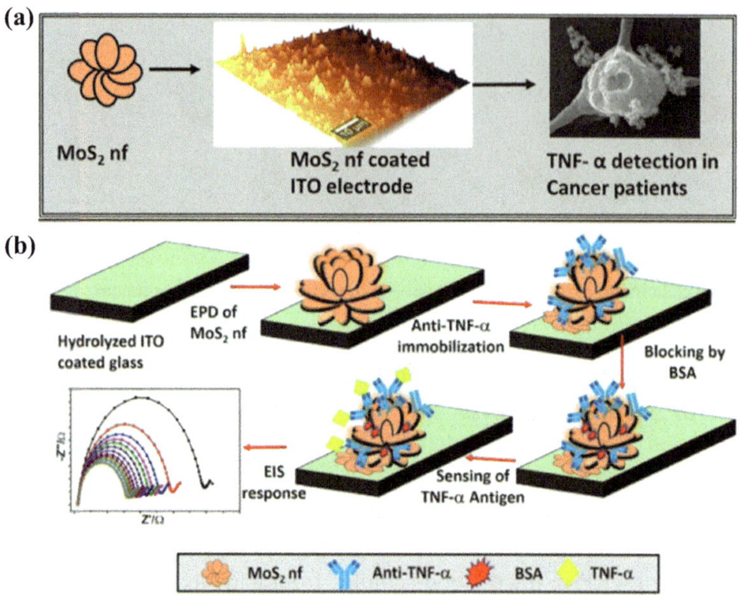

Fig. 10.8 **a** MoS_2 nanoflower sensor designed to monitor TNF-α in patients with cancer; **b** Schematic presentation of the biosensor fabrication [70]

magnetic particle-modified capacitive sensor for the measurement of cancer markers including CRP [71].

10.4.3 Analysis and Application in Cardiovascular Disease

The invention of electrochemical biosensors for quick and super-sensitive recognition of cardiovascular risk markers from patient blood samples will help improve the identification of patients at high risk for cardiovascular disease and help guide treatment. Elevated levels of interleukins and TNF-α as potential cardiovascular risk markers in humans are responsible for the risk of cardiovascular disease [72]. Hou et al. produced a hydrogel electrochemical immunosensor and used it to measure TNF-α in clinical serum samples [73]. The immunosensor demonstrated excellent sensitivity, a good linearity range, satisfactory selectivity, and acceptable reproducibility. Aydın developed a novel electrochemical immunosensor for the high-precision detection of IL 6 in clinical samples using disposable indium tin oxide electrodes [74]. Liu et al. designed a dual-response sandwich immunosensor for IL-6 detection based on two TiO_2 mesocrystal nanoarchitectures combined with electrochemiluminescence and electrochemical detection [75]. The developed sensor has good selectivity and reproducibility and shows potential for application in clinical detection and diagnosis. (Fig. 10.9).

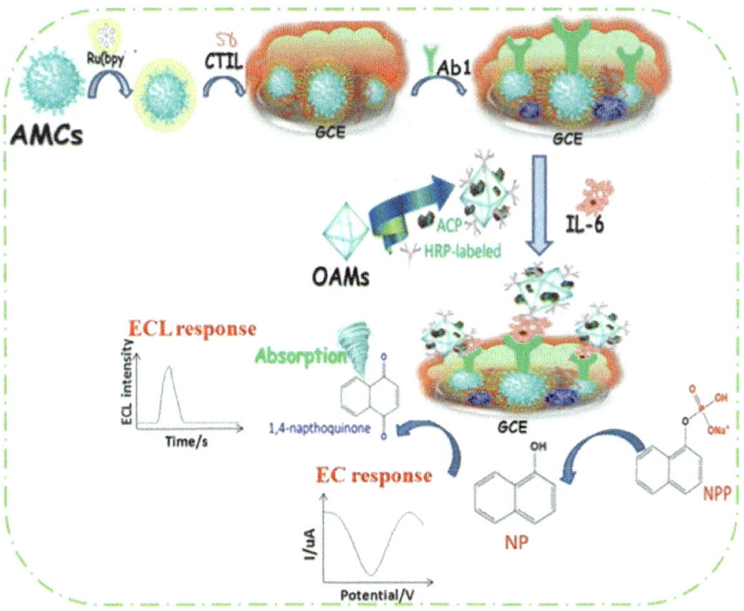

Fig. 10.9 Preparation of IL-6 dual response immunosensor [75]

10.4.4 Analysis and Application in Autoimmune Diseases

Interferon-γ (IFN-γ) is an inflammatory cytokine secreted by immune cells, and its production is associated with effective immune defense. Elevated IFN-γ levels are early indicators of many diseases and are important in the evaluation, treatment progression, and follow-up of autoimmune diseases [76]. Zhou et al. described an electrochemical aptamer sensor for the recognition of IFN-γ [77]. The demonstrated sensor exhibits good stability, repeatability, and reproducibility and responded to IFN-γ down to 1.7 pg/mL. This work was validated to be applicable to the detection of cellular secretion of IFN-γ (Fig. 10.10). Chu et al. constructed an electrochemical sensor composed of TiO_2 nanoparticles/glassy carbon electrode and IFNγ monoclonal antibody, an outstanding strategy for simple, specific and sensitive sensing of IFNγ [78].

10.4.5 Analysis and Application in the Area of Organ Transplantation

There is a persistent state of systemic inflammatory response after organ transplantation, and increased inflammatory markers can be detected in the serum of transplant recipients [79]. PCT is a sensitive and dynamic marker of inflammation, providing

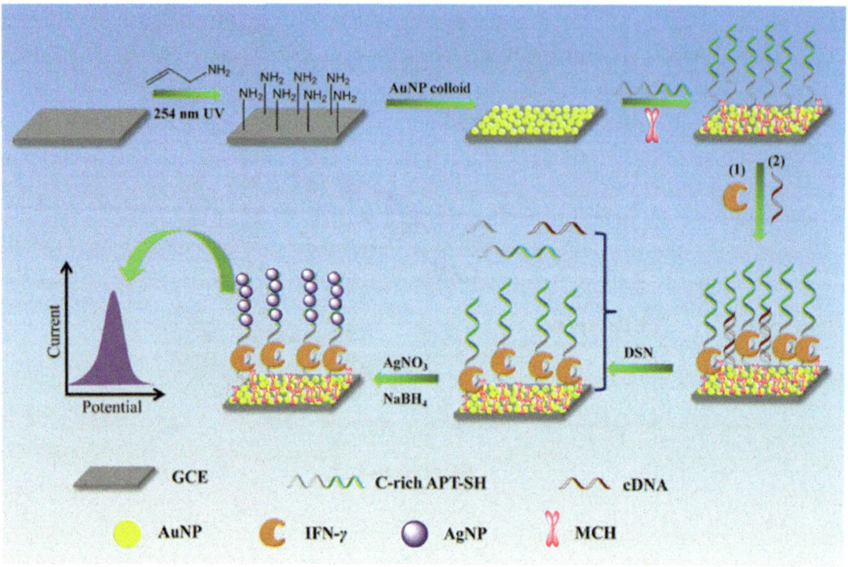

Fig. 10.10 Aptasensing strategies for assaying IFN-γ [77]

crucial parameters for early identification of symptoms and clinical management of infection or rejection, and a rapid decrease in PCT after clinical treatment is highly suggestive of treatment efficacy [80]. Li et al. presented a dual-mode electrochemical immunoassay solution for the effective determination of PCT [81]. The designed assay demonstrated superior performance for PCT protein (Fig. 10.11). Moreover, the method possesses great potential for research in clinical diagnosis and treatment. Huang et al. reported an electrochemical immunosensor using 3D wood as a matrix, which can achieve high sensitivity and accurate recognition of PCT [82].

10.4.6 Analysis and Application in Obstetrics and Gynecology

Gynecological cancer is one of the major causes of serious threats to women's health. Serum inflammatory factors have been demonstrated to be used as an adjunctive diagnosis for gynecologic tumors. IL-10 is a key biomarker for the correct diagnosis and rapid treatment of adverse preeclampsia [83]. IL-6 values were observed at a markedly higher level in women with pelvic inflammatory disease. Arkusz et al. established an electrochemical impedance sensor with titanium dioxide nanotubes and IL-6 antibodies for ultrasensitive sensing of IL-6 [84]. Errachid et al. reported a highly sensitive impedimetric determination for the inflammatory marker IL-10 [85]. It mainly utilized pyrene carboxylic acid and graphene foam flexible electrodes

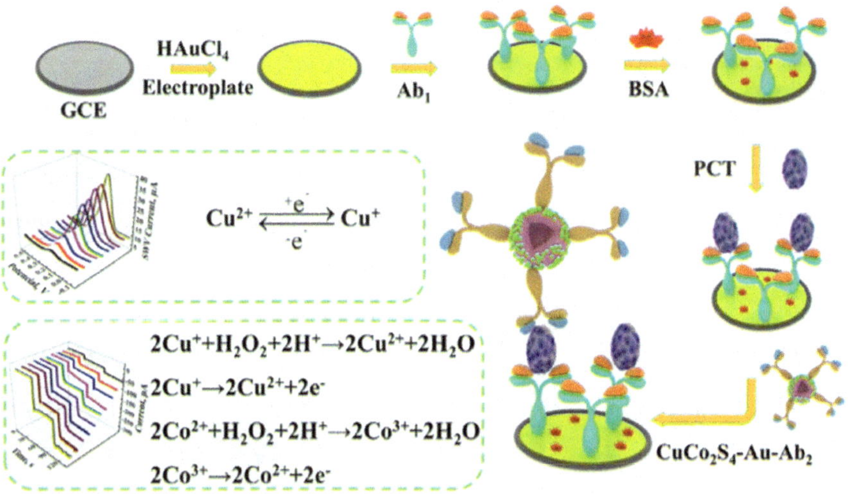

Fig. 10.11 Schematic presentation for the fabrication of the dual-mode PCT electrochemical immunosensor [81]

as the functional components. The π-π non-covalent functionalization improved the stability of the device. The proposed electrochemical sensor provided superior specificity, reproducibility, and stability (Fig. 10.12).

10.5 Conclusions and Perspectives

The detection of inflammatory biomarkers by electrochemical biosensors is expected to play a key role in analytical science in the near future. The rapid development of electrochemical biosensors highlights their future applications in different scientific fields, especially in the early detection of diseases, gene mutations, and biological targets. The detection of inflammatory biomarkers can significantly improve the detection sensitivity, reduce the detection cost, and has certain evaluation value and clinical significance for the early prediction, auxiliary diagnosis, curative effect observation, and prognosis of inflammation-related diseases. In this paper, we summarized the detection of inflammatory biomarkers by electrochemical biosensors, especially the detection and application of inflammatory biomarkers in blood. Next, the detection of inflammatory biomarkers in whole blood by electrochemical sensors is introduced in detail. Finally, the application of electrochemical sensors in life analysis and clinical application is summarized.

Although progress in researches have been made in electrochemical biosensors for inflammatory biomarkers detection, several challenges still exist: (1) There are some limitations in the development of strategies to improve sensitivity and practical application. (2) Now, rapid diagnosis has become the goal pursued by the functional

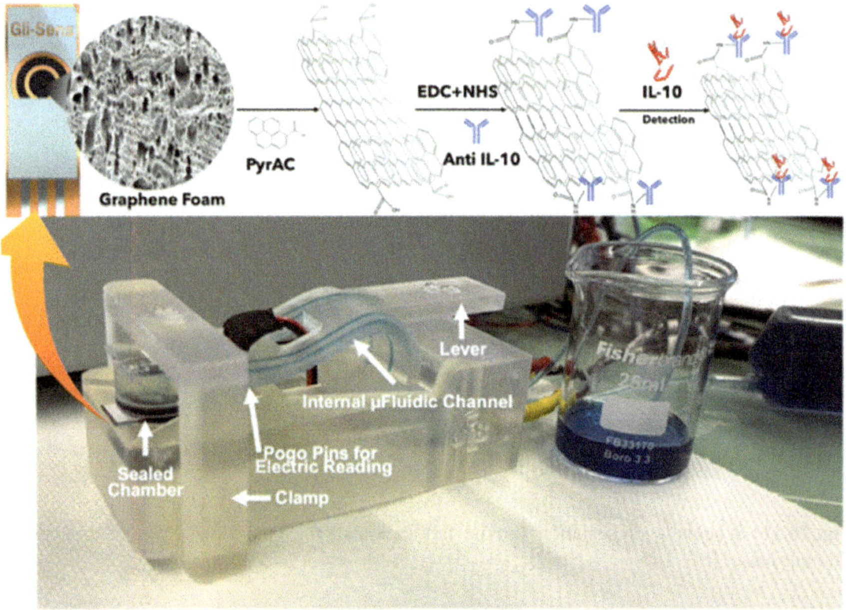

Fig. 10.12 Diagram for fixation of anti-IL-10 and analysis of IL-10 [85]

diagnosis system. Because rapid diagnosis is directly related to the treatment decision, the previous electrochemical analysis methods generally require a very cumbersome operation, while the current electrochemical sensors also require immediate use while pursuing stability, so some portable electrochemical methods for rapid detection of inflammatory factors in serum urgently need to be developed. (3) Each disease corresponds to a variety of inflammatory biomarkers, and there is also an inflammatory biomarker involved in many diseases. Therefore, it is necessary to develop electrochemical methods for the detection of multi-target inflammatory biomarkers to improve the accuracy and reaction rate of the detection results.

Compared with other detection methods, the detection of inflammatory biomarkers can significantly improve the detection sensitivity, reduce the detection cost, and provide a good perspective for clinical diagnosis. Encouraged by the current development of electrochemical biosensors for inflammatory biomarkers detection, we believe that all the efforts and research in this field will overcome the present difficulties and obstacles, integrated with other technologies, such as nanomaterials and microfluidic systems to promote the development of the entire electrochemical filed.

References

1. Chakraborty C, Sharma AR, Sharma G, Lee SS (2020) The Interplay among miRNAs, major cytokines, and cancer-related inflammation. Mol Ther Nucl Acids 20:606–620
2. Macovei DG, Irimes MB, Hosu O, Cristea C, Tertis M (2023) Point-of-care electrochemical testing of biomarkers involved in inflammatory and inflammatory-associated medical conditions. Anal Bioanal Chem 415:1033–1063
3. Wilson JL, Altman RB (2018) Biomarkers: delivering on the expectation of molecularly driven, quantitative health. Exp Biol Med 243:313–322
4. Khusnurrokhman G, Wati FF (2022) Tumor-promoting inflammation in lung cancer: a literature review. Ann Med Surg 79:104022
5. Marin M, Moya C, Manez S (2019) Mutual influences between nitric oxide and paraoxonase 1. Antioxidants 8:619
6. Sun S, Luo J, Zhu Y, Kong F, Mao G, Ming T, Xing Y, Liu J, Dai Y, Yan S et al (2022) Multifunctional self-driven origami paper-based integrated microfluidic chip to detect CRP and PAB in whole blood. Biosens Bioelectron 208:114225
7. Sivertsen J, Braathen H, Lunde THF, Kristoffersen EK, Hervig T, Strandenes G, Apelseth TO (2020) Cold-stored leukoreduced CPDA-1 whole blood: in vitro quality and hemostatic properties. Transfusion 60:1042–1049
8. Amory JH, Hitti J, Lawler R, Eschenbach DA (2001) Increased tumor necrosis factor-alpha production after lipopolysaccharide stimulation of whole blood in patients with previous preterm delivery complicated by intra-amniotic infection or inflammation. Am J Obstet Gynecol 185:1064–1067
9. Isiksacan Z, Asghari M, Elbuken C (2017) A microfluidic erythrocyte sedimentation rate analyzer using rouleaux formation kinetics. Microfluid Nanofluid 21:44
10. Zhang M, Hu X, Mei L, Zhang L, Wang X, Liao X, Qiao X, Hong C (2021) PSA detection electrochemical immunosensor based on MOF-235 nanomaterial adsorption aggregation signal amplification strategy. Microchem J 171:106870
11. Liu Y, Xu Q, Zhang Y, Ren B, Huang L, Cai H, Xu T, Liu Q, Zhang X (2021) An electrochemical aptasensor based on AuPt alloy nanoparticles for ultrasensitive detection of amyloid-beta oligomers. Talanta 231:122360
12. Zhao H, Liu F, Xie W, Zhou TC, OuYang J, Jin L, Li H, Zhao CY, Zhang L, Wei J, et al (2021) Ultrasensitive supersandwich-type electrochemical sensor for SARS-CoV-2 from the infected COVID-19 patients using a smartphone. Sens Actuat B Chem 327:128899
13. Weyrich AS, Lindemann S, Zimmerman GA (2003) The evolving role of platelets in inflammation. J Thromb Haemost 1:1897–1905
14. Denis MM, Tolley ND, Bunting M, Schwertz H, Jiang H, Lindemann S, Yost CC, Rubner FJ, Albertine KH, Swoboda KJ et al (2005) Escaping the nuclear confines: signal-dependent pre-mRNA splicing in anucleate platelets. Cell 122:379–391
15. Dinarello CA (1996) Biologic Basis for Interleukin-l in disease. Blood 87:2095–2147
16. Alkire RC, Gogotsi Y, Simon P (2008) Nanostructured materials in electrochemistry (John Wiley & Sons)
17. Attar A, Mandli J, Ennaji MM, Amine A (2016) Label-free electrochemical impedance detection of rotavirus based on immobilized antibodies on gold sononanoparticles. Electroanalysis 28:1839–1846
18. Wang Y, Ping J, Ye Z, Wu J, Ying Y (2013) Impedimetric immunosensor based on gold nanoparticles modified graphene paper for label-free detection of Escherichia coli O157: H7. Biosens Bioelectron 49:492–498
19. De Luna P, Mahshid SS, Das J, Luan B, Sargent EH, Kelley SO, Zhou R (2017) High-curvature nanostructuring enhances probe display for biomolecular detection. Nano Lett 17:1289–1295
20. Mahshid S, Mepham AH, Mahshid SS, Burgess IB, Saberi Safaei T, Sargent EH, Kelley SO (2016) Mechanistic control of the growth of three-dimensional gold sensors. J Phys Chem C 120:21123–21132

21. Patris S, Vandeput M, Kauffmann J-M (2016) Antibodies as target for affinity biosensors. TrAC Trends Anal Chem 79:239–246
22. Oue S, Okamoto A, Yano T, Kagamiyama H (1999) Redesigning the substrate specificity of an enzyme by cumulative effects of the mutations of non-active site residues. J Biol Chem 274:2344–2349
23. Song K-M, Lee S, Ban C (2012) Aptamers and their biological applications. Sensors 12:612–631
24. Crapnell RD, Dempsey-Hibbert NC, Peeters M, Tridente A, Banks CE (2020) Molecularly imprinted polymer based electrochemical biosensors: overcoming the challenges of detecting vital biomarkers and speeding up diagnosis. Talanta Open 2:100018
25. Betlem K, Mahmood I, Seixas R, Sadiki I, Raimbault R, Foster C, Crapnell R, Tedesco S, Banks C, Gruber J (2019) Evaluating the temperature dependence of heat-transfer based detection: a case study with caffeine and Molecularly Imprinted Polymers as synthetic receptors. Chem Eng J 359:505–517
26. Jamieson O, Soares TC, de Faria BA, Hudson A, Mecozzi F, Rowley-Neale SJ, Banks CE, Gruber J, Novakovic K, Peeters M (2019) Screen printed electrode based detection systems for the antibiotic amoxicillin in aqueous samples utilising molecularly imprinted polymers as synthetic receptors. Chemosensors 8:5
27. Crapnell RD, Hudson A, Foster CW, Eersels K, Grinsven BV, Cleij TJ, Banks CE, Peeters M (2019) Recent advances in electrosynthesized molecularly imprinted polymer sensing platforms for bioanalyte detection. Sensors 19:1204
28. Kim YJ, Jin YH, Salieb-Beugelaar GB, Nam CH, Stieglitz T (2014) Genetically engineered bacteriophage delivers a tumor necrosis factor alpha antagonist coating on neural electrodes. Biomed Mater 9:015009
29. Thaler M, Luppa PB (2019) Highly sensitive immunodiagnostics at the point of care employing alternative recognition elements and smartphones: hype, trend, or revolution? Anal Bioanal Chem 411:7623–7635
30. Justino CI, Freitas AC, Pereira R, Duarte AC, Santos TAR (2015) Recent developments in recognition elements for chemical sensors and biosensors. TrAC Trends Anal Chem 68:2–17
31. Molinero-Fernandez A, Lopez MA, Escarpa A (2020) Electrochemical microfluidic micromotors-based immunoassay for C-reactive protein petermination in preterm neonatal samples with sepsis suspicion. Anal Chem 92:5048–5054
32. Bard AJ, Faulkner LR, White HS (2022) Electrochemical methods: fundamentals and applications (John Wiley & Sons)
33. Amiri M, Bezaatpour A, Jafari H, Boukherroub R, Szunerits S (2018) Electrochemical methodologies for the detection of pathogens. ACS Sens 3:1069–1086
34. Hao Z, Pan Y, Huang C, Wang Z, Lin Q, Zhao X, Liu S (2020) Modulating the linker immobilization density on aptameric graphene field effect transistors using an electric field. ACS Sens 5:2503–2513
35. Geetha Bai R, Muthoosamy K, Zhou M, Ashokkumar M, Huang NM, Manickam S (2017) Sonochemical and sustainable synthesis of graphene-gold (G-Au) nanocomposites for enzymeless and selective electrochemical detection of nitric oxide. Biosens Bioelectron 87:622–629
36. Molina Á, González J, Molina Á, González J (2016) Some fundamental concepts. In: Pulse voltammetry in physical electrochemistry and electroanalysis, pp 1–66
37. Fujii K, Nakano T, Kanazawa M, Akimoto S, Hirano T, Kato H, Nishimura T (2005) Clinical-scale high-throughput human plasma proteome analysis: lung adenocarcinoma. Proteomics 5:1150–1159
38. Borgia JA, Basu S, Faber LP, Kim AW, Coon JS, Kaiser-Walters KA, Fhied C, Thomas S, Rouhi O, Warren WH et al (2009) Establishment of a multi-analyte serum biomarker panel to identify lymph node metastases in non-small cell lung cancer. J Thorac Oncol 4:338–347
39. Nolen BM, Marks JR, Ta'san S, Rand A, Luong TM, Wang Y, Blackwell K, Lokshin AE (2008) Serum biomarker profiles and response to neoadjuvant chemotherapy for locally advanced breast cancer. Breast Cancer Res 10:R45

40. Stern E, Vacic A, Rajan NK, Criscione JM, Park J, Ilic BR, Mooney DJ, Reed MA, Fahmy TM (2010) Label-free biomarker detection from whole blood. Nat Nanotechnol 5:138–142
41. Dimeski G, Badrick T, John AS (2010) Ion selective electrodes (ISEs) and interferences–a review. Clin Chim Acta 411:309–317
42. Wang D, Kou R, Gil MP, Jakobson HP, Tang J, Yu D, Lu Y (2005) Templated synthesis, characterization, and sensing application of macroscopic platinum nanowire network electrodes. J Nanosci Nanotechnol 5:1904–1909
43. Ibupoto ZH, Jamal N, Khun K, Willander M (2012) Development of a disposable potentiometric antibody immobilized ZnO nanotubes based sensor for the detection of C-reactive protein. Sens Actuat B Chem 166–167:809–814
44. Memoli B (1999) Cytokine production in haemodialysis. Blood Purif 17:149–158
45. Panichi V, Migliori M, De Pietro S, Metelli MR, Taccola D, Perez R, Palla R, Rindi P, Cristofani R, Tetta C (2000) Plasma C-reactive protein in hemodialysis patients: a cross-sectional, longitudinal clinical survey. Blood Purif 18:30–36
46. Maes M, Galecki P, Chang YS, Berk M (2011) A review on the oxidative and nitrosative stress (O&NS) pathways in major depression and their possible contribution to the (neuro)degenerative processes in that illness. Prog Neuropsychopharm Biol Psychiatry 35:676–692
47. Behr GA, Moreira JC, Frey BN (2012) Preclinical and clinical evidence of antioxidant effects of antidepressant agents: implications for the pathophysiology of major depressive disorder. Oxid Med Cell Longev 2012:609421
48. Zhang M, Mullens C, Gorski W (2005) Insulin oxidation and determination at carbon electrodes. Anal Chem 77:6396–6401
49. Kim SH, Thoa TTT, Gu MB (2019) Aptasensors for environmental monitoring of contaminants in water and soil. Curr Opin Environ Sci Health 10:9–21
50. Zupančič U, Jolly P, Estrela P, Moschou D, Ingber DE (2021) Graphene enabled low–noise surface chemistry for multiplexed sepsis biomarker detection in whole blood. Adv Funct Mater 31
51. Boonyasit Y, Chailapakul O, Laiwattanapaisal W (2016) A multiplexed three-dimensional paper-based electrochemical impedance device for simultaneous label-free affinity sensing of total and glycated haemoglobin: the potential of using a specific single-frequency value for analysis. Anal Chim Acta 936:1–11
52. Fairchild AB, McAferty K, Demirok UK, La Belle JT (2009) A label-free, rapid multimarker protein impedance-based immunosensor. Icme Int Conf Complex Med Eng 2009:170–174
53. Elshafey R, Tavares AC, Siaj M, Zourob M (2013) Electrochemical impedance immunosensor based on gold nanoparticles-protein G for the detection of cancer marker epidermal growth factor receptor in human plasma and brain tissue. Biosens Bioelectron 50:143–149
54. Periyakaruppan A, Gandhiraman RP, Meyyappan M, Koehne JE (2013) Label-free detection of cardiac troponin-I using carbon nanofiber based nanoelectrode arrays. Anal Chem 85:3858–3863
55. Lasseter TL, Cai W, Hamers RJ (2004) Frequency-dependent electrical detection of protein binding events. Analyst 129:3–8
56. Kanyong P, Catli C, Davis JJ (2020) Ultrasensitive impedimetric immunosensor for the detection of C-reactive protein in blood at surface-initiated-reversible addition-fragmentation chain transfer generated poly(2-hydroxyethyl methacrylate) brushes. Anal Chem 92:4707–4710
57. Bryan T, Luo X, Bueno PR, Davis JJ (2013) An optimised electrochemical biosensor for the label-free detection of C-reactive protein in blood. Biosens Bioelectron 39:94–98
58. Boonyasit Y, Chailapakul O, Laiwattanapaisal W (2019) A folding affinity paper-based electrochemical impedance device for cardiovascular risk assessment. Biosens Bioelectron 130:389–396
59. Symanski JS, Martinchek GA, Bruckenstein S (1983) Conductometric sensor for atmospheric carbon-dioxide determination. Anal Chem 55:1152–1156
60. Chua JH, Chee RE, Agarwal A, Wong SM, Zhang GJ (2009) Label-free electrical detection of cardiac biomarker with complementary metal-oxide semiconductor-compatible silicon nanowire sensor arrays. Anal Chem 81:6266–6271

61. Ronkainen-Matsuno NJ, Thomas JH, Halsall HB, Heineman WR (2002) Electrochemical immunoassay moving into the fast lane. TrAC Trends Anal Chem 21:213–225
62. Heineman WR, Halsall HB, Pesce AJ, Jenkins SH, Yao H (1993) Electrochemical homogeneous enzyme immunoassay of theophylline in hemolyzed, icteric, and lipemic samples. Clin Chem 39:1432–1434
63. Rao L, Zhang K, Luo H, He S, Li Y, Liu C, Yang Y, Wang D (2022) The ability of inflammatory markers to recognize infection in cancer patients with fever at admission. Immunol Res 70(5):667–677
64. Yoon J, Cho HY, Shin M, Choi HK, Lee T, Choi JW (2020) Flexible electrochemical biosensors for healthcare monitoring. J Mater Chem B 8:7303–7318
65. Balayan S, Chauhan N, Chandra R, Jain U (2022) Molecular imprinting based electrochemical biosensor for identification of serum amyloid A (SAA), a neonatal sepsis biomarker. Int J Biol Macromol 195:589–597
66. Medetalibeyoglu H, Beytur M, Akyıldırım O, Atar N, Yola ML (2020) Validated electrochemical immunosensor for ultra-sensitive procalcitonin detection: carbon electrode modified with gold nanoparticles functionalized sulfur doped MXene as sensor platform and carboxylated graphitic carbon nitride as signal amplification. Sens Actuat B Chem 319:128195
67. Li G, Zhang H, Zhang L, Liu H, Li S, Wang Y, Deng X (2022) Serum markers CA125, CA153, and CEA along with Inflammatory cytokines in the early detection of lung cancer in high-risk populations. Biomed Res Int 2022:1394042
68. Deng Y, Qiu T, Patel N, Zhou S, Xue T, Zhang H (2019) Clinical management of risk of radiation pneumonia with serum markers during the radiotherapy for patients with thoracic malignant tumors. Cancer Manag Res 11:10249–10256
69. Valverde A, Serafín V, Garoz J, Montero-Calle A, González-Cortés A, Arenas M, Camps J, Barderas R, Yáñez-Sedeño P, Campuzano S, Pingarrón JM (2020) Electrochemical immunoplatform to improve the reliability of breast cancer diagnosis through the simultaneous determination of RANKL and TNF in serum. Sens Actuat B Chem 314:128096
70. Sri S, Chauhan D, Lakshmi GBVS, Thakar A, Solanki PR (2022) MoS_2 nanoflower based electrochemical biosensor for TNF alpha detection in cancer patients. Electrochim Acta 405:139736
71. Altintas Z, Kallempudi SS, Sezerman U, Gurbuz Y (2012) A novel magnetic particle-modified electrochemical sensor for immunosensor applications. Sens Actuat B Chem 174:187–194
72. Bakirhan NK, Ozcelikay G, Ozkan SA (2018) Recent progress on the sensitive detection of cardiovascular disease markers by electrochemical-based biosensors. J Pharm Biomed Anal 159:406–424
73. Hou Y, Li T, Huang H, Quan H, Miao X, Yang M (2013) Electrochemical immunosensor for the detection of tumor necrosis factor α based on hydrogel prepared from ferrocene modified amino acid. Sens Actuat B Chem 182:605–609
74. Aydın EB, Aydın M, Sezgintürk MK (2021) A novel electrochemical immunosensor based on acetylene black/epoxy-substituted-polypyrrole polymer composite for the highly sensitive and selective detection of interleukin 6. Talanta 222:121596
75. Liu N, Yi H, Lin Y, Zheng H, Zheng X, Lin D, Dai H (2018) Combined electrochemiluminescent and electrochemical immunoassay for interleukin 6 based on the use of TiO_2 mesocrystal nanoarchitectures. Microchim Acta 185:277
76. Yerrapragada RM, Mampallil D (2022) Interferon-γ detection in point of care diagnostics: short review. Talanta 245:123428
77. Zhou Y, Liu J, Dong H, Liu Z, Wang L, Li Q, Ren J, Zhang Y, Xu M (2022) Target-induced silver nanocluster generation for highly sensitive electrochemical aptasensor towards cell-secreted interferon-γ. Biosens Bioelectron 203:114042
78. Chu Z (2017) Development of a semiconductor-based electrochemical sensor for interferon-γ detection. Int J Electrochem Sci 9141–9149
79. Graham ML, Ramachandran S, Singh A, Moore MEG, Flanagan EB, Azimzadeh A, Burlak C, Mueller KR, Martins K, Anazawa T et al (2022) Clinically available immunosuppression averts rejection but not systemic inflammation after porcine islet xenotransplant in cynomolgus macaques. Am J Transplant 22:745–760

80. Sandkovsky U, Kalil AC, Florescu DF (2015) The use and value of procalcitonin in solid organ transplantation. Clin Transplant 29:689–696
81. Li Y, Liu L, Liu X, Ren Y, Xu K, Zhang N, Sun X, Yang X, Ren X, Wei Q (2020) A dual-mode PCT electrochemical immunosensor with $CuCo_2S_4$ bimetallic sulfides as enhancer. Biosens Bioelectron 163:112280
82. Huang J, Cheng W, Li Y (2022) 3D carbonized wood-based integrated electrochemical immunosensor for ultrasensitive detection of procalcitonin antigen. Talanta 238:122991
83. Zhao F, Ai F, Wu J, Dong X (2020) Changes and clinical significance of serum inflammatory factors in the treatment of pregnancy hypertension syndrome with magnesium sulfate combined with nifedipine. Exp Ther Med 20:1796–1802
84. Arkusz K, Paradowska E (2020) Impedimetric detection of femtomolar levels of interleukin 6, interleukin 8, and tumor necrosis factor alpha based on thermally modified nanotubular titanium dioxide arrays. Nanomaterials 10:2399
85. Frias IAM, Zine N, Sigaud M, Lozano-Sanchez P, Caffio M, Errachid A (2023) Non-covalent π–π functionalized Gii-sense® graphene foam for interleukin 10 impedimetric detection. Biosens Bioelectron 222:114954

Chapter 11
Electrochemical Biosensors for Tumor Biomarkers Detection

Linfeng Chen and Yanan Li

Abstract Cancer is a big public health challenge and a leading cause of death worldwide, leading to millions of deaths each year. Developing efficient strategies to detect cancer at early stages is critically important as it will be able to provide appropriate treatment and effectively improve the life quality of patients. Cancer is typically featured by released biomarkers (e.g., proteins, nucleic acids, and cells), which could be commonly found in biological fluids and have been used to indicate the presence of cancer tumors. Electrochemical biosensors have attracted tremendous attention due to the superiorities of the techniques, such as non-invasiveness, easy operation, high sensitivity, and specificity. This chapter is focused on the recent advance of electrochemical biosensors for tumor biomarker detection. The tumor biomarkers and typical electrochemical techniques are also briefly introduced.

Keywords Tumor biomarker · Cancer diagnostic · Electrochemical biosensor · Blood testing

11.1 Introduction

Cancer is referred to a large group of serious diseases that could rapidly grow and even spread to any part of the body. Cancer has long been a primary cause of death globally. According to the statistical report by World Health Organization (WHO), nearly 10 million people died due to cancer in 2020 (Fig. 11.1). In more detail, the top five prevalent cancers in 2020 in terms of new cases were followed by breast (2.26 million), lung (2.21 million), colon and rectum (1.93 million), prostate (1.41 million), and stomach (1.09 million), while the top cancers causing most deaths were ordered by lung (1.80 million deaths), colon and rectum (0.94 million deaths), liver (0.83 million deaths), stomach (0.77 million deaths), and breast (0.68 million deaths) (Fig. 11.1). No matter the cancers result in patients' death or not, they have a

L. Chen (✉) · Y. Li
State Key Laboratory of Biogeology and Environmental Geology, Faculty of Materials Science and Chemistry, China University of Geosciences, Wuhan 430074, P. R. China
e-mail: chenlinfeng@cug.edu.cn

F. Xia et al. (eds.), *Electrochemical Biosensors for Whole Blood Analysis*,
https://doi.org/10.1007/978-981-99-5644-9_11

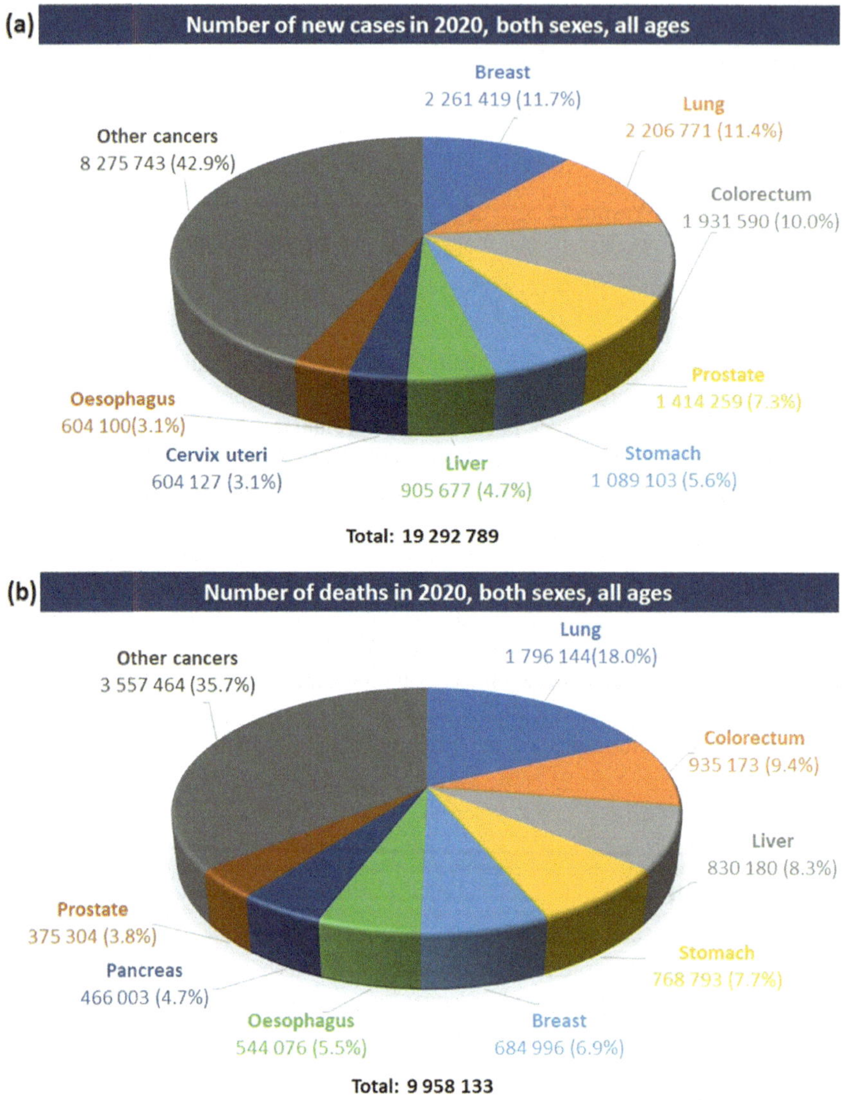

Fig. 11.1 Statistics of new cases and cancer death in 2020 worldwide (World Health Organization)

significantly negative impact in patients' life, such as reducing patients' life quality, and requiring long-term and challenging treatment.

Cancer develops from malignant tumors whose growth is abnormally rapid. In addition, malignant tumors often transfer to other organs through the blood and/or lymphatic circulation system, which has more serious health consequences. For example, colon cancer tumors frequently spread to the lung [1]. Until now, there has been no effective method to solve the problem but to improve the life quality and time of cancer patients. In order to provide proper treatment, it is critical to detect the cancer at its early stages. Cancer cells usually generate featured species (i.e., tumor biomarkers) which could be detected in biological fluids, such as blood, urine, saliva, and sweat. Thus, detection of tumor biomarkers will provide information about the physiological state of the cancer disease. However, to detect the biomarkers of cancer tumors in the early stages is quite challenging as the biomarker concentration existing in the tumor cells and the body fluids is rather low. Thus, to develop reliable and highly sensitive analysis methods for trace-level determination of tumor biomarkers becomes critically important.

Traditional clinical methods used for early tumor diagnosis are based on imaging or morphological analysis of tissues, mainly including X-ray, magnetic resonance imaging (MRI), computed tomography (CT), and ultrasound, which are limited to detecting the cancer when a visible change to the tissue could be observed [2]. Although more quantitatively analytical techniques have been developed for the detection of tumor biomarkers, e.g., radioimmunoassay, tissue biopsy, and enzyme-linked immunosorbent assay (ELISA), they are time-consuming, operation-complex, and highly dependent on sophisticated instrumentation [3]. In addition, optical biosensors often suffer from interference from other luminescence molecules. In this context, electrochemical biosensors have attracted considerable attention as effective strategies for the determination of biomarkers due to their unique advantages of easy operation, fast response, high specificity, and sensitivity [4]. In this chapter, we will focus on the electrochemical biosensors for the detection of tumor biomarkers, which have been rapidly developed in the past decades.

11.2 Tumor Biomarkers and Cancers

According to the National Cancer Institute (NCI), a tumor biomarker is referred to one specific substance that is closely correlated with cancers and could be commonly used as an indicator to provide information about cancer [5]. Tumor biomarkers can be mainly classified into three types, including nucleic acids (e.g., messenger RNA, microRNA, DNA), proteins (e.g., enzymes, hormones, antibodies, and receptors), and cancer cells, which may be found in tumor tissues, body fluids or excretions/secretions, such as blood, urine, and saliva [6, 7]. It's worth noting that, among the FDA-approved biomarkers, more than 90% are proteins and nucleic acids [7].

Many tumor biomarkers have already been widely used in clinical practice. However the relationship between the biomarkers and the cancer is not simple, as while some biomarkers are only found in one type of cancer, some others are detected in different tumor types. For example, prostate-specific antigens (PSAs) are a specific type of biomarker found in prostate cancer. Carcinoembryonic antigen (CEA), which is an oncofetal glycoprotein, has been identified as a multi-tumor biomarker for the detection of lung, breast, colon, gastric, and pancreatic tumors [8] (Table 11.1). Thus, CEA is often combined with cytokeratin 19 fragment (CYFRA 21–1) to diagnose lung tumors [9]. Representative cancers and the main tumor biomarkers found in the corresponding cancers are shown in Table 11.1.

The concentration of tumor biomarkers in a cancer patient is different from that in a healthy person, and the change in biomarker concentration can be used as an indicator of the presence of cancer. For example, the CEA concentration in the blood of healthy persons is in the range of 3~5 ng mL^{-1}; thus, CEA concentration >20 ng mL^{-1} strongly indicates the presence of metastatic cancer [23]. Alpha-fetoprotein (AFP) is a featured glycoprotein mainly made by liver cells, and has been extensively used as a biomarker for the test of liver cancer (LC). The AFP level in human serum of the normal adult is generally <20 ng mL^{-1}. However, AFP concentration will significantly increase and even be higher than 400 ng mL^{-1} when liver cells are damaged or LC exists. Therefore, the detection of AFP in human serum is of great importance for the early diagnosis of LC [24]. Cancer antigen (CA125 or MUC16)

Table 11.1 Cancer types and the associated tumor biomarkers

Cancer	Associated tumor biomarkers	Ref
Lung cancer	CEA,[a] CA19-9,[b] SCCA,[c] CYFRA21-1,[d] NSE,[e] EGFR 2 (HER2),[f] VEGF,[g] CA 125 (MUC16),[h] TPA[i]	[4, 9– 11]
Breast cancer	CA15-3 (MUC1),[j] CA 125 (MUC16), CEA, ER,[k] PR,[l] EGFR 2 (HER2), MCF-7,[m] VEGF, BRCA1,[n] BRCA2,[n] microRNA, ctDNA,[o] ccfDNA[p]	[1, 11, 12–14
Prostate cancer	PSA,[q] Pro2PSA,[r] Sarcosine	[11, 15, 16]
Colon cancer	CEA, CA19-9, CA15-3 (MUC1), p53,[s] oncomiRNA-21, CA72-4[t]	[6, 11, 17–19
Liver cancer	AFP[u]	[20]
Gastric cancer	CEA, CA19-9	[11, 21]
Pancreatic cancer	CA19-9, CEA	[11]
Esophagus cancer	ESCC[v]	[22]
Ovarian cancer	CA19-9, CA 125 (MUC16), HE4[w]	[11]

[a] CEA: carcinoembryonic antigen. [b] CA19-9: carbohydrate antigen 19–9. [c] SCCA: squamous cell carcinoma antigen. [d] CYFRA21-1: cytokeratin 19 fragment. [e] NSE: neuron specific enolase. [f] EGFR 2 (HER2): human epidermal growth factor receptor 2. [g] VEGF: vascular endothelial growth factor. [h] CA 125 (MUC16): a 200 kDa membrane mucin-like glycoprotein. [i] TPA: tissue polypeptide antigen. [j] MUC1 (CA15-3): mucin 1, a transmembrane mucin glycoprotein. [k] ER: estrogen receptor. [l] PR: progesterone receptor, a hormone receptor. [m] MCF-7: human breast cancer cells. [n] BRCA1/ BRCA2: breast cancer susceptibility gene. [o] ctDNA: circulating tumor DNA. [p] ccfDNA: circulating cell-free DNA. [q] PSA: prostate-specific antigen. [r] Pro2PSA: zymogen precursor of PSA. [s] p53 (TP53): a tumor suppressor protein. [t] CA72-4: carbohydrate antigen 72–4. [u] AFP: alpha-fetoprotein. [v] ESCC: esophageal squamous cell carcinoma. [w] HE4: human epididymis protein 4

Table 11.2 Normal concentration level of typical tumor biomarkers (Reproduced from [26] with modification)

Tumor biomarkers	Thresholds
NSE	12.5 μg mL^{-1}
PSA	4 ng mL^{-1}
GST	3.2 mU mL^{-1}
ALP	~0
CT	100 pg mL^{-1}
SCCA	1.5 ng mL^{-1}
Ferritin	250 ng mL^{-1}
hCG	5.0 IU L^{-1}
AFP	20 ng mL^{-1} [24]
CEA	3.0~5.0 ng mL^{-1} [23]
CA125	35 U mL^{-1} [27]
CA153	25 U mL^{-1}
CA27-29	36.4 U mL^{-1}
CA549	11 U mL^{-1}
CA19-9	37 U mL^{-1}
CA50	14~20 U mL^{-1}
CA242	20 U mL^{-1}
CA72-4	6 U mL^{-1}

is a 200 kDa cell surface mucin glycoprotein and could be used as a sign of ovarian cancer. The normal CA125 concentration in humans is generally lower than 35 units per milliliter (U mL^{-1}). The elevated levels of CA-125 in the blood indicate a high possibility of ovarian cancer [25]. The normal level of typical tumor biomarkers is presented in Table 11.2.

11.3 Electrochemical Biosensors for the Detection of Tumor Markers

11.3.1 Electrochemical Biosensor Principles

Electrochemical biosensors are sensing devices used for quantitative and semi-quantitative analysis based on biological recognition. Owing to the fascinating superiorities, such as rapid response, easy operation, low cost, high sensitivity, and specificity, the electrochemical platforms show tremendous potential for the diagnosis of cancer. To construct an electrochemical biosensor, three integrated parts are usually needed (Fig. 11.2): (a) a biorecognition element that has a specific interaction with

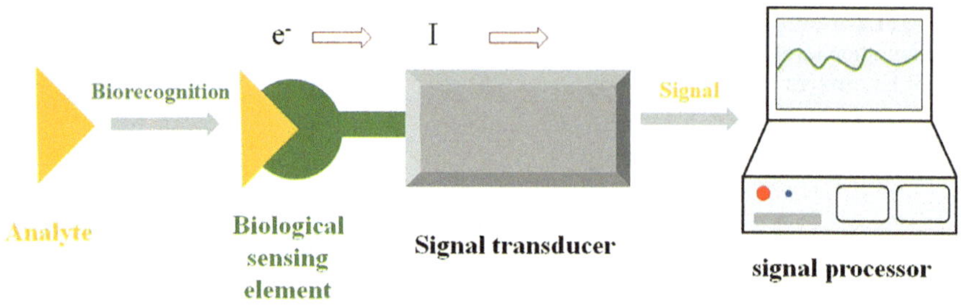

Fig. 11.2 Basic schematics of an electrochemical biosensor. There are three integrated components: biological sensing element, signal transducer, and signal processor (Reproduced from [31])

the bio-target. Biorecognition elements are a crucial component of electrochemical biosensors, and enzyme, aptamers, and antibodies are typical examples. (b) a signal transducer that can convert the specific interaction to a measurable signal (e.g., current, potential, conductance, or impedance), which is proportionally correlated to the target concentration. (c) an electronic system for data analysis [28].

Based on the analytical signal, electrochemical techniques mainly include potentiometry, voltammetry, electrochemical impedance spectroscopy (EIS), and electrochemiluminescence [29]. Potentiometric biosensors are traditionally defined as a zero-current technique that measures the potential across an interface [30]. Voltammetry, which involves measuring the current versus the applied voltage during a biological recognition event, is the most accepted method in electrochemical analysis. Voltammetry can be further divided into linear-sweep voltammetry, differential-pulse voltammetry, cyclic voltammetry, and square-wave voltammetry according to different modes of potential control [31]. The principle of EIS is to measure the impedance of the circuit as biological recognition events occur at the electrode interface by applying an alternating current (AC) potential. Usually, the AC potential is low (2–10 mV) while the AC frequency is varied in a wide range from 100 kHz to 1 MHz [32]. Electrochemiluminescence is a kind of analytic technique that involves the luminescence reaction induced by the substances produced by the electrochemical process at the electrode surface. The chemiluminescence intensity is correlated with the concentration of the biotargets. This method exhibits the advantages of high sensitivity, good producibility, easy operation, and easy control as it combines both the electrochemistry and chemiluminescence process [33].

It is worth mentioning that electrochemical biosensors achieve sensing by relying on the electrochemical process occurring on the surface of an electrode. Thus, the electrode plays a key role in an electrochemical biosensor. The electrode not only acts as a surface for immobilizing biorecognition elements (e.g., enzymes, aptamers, or antibodies), but also conducts electrons (i.e., electrical signal generation). Therefore, the performance (e.g., sensitivity and detection limitation) of electrochemical biosensors could be effectively improved by either selecting appropriate conductive materials as the electrodes or by modifying the electrode with functional nanomaterials. Various types of materials are reported to be used as the electrode for biosensing. One typical kind of material are intrinsically conductive polymers, such as polyaniline, polypyrrole, and polythiophene, which have been extensively utilized as the electrode and biological probes could be immobilized on the electrode surface for sensing. These polymers could also be doped with other nanomaterials to form a polymer composite matrix with improved conductivity or mechanical properties [34]. The other type of widely used conductive materials is carbon-based nanomaterials, noble metal nanoparticles, and their mixed composites. Particularly, gold nanoparticles (AuNPs) have been widely used as the electrode due to their high electrical conductivity, easy modification with biorecognition molecules, and biocompatibility [35, 36]. In recent years, carbon-based nanomaterials (e.g., glassy carbon, carbon nanotubes, and graphene) have received considerable attention and have been frequently utilized as electrodes in the construction of electrochemical biosensors, as carbon-based nanomaterials have superior electric and mechanical properties, and are inexpensive [30, 37].

11.3.2 Electrochemical Biosensors for Tumor Biomarker Detection

As a human blood sample contains different tumor biomarkers, including enzymes, exosomes, antigens, tumor cells, and nucleic acid, it's able to obtain sufficient information about the cancer by detecting the biomarkers in the biological fluids. Additionally, a blood-based test is a minimally invasive method when taking samples [38, 39]. Thus, in this section, the attention is paid to electrochemical biosensors for the tumor biomarker detection with blood samples.

11.3.2.1 Detection of Lung Tumor Biomarkers

Lung cancer is one of the most common forms of cancer. According to the statistical report in 2020, about 11.4% of the total new cancer cases are lung cancer, which is only slightly lower than breast cancer; the death percentage of lung cancer is about 18.0%, which is the highest (Fig. 11.1). The disease is not easily treated as the lung tumor will always spread in the early stages to other organs, e.g., the liver, brain, and bones, which results in dangerous outcome. Conventional methods commonly used to test lung cancer tumor include ultrasound [40], magnetic resonance imaging (MRI) [41], computerized tomography (CT) or computerized axial tomography (CAT) scan [42]. However, these methods are invasive, and expensive and their resolution is not enough to detect small-sized tumors in their early stages. Currently, major trouble is the late diagnosis of lung cancer, which leads to a difficult prognosis of the disease [43].

Developing techniques for early detection of lung cancer thus becomes critically important, which will allow clinicians to choose appropriate treatments for the patients and improve the survival time. As mentioned above, when cancer tumors occur in the body, special biomarkers will be released by the cancer cells. The amount of the biomarkers in the biological fluids is different with the growth of the cancer. Therefore, detection of the lung cancer biomarkers becomes an effective way for clinical diagnosis of the disease. Numerous biomarkers are identified for lung cancer, e.g., CEA, CA19-9, SCCA, CYFRA21-1, NSE, EGFR 2 (HER2), VEGF, CA 125 (MUC16), and TPA (Table 11.1) [44]. Quantification of such tumor biomarkers in human blood samples by electrochemical biosensors may indicate the stages of lung cancer [45, 46].

CEA, a 180 kDa molecular weight glycoprotein, was previously identified as a multi-tumor biomarker that could be found in various cancers, such as colon cancer, breast cancer, ovarian cancer, as well as lung cancers [26]. The abnormal increase of CEA concentration in blood may indicate the presence of lung cancer. Huang et al. developed a simple and sensitive electrochemical biosensor for the detection of CEA [47]. The glass carbon was selected as the electrode (GCE), and they coated the GCE with Au-F127 strawberry-like nanospheres. Here, the specific Au-F127 nanostructures had a large surface area and could be effectively immobilized with anti-CEA (Fig. 11.3a). In addition, the noble gold could accelerate electron transfer at the interface and improve the signal sensitivity. DPV test was then performed to evaluate the performance of the biosensor for the CEA detection. As a result, in a wide CEA concentration ranging from 0.01 and 80 ng/mL, the biosensor exhibited a linear response. And a low detection limit of 0.24 pg mL^{-1} at a signal-to-noise (S/N) ratio of 3 was achieved. This biosensor was further used to analyze CEA in human serum samples, which was consistent with the reference values obtained from the hospital. In another example, a label-free electrochemical biosensor was constructed with Prussian blue nanoparticles-decorated molybdenum disulfide nanocomposite. When it was used for the CEA detection with a concentration down to 0.005–10 ng mL^{-1}, the biosensor still showed a linear response. The detection limit was down to about 0.54 pg mL^{-1} [48].

There are mainly two histological types of lung cancers depending on the cell size. If the cell size is small, the cancer is defined as small cell lung cancer (SCLC); otherwise, it's non-small cell lung cancer (NSCLC). About 20% of lung cancer falls into the SCLC category, which spreads quickly and should be determined as early as possible [49], 4]. One effective way is to detect neuron-specific enolase (NSE), which is a sensitive and featured tumor biomarker for SCLC [50]. NSE is composed of two nearly identical polypeptide chains with each molecular weight of 39 kDa, and is found to be abnormally high in patients with SCLC. Thus, it's of great importance to test the NSE in order to determine and monitor the SCLC state [51]. Chen et al. developed an electrochemical biosensor for the detection of NSE (Fig. 11.3b) [52]. They fabricated a microporous reduced graphene oxide/polyaniline (rGO/PANI) film, which had a 3D porous structure and could provide a large surface area for immobilizing the antibodies. In addition, the composite exhibited enhanced conductivity, which may contribute to the sensitivity. Under the optimal conditions, the biosensor was used to test the NSE solutions, and a DPV curve was collected. In the NSE concentration ranging from 0.5 pg mL^{-1} to 10.0 ng mL^{-1}, a linear

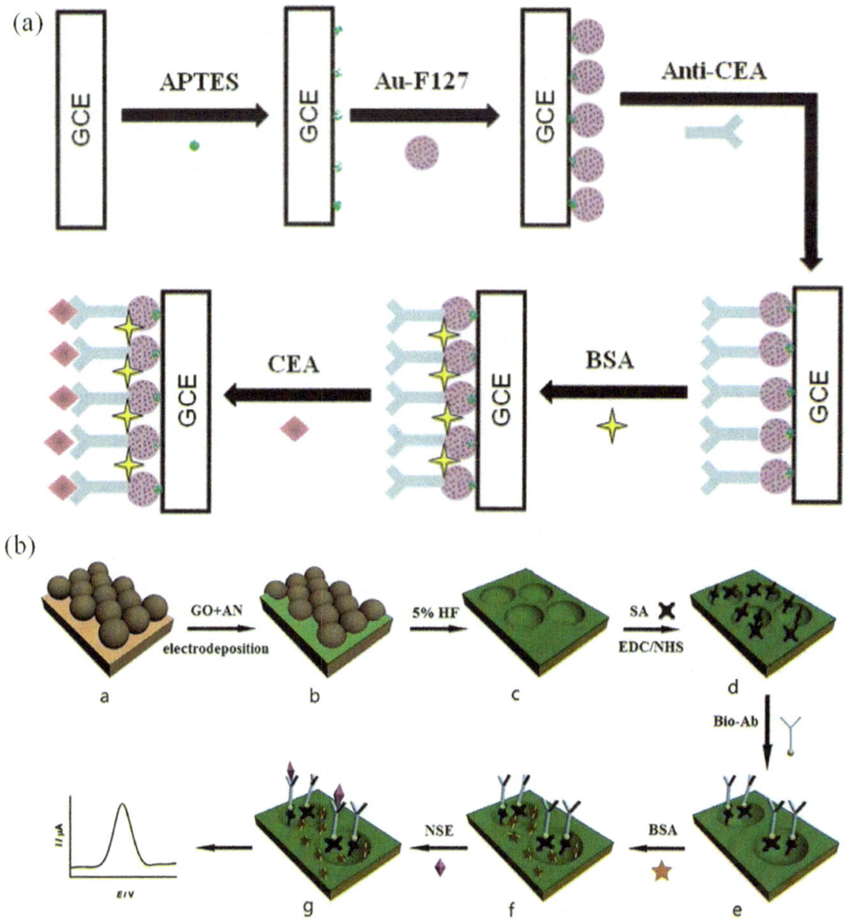

Fig. 11.3 Electrochemical biosensors for detection of lung tumor biomarker CEA **a** (Reproduced from [47]) and NSE **b** (Reproduced from [52])

current response was obtained. The NSE detection limit of 0.1 pg mL^{-1} was finally achieved. The biosensor was proposed to detect NSE in serum samples due to its good analytical performance, such as high selectivity, good stability, and sufficient reproducibility.

11.3.2.2 Detection of Breast Cancer Biomarkers

Breast cancer (BC) is a specific threat to women. In 2020, BC is reported to occur in women with the highest incidence rate (11.7%), which is still increasing. Globally, about 684 996 deaths (6.9%) were caused by BC in 2020 (Fig. 11.1). Only in the

United States, it's estimated that there will be 297,790 new BC cases in women in 2023 [53]. To improve cancer survival rates, diagnosis of the BC disease at early stages is critical, which will enable early intervention and proper treatment. Previous reports have indicated BC survival rates could be significantly improved in the long term when suitable treatment is applied. Traditional diagnostic techniques which are very expensive and time-consuming could not effectively determine the BC tumor in the early stage [54].

Recent electrochemical diagnostic platforms were reported as a more efficient and less costly way for the determination of BC by detecting various BC biomarkers [1]. FDA-approved diverse BC biomarkers could be classified as nucleic acid, protein, tumor cell, and exosomes [1]. For example, vital protein biomarkers include human epidermal growth factor receptor-2 (HER-2), estrogen receptor (ER), and progesterone receptor (PR) [55]. The other typical biomarkers include breast cancer susceptibility gene (BRCA1, BRCA2), microRNA, CA15-3, CEA, CA19-9, CA125, and circulating tumor cells (CTC), et al. (Table 11.1) [55].

BRCA1 and/or BRCA2 are important tumor biomarkers and could be used for diagnosis of BC. The change of BRCA1 and/or BRCA2 strongly indicates the presence of BC [56, 57]. Recently, a DPV electrochemical biosensor was constructed for the detection of BRCA1 (Fig. 11.4a) [58]. A selected GCE surface was coated with polypeptide-doped poly(3,4-ethylenedioxythiophene), followed by further modification with a capture of DNA probe and methylene blue (MB). Here, MB was added to improve the DPV signal. When the biosensor was used to detect BRCA1, which was attached to the electrode due to the specific interaction with the DNA probe, MB would be released and generated a large change in the DPV signal. The peak current change was linearly related to the BRCA1 concentration in the range of 0.01 pM to 1 nM. The BRCA1 concentration as low as 0.0034 pM was promising to be detected (S/N = 3). Usually, it's a challenge for electrochemical biosensors to be applied for real clinical applications because biomolecules in serum samples will attach to the electrode surface and produce troublesome background signals. The DPV biosensor in this work was highly antifouling, and exhibited good analytical performance in detecting BRCA1 in human serum samples.

Another typical BC biomarker is HER2, which is a transmembrane protein with a molecular weight of 185 kDa [59]. The protein could promote tumor growth, which in turn enhances cell proliferation, invasion, and spread to other tissues [60]. HER2 is found to over-express in 15–20% of invasive BC, and has been widely studied [12]. Marques et al. reported a carbon electrode-based electrochemical biosensor for the quantitative analysis of HER2 extracellular domain (ECD) (HER2 ECD) in human serum sample (Fig. 11.4b) [61]. The detection process was relatively complex. The biosensor electrode was fabricated by screen-printing and coated with gold nanoparticles, followed by modification with HER2 antibodies. After the biomarker HER2 ECD was captured, the biotin-modified HER2 antibodies were then added to connect with the biomarker. Subsequently, a streptavidin–alkaline phosphatase conjugate was utilized and attached to the biotin to label the detection HER2 antibodies. Finally, an enzymatic substrate (3-indoxyl phosphate) and silver ions were used. Metal silver would be obtained in the enzymatic reaction, and analyzed using

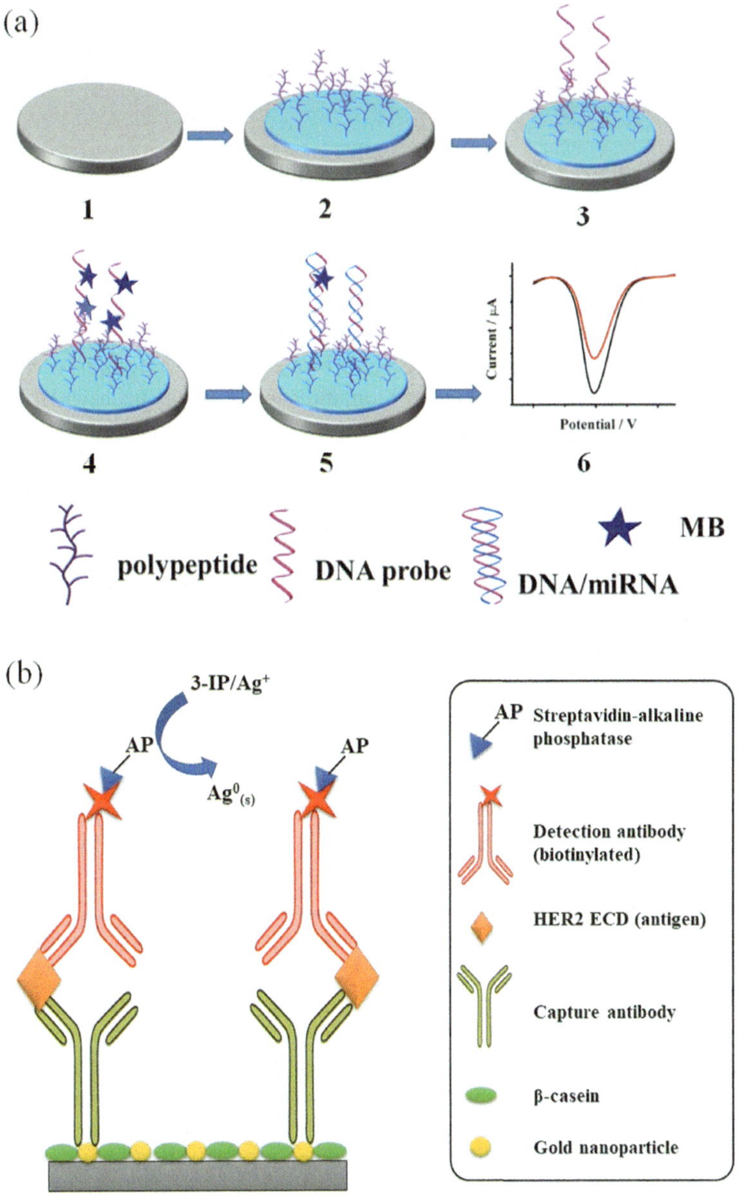

Fig. 11.4 Electrochemical biosensors for detection of breast tumor biomarker BRCA1 **a** (Reproduced from [58]), and HER2 ECD **b** (Reproduced from [61])

linear sweep voltammetry. The biosensor could detect the HER2 ECD in human serum with a linear response in the concentration range of 15–100 ng mL^{-1}. The detection limit of 4.4 ng mL^{-1} was attained.

11.3.2.3 Detection of Prostate Cancer Biomarkers

Prostate cancer is one of the biggest threats to men. In 2020, prostate cancer resulted in 375, 304 deaths (Fig. 11.1). In order to improve the survival rate, it's necessary to monitor the status of prostate cancer in the long term. The amount of prostate-specific antigen (PSA) in the blood samples is higher in prostate cancer patients than in normal persons [16]. Thus, in the case of prostate cancer testing in its early stages, PSA, which was a US FDA-approved biomarker in 1986, has been widely used to screen for prostate cancer in men. With the development of nanotechnologies and the demand for more sensitive detection, electrochemical sensors for PSA detection have attracted increasing attention. For example, Pothipor et al. designed and developed a highly sensitive electrochemical biosensor for the detection of PSA [62]. The electrode of the biosensor was modified with a conductive polymer composite doped with graphene–poly(3-aminobenzoic acid) (GP-P3ABA) nanomaterial. Then porous-hollowed-Ag-AuNP core–shell nanoparticles (PHSGNP) were synthesized as labeling for the detection of PSA. In the presence of PSA, PHSGNP labeling was combined to the electrode modified with GP-P3ABA, which exhibited significantly enhanced sensing efficiency. The detect limit was 0.13 pg mL^{-1}.

As PSA is not specific, PSA detection is not sufficiently sensitive (limited sensitivity of ~80%). In practical use, false-positive results often happen, which result in unnecessary biopsy and overtreatment for many patients [16, 63]. In addition, increased PSA concentration can also be found in normal persons, thus the PSA-based screening is not always correct [64]. Therefore, selecting new biomarkers or developing strategies for multiple tumor biomarkers detection is becoming more and more important for accurate early-stage prostate cancer diagnosis. Previous studies showed that sarcosine has a close correlation with the prostate cancer. Sarcosine's concentration is significantly increased in cancer patients, and thus could be used as a combined biomarker [65]. Recently, Lu et al. elaborately designed an electrochemical aptasensor for simultaneous detection of PSA and sarcosine [66]. They first prepared hierarchical MoS_2 nanostructures on the surface of GCE (Fig. 11.5). The flower-like structure could enhance surface area and improve DNA hybridization efficiency. Then two functional SiO_2 nanoprobes modified with both numerous electroactive redox indicators and a carboxyl-modified DNA were designed and attached to the MoS_2 surface by DNA hybridization. In the presence of PSA and sarcosine, SiO_2 nanoprobes were detached from the MoS_2 surface, and the electrochemical signal was dramatically reduced. The biosensor could successfully detect both PSA and sarcosine biomarkers in human serum with corresponding detect limits of 2.5 and 14.4 fg mL^{-1}, respectively. Furthermore, cancer patients and healthy ones could be distinguished.

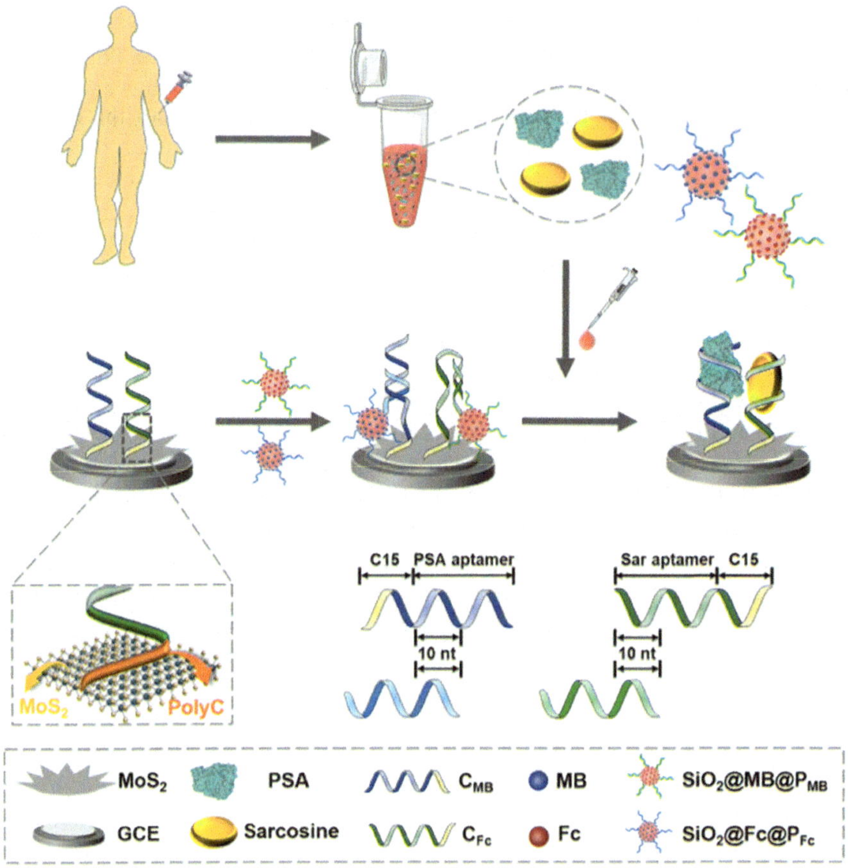

Fig. 11.5 Electrochemical biosensors for the detection of prostate tumor biomarker PSA (Reproduced from [66])

11.3.2.4 Detection of Colon Tumor Biomarkers

Colon cancer is usually regarded as an old age disease. It is the third most frequent cancer type, and the second leading cause of death worldwide (Fig. 11.1). According to a recent report in 2020, there were 1, 931,590 new cases of colon cancer and 935,173 deaths occurred due to colon cancer globally. The exact reason for colon cancer is still unclear. Colon cancer-related biomarkers, such as nuclei acids (e.g., specific genes, mutant DNA, specific microRNA), proteins (e.g., CEA, p53), and circulating tumor cells, have been utilized for cancer detection [67].

miRNA-21 is a microRNA acting as an oncogene, and has been proposed as a biomarker for colon cancer monitoring. For example, the electrochemical biosensor was reported for the detection of miRNA-21 extracted from a colon cancer cell sample [18]. Commercially available magnetic particles labeled with streptavidin were used and modified further with the biotinylated capture probe. The miRNA-21 extracted from exosomes was initially captured by the magnetic particles, and then was further collected from the bioconjugates after heating and magnetic separation. When miRNA-21 was finally adsorbed onto the electrode surface, DPV was performed to successfully detect its concentration in the presence of the $[Fe(CN)_6]^{4-/3-}$ redox system. The reported method exhibited a detection sensitivity of 1.0 pM for the serum sample with a detection limit of 1.0 pM.

p53 is a kind of tumor-associated antigen and plays a significant role in cellular functions [68]. It's an important colon tumor biomarker and its concentration in the serum samples will increase to 1.03 ng mL^{-1} when the cancer is present, compared to the normal level of 0.52 ng mL^{-1} [69]. In an example reported by Garranzo-Asensio et al., an electrochemical biosensor was designed for the detection of p53-specific autoantibodies (Fig. 11.6) [70]. The functional magnetic beads (MBs) were first prepared by covalently modified with HaloTag fusion p53 protein which had specific interaction with bio-targets. The p53-autoantibodies were then selectively captured by the MBs, which were subsequently conjugated with immunocomplexes anti-human IgG-HRP, and added onto carbon-based electrodes. Amperometric test under the enzymatic reduction of hydroquinone/H_2O_2 was used to determine the concentration of p53 autoantibodies in the serum sample. The proposed biosensor could accurately determine the p53 in serum samples from patients with a high possibility of developing colon cancer and from patients with diagnosed ovarian and colon cancer.

11.3.2.5 Detection of Liver Tumor Biomarkers

LC is the sixth most prevalent malignant tumor worldwide, and there are 905,677 new cases of LC in 2020. LC is ranked the third leading cause of cancer-related death in 2020, and about 830,180 deaths are recorded (Fig. 11.1). The highly malignant LC develops rapidly in a short time but there is no effective prognosis currently. More seriously, no obviously abnormal symptoms could be seen in LC patients at the early stages. Thus, in most cases the diagnosis is relatively late and the limited treatment is rather poor, resulting in a high mortality rate is high [71]. According to the Barcelona Clinic Liver Cancer (BCLC) system, LC can be classified into five stages, that is very early disease (stage 0), early disease (stage A), intermediate disease (stage B), advanced disease (stage C), and terminal disease (stage D) [72]. In the early stage (0 or A), survival at five years is in high possibility, which significantly reduces in the advanced stage [73]. Therefore, early diagnosis and interference of LC seem quite important to improve the survival time and quality of patients. The electrochemical detection of LC-related biomarkers is an effective way and has received great attention.

AFP is a featured glycoprotein mainly produced by liver cells, and has been commonly used as a clinical tumor biomarker for the diagnosis of liver cancer tumor [24]. The normal AFP concentration in human blood is generally lower than 20 ng ml^{-1}, but will increase significantly to be higher than 400 ng mL^{-1} when the

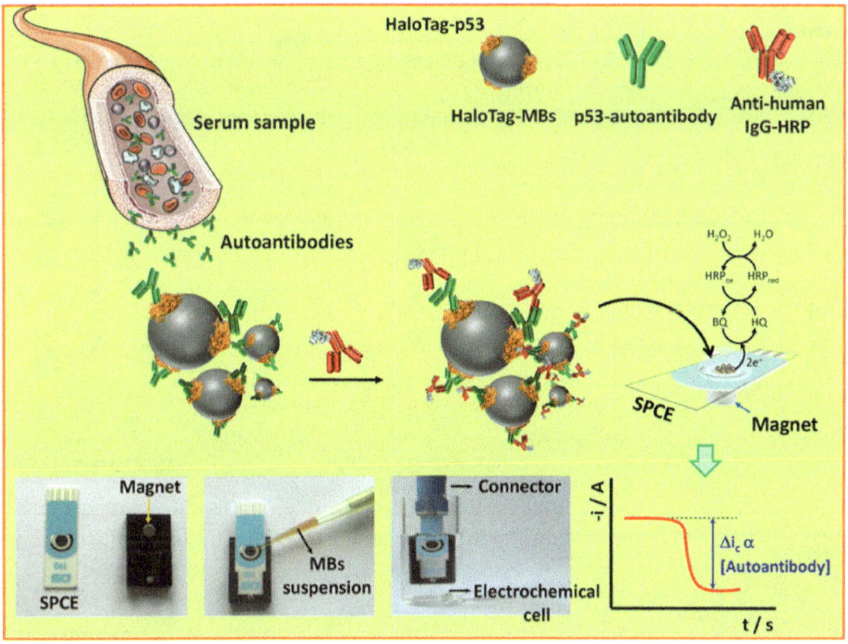

Fig. 11.6 Electrochemical biosensors for detection of colon tumor biomarker p53. (Reproduced from [70])

liver cells are damaged or in the presence of a liver tumor. Therefore, the detection of AFP in the blood may indicate the presence of a liver tumor, and then proper treatment could be used [24]. Recently, highly sensitive electrochemical detection of AFP was reported by using a gold-based electrode [74]. In order to improve the electroanalytical performance, a bare gold electrode was modified by depositing gold nanowires on the surface, which was then coated with polydopamine (PDA) nanofilm. The as-prepared electrode had a structured surface with significantly increased active surface area, which improved both the electron transfer and the immobilization of antibodies for AFP capture. The new electrochemical biosensor could detect AFP in the concentration range of 0.1 pg mL^{-1} to 50 ng mL^{-1} with a linear response. The detection limit reached 0.01 pg mL^{-1}. In another study, Wang et al. constructed a sandwich-type electrochemical biosensor for AFP detection with a high selectivity (Fig. 11.7) [75]. The biosensor was fabricated on a gold electrode. Two recognition elements, i.e., Ab1 and Ab2 were used. Ab1 was directly modified to the electrode and Ab2 assembled with Ab1 to form the sandwich structure through the AFP connection. Ab2 and electroactive redox probe of aminoferrocene (AF) were also attached to PDA nanoparticles (AB2-PDA-AF). The electrochemical signal of the biosensor was first amplified by the oxidation of the AF on Ab2-PDA-AF, and then by the reaction between AF/AF^{+} and the catechol groups of PDA nanoparticles. This immunosensor exhibited good performance of good selectivity, stability, and reproducibility, and the

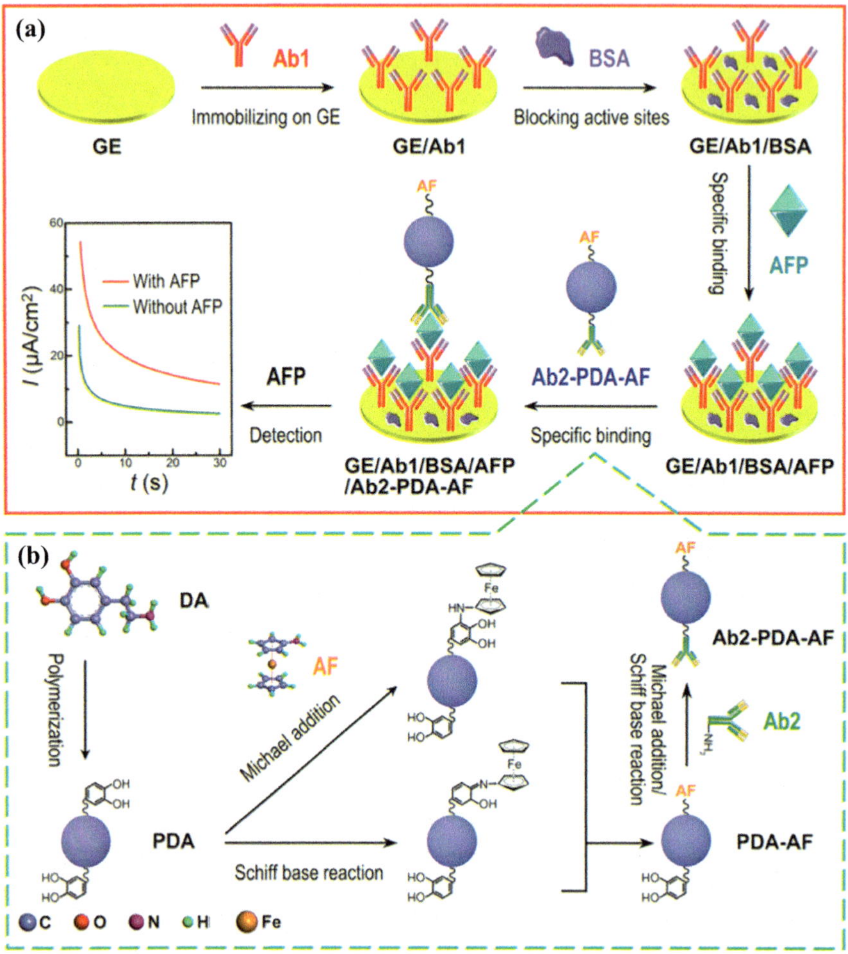

Fig. 11.7 Electrochemical biosensors for detection of liver tumor biomarker AFP (Reproduced from [75])

detection of AFP in human serum samples was well achieved with a detection limit of ca. 0.01 pg mL^{-1}.

11.4 Conclusion and Outlook

With continuous efforts devoted to the early and precise diagnosis of cancer, significant progress has been made in the construction and development of various electrochemical biosensors for tumor biomarker detection in the past decade. Particularly, owing to the rapid development of new nanomaterials and nanotechnology, novel electrochemical biosensors with ultrahigh sensitivity, selectivity, and simplicity have been achieved, which make them quite promising to be applied in the diagnostic test at the molecular and cellular level and using small amount blood samples.

On the other hand, towards practical utilization, many challenges remain. There is a need for the enhancement of electrochemical biosensors for multiple tumor biomarker detection as some tumors cannot be determined by a single biomarker. For example, CEA is a common tumor biomarker found in lung cancer, breast cancer, colon cancer, and gastric cancer, and the detection of CEA could not provide sufficient information about cancer. Achieving successful clinical tests also requires further improvement of the electrochemical biosensors in the aspects of stability, repeatability, and interference immunity. To develop cheap and portable biosensors for real-time analysis is also important, which will enable the techniques to be widely utilized at home.

References

1. Mostafa IM, Tian Y, Anjum S et al (2022) Comprehensive review on the electrochemical biosensors of different breast cancer biomarkers. Sens Actuat B Chem 365:20
2. Zhang Y, Li MY, Gao XM et al (2019) Nanotechnology in cancer diagnosis: progress, challenges and opportunities. J Hematol Oncol 12:137
3. Ahlawat S, Singh KP, Nehra A et al (2022) Ultra rapid biosensors for detecting carcinoma biomarkers. Mater Express 12:179–201
4. Sharifianjazi F, Jafari RA, Esmaeilkhanian A et al (2021) Biosensors and nanotechnology for cancer diagnosis (lung and bronchus, breast, prostate, and colon): a systematic review. Biomed Mater 17:012002
5. National Cancer Institute (2023) Tumor markers. https://www.cancer.gov/about-cancer/diagnosis-staging/diagnosis/tumormarkers-fact-sheet
6. Duffy MJ (2001) Clinical uses of tumor markers: a critical review. Crit Rev Clin Lab Sci 38:225–262
7. Zhang ZJ, Sen P, Adhikari BR et al (2022) Development of nucleic acid-based electrochemical biosensors for clinical applications. Angew Chem Int Ed 61:e202212496
8. Öbrink B (1997) CEA adhesion molecules: multifunctional proteins with signal-regulatory properties. Curr Opin Cell Biol 9:616–626
9. Okamura K, Takayama K, Izumi M et al (2013) Diagnostic value of CEA and CYFRA 21–1 tumor markers in primary lung cancer. Lung Cancer 80:45–49
10. Oremek GM, Sauer-Eppel H, Bruzdziak TH (2007) Value of tumour and inflammatory markers in lung cancer. Anticancer Res 27:1911–1915
11. Kirwan A, Utratna M, O'Dwyer ME et al (2015) Glycosylation-based serum biomarkers for cancer diagnostics and prognostics. BioMed Res Int 2015:490531
12. Loibl S, Gianni L (2017) HER2-positive breast cancer. Lancet 389:2415–2429
13. Wang K, He M, Zhai F et al (2017) A novel electrochemical biosensor based on polyadenine modified aptamer for label-free and ultrasensitive detection of human breast cancer cells. Talanta 166:87–92
14. Yadav BS, Chanana P, Jhamb S et al (2015) Biomarkers in triple negative breast cancer: a review. J Clin Oncol 6:252–263
15. Sreekumar A, Poisson ML, Rajendiran MT et al (2009) Metabolomic profiles delineate potential role for sarcosine in prostate cancer progression. Nature 457:910–914

16. Numan A, Singh S, Zhan YQ et al (2021) Advanced nanoengineered-customized point-of-care tools for prostate-specific antigen. Microchim Acta 189:27
17. Swiderska M, Choromańska B, Dąbrowska E et al (2014) The diagnostics of colorectal cancer. Contemporary Oncol (Poznan, Poland) 18:1–6
18. Boriachek K, Umer M, Islam MN et al (2018) An amplification-free electrochemical detection of exosomal miRNA-21 in serum samples. Analyst 143:1662–1669
19. Nikolaou S, Qiu SY, Fiorentino F et al (2018) Systematic review of blood diagnostic markers in colorectal cancer. Tech Coloproctol 22:481–498
20. Mohammadinejad A, Oskuee KR, Eivazzadeh-Keihan R et al (2020) Development of biosensors for detection of alpha-fetoprotein: as a major biomarker for hepatocellular carcinoma. Trends Environ Anal Chem 130:115961
21. Adamczyk B, Tharmalingam T, Rudd MP (2012) Glycans as cancer biomarkers. Biochim Biophys Acta Gen Subj 1820:1347–1353
22. Pawar H, Kashyap MK, Sahasrabuddhe NA et al (2011) Quantitative tissue proteomics of esophageal squamous cell carcinoma for novel biomarker discovery. Cancer Biol Ther 12:510–522
23. Dnistrian AM, Schwartz MK, Greenberg EJ et al (1991) CA 15–3 and carcinoembryonic antigen in the clinical evaluation of breast cancer. Clin Chim Acta 200:81–93
24. Zhang QY, Ou LB (2022) Recent progress in electrochemical biosensors for the detection of alpha-fetoprotein as the biomarker of liver cancer. Int J Electrochem Sci 17:15
25. Gupta D, Lis CG (2009) Role of CA125 in predicting ovarian cancer survival-a review of the epidemiological literature. J Adv Res 2:13
26. Wu J, Fu Z, Yan F et al (2007) Biomedical and clinical applications of immunoassays and immunosensors for tumor markers. Trends Anal Chem 26:679–688
27. Jafari M, Hasanzadeh M, Solhi E et al (2019) Ultrasensitive bioassay of epitope of Mucin-16 protein (CA 125) in human plasma samples by using of novel immunoassay: a new platform in early stage diagnosis of ovarian cancer and efficient management. J Biol Macromol 126:1255–1265
28. Grieshaber D, MacKenzie R, Vörös J, et al (2008) Electrochemical biosensors-sensor principles and architectures. Sensors 8:1400–1458
29. Li MT, Jiang F, Xue LY et al (2022) Recent progress in biosensors for detection of tumor biomarkers. Molecules 27:7327
30. Kimmel DW, Leblanc G, Meschievitz ME et al (2012) Electrochemical sensors and biosensors. Anal Chem 84:685–707
31. Ronkainen NJ, Halsall BH, Heineman RW (2010) Electrochemical biosensors. Chem Soc Rev 39:1747–1763
32. Hammond JL, Formisano N, Estrela P (2016) Electrochemical biosensors and nanobiosensors. Essays Biochem 60:69–80
33. Ning QH, Feng SQ, Cheng YM et al (2022) Point-of-care biochemical assays using electrochemical technologies: approaches, applications, and opportunities. Mikrochim Acta 189:310–310
34. Shrivastava S, Jadon N, Jain R (2016) Next-generation polymer nanocomposite-based electrochemical sensors and biosensors: a review. Trends Anal Chem 82:55–67
35. Hasanzadeh M, Sahmani R, Solhi E et al (2018) Ultrasensitive immunoassay of carcinoma antigen 125 in untreated human plasma samples using gold nanoparticles with flower like morphology: a new platform in early stage diagnosis of ovarian cancer and efficient management. J Biol Macromol 119:913–925
36. Shekari Z, Zare HR, Falahati A (2021) Dual assaying of breast cancer biomarkers by using a sandwich–type electrochemical aptasensor based on a gold nanoparticles-3D graphene hydrogel nanocomposite and redox probes labeled aptamers. Sens Actuat B Chem 332:129515
37. Farzin L, Shamsipur M, Samandari L et al (2018) Recent advances in designing nanomaterial based biointerfaces for electrochemical biosensing cardiovascular biomarkers. J Pharm Anal 161:344–376

38. Worm Ørntoft MB (2018) Review of blood-based colorectal cancer screening: how far are circulating cell-free DNA methylation markers from clinical implementation? Clin Colorectal Cancer 17:e415–e433
39. Yörüker EE, Holdenrieder S, Gezer U (2016) Blood-based biomarkers for diagnosis, prognosis and treatment of colorectal cancer. Clin Chim Acta 455:26–32
40. Um S, Kim KH, Jung S et al (2015) Endobronchial ultrasound versus mediastinoscopy for mediastinal nodal staging of non-small-cell lung cancer. J Thorac Oncol 10:331–337
41. Deng Y, Li X, Lei Y et al (2016) Use of diffusion-weighted magnetic resonance imaging to distinguish between lung cancer and focal inflammatory lesions: a comparison of intravoxel incoherent motion derived parameters and apparent diffusion coefficient. Acta Radiol 57:1310–1317
42. Cunliffe A, Armato GS, Castillo R et al (2015) Lung texture in serial thoracic computed tomography scans: correlation of radiomics-based features with radiation therapy dose and radiation pneumonitis development. Int J Radiat Biol 91:1048–1056
43. Tang RX, Liang L, Luo DZ et al (2015) Downregulation of MiR-30a is associated with poor prognosis in lung cancer. Med Sci Monit 21:2514–2520
44. Schneider J (2006) Tumor markers in detection of lung cancer. Adv Clin Chem 42:1–41
45. Atapour A, Khajehzadeh H, Shafie M et al (2022) Gold nanoparticle-based aptasensors: a promising perspective for early-stage detection of cancer biomarkers. Mater. Today Commun 30:20
46. Khanmohammadi A, Aghaie A, Vahedi E et al (2020) Electrochemical biosensors for the detection of lung cancer biomarkers: a review. Talanta 206:120251
47. Li J, Xie H, Liu Y et al (2015) Au-F127 strawberry-like nanospheres as an electrochemical interface for sensitive detection of carcinoembryonic antigen in real sample. Talanta 144:404–410
48. Su S, Han X, Lu Z et al (2017) Facile synthesis of a MoS_2-prussian blue nanocube nanohybrid-based electrochemical sensing platform for hydrogen peroxide and carcinoembryonic antigen detection. ACS Appl Mater Interfaces 9:12773–12781
49. Rankin NM, McGregor D, Stone E et al (2018) Evidence-practice gaps in lung cancer: a scoping review. Eur J Cancer Care (Engl) 27:e12588
50. Yang G, Xiao Z, Tang C et al (2019) Recent advances in biosensor for detection of lung cancer biomarkers. Biosens Bioelectron 141:11
51. Qi W, Li X, Kang J (2014) Advances in the study of serum tumor markers of lung cancer. J Cancer Res Ther 10:C95–C101
52. Zhang Q, Li X, Qian C et al (2018) Label-free electrochemical immunoassay for neuron specific enolase based on 3D macroporous reduced graphene oxide/polyaniline film. Anal Biochem 540–541:1–8
53. Siegel RL, Miller KD, Wagle NS et al (2023) Cancer statistics, 2023. CA Cancer J Clin 73:17–48
54. Mittal S, Kaur H, Gautam N et al (2017) Biosensors for breast cancer diagnosis: a review of bioreceptors, biotransducers and signal amplification strategies. Biosens Bioelectron 88:217–231
55. Hong R, Sun HY, Li DJ et al (2022) A review of biosensors for detecting tumor markers in breast cancer. Life-Basel 12:23
56. Donepudi SM, Kondapalli K, Amos JS et al (2014) Breast cancer statistics and markers. Cancer Res 10:506
57. Friedenson B (2007) The BRCA1/2 pathway prevents hematologic cancers in addition to breast and ovarian cancers. BMC Cancer 7:152
58. Wang J, Wang D, Hui N (2020) A low fouling electrochemical biosensor based on the zwitterionic polypeptide doped conducting polymer PEDOT for breast cancer marker BRCA1 detection. Bioelectrochemistry 136:107595
59. Tsé C, Gauchez A, Jacot W et al (2012) HER2 shedding and serum HER2 extracellular domain: biology and clinical utility in breast cancer. Cancer Treat Rev 38:133–142
60. Nicolini A, Ferrari P, Duffy JM (2018) Prognostic and predictive biomarkers in breast cancer: past, present and future. Semin Cancer Biol 52:56–73

61. Marques CR, Viswanathan S, Nouws PH et al (2014) Electrochemical immunosensor for the analysis of the breast cancer biomarker HER2 ECD. Talanta 129:594–599
62. Pothipor C, Wiriyakun N, Putnin T et al (2019) Highly sensitive biosensor based on graphene–poly (3-aminobenzoic acid) modified electrodes and porous-hollowed-silver-gold nanoparticle labelling for prostate cancer detection. Sens Actuat B Chem 296:126657
63. Nam J, Thaxton SC, Mirkin AC (2003) Nanoparticle-based bio-bar codes for the ultrasensitive detection of proteins. Science 301:1884–1886
64. Sohn E (2015) Screening: diagnostic dilemma. Nature 528:S120-S122
65. Jentzmik F, Stephan C, Miller K et al (2010) Sarcosine in urine after digital rectal examination fails as a marker in prostate cancer detection and identification of aggressive tumours. Eur Urol 58:12–18
66. Yan RH, Lu N, Han SP, et al (2022) Simultaneous detection of dual biomarkers using hierarchical MoS2 nanostructuring and nano-signal amplification-based electrochemical aptasensor toward accurate diagnosis of prostate cancer. Biosens Bioelectron 197:113797–113797
67. Quinchia J, Echeverri D, Cruz-Pacheco FA et al (2020) Electrochemical biosensors for determination of colorectal tumor biomarkers. Micromachines 11:46
68. Aydın M, Aydın BE, Sezgintürk KM (2018) A disposable immunosensor using ITO based electrode modified by a star-shaped polymer for analysis of tumor suppressor protein p53 in human serum. Biosens Bioelectron 107:1–9
69. Attallah AM, Abdel-Aziz MM, El-Sayed AM et al (2003) Detection of serum p53 protein in patients with different gastrointestinal cancers. Cancer Prev Res (Phila) 27:127–131
70. Garranzo-Asensio M, Guzman-Aranguez A, Poves C et al (2016) Toward liquid biopsy: determination of the humoral immune response in cancer patients using HaloTag fusion protein-modified electrochemical bioplatforms. J Anal Chem 88:12339–12345
71. Vogel A, Saborowski A (2020) Current strategies for the treatment of intermediate and advanced hepatocellular carcinoma. Cancer Treat Rev 82:101946
72. Llovet JM, Kelley RK, Villanueva A, et al (2021) Hepatocellular carcinoma. Nat Rev Dis Primers 7:6
73. Villanueva A (2019) Hepatocellular carcinoma. N Engl J Med 380:1450–1462
74. Hong YX, Wang YL, Zhu YC (2020) Highly sensitive immunosensor based on polydopamine-nanofilm modified 3D gold nanoelectrode for α-fetoprotein detection. Electrochim Acta 364:137328
75. Wang Y, Hong Y, Wang M et al (2022) Multifunctional nanolabels based on polydopamine nanospheres for sensitive alpha fetoprotein electrochemical detection. ACS Appl Nano Mater 5:1588–1599

Chapter 12
Electrochemical Biosensors for Virus Detection

Zhi Zheng, Haiyang Li, Hui Li, and Fan Xia

Abstract Viruses could be copied with millions of new particles in the host cell, making long-term threats to public health due to their stable gene and fast spreading capacity. Rapid and accurate detection of viruses could prevent proliferation and is very important for global health. Conventional strategies for detecting viruses encounter difficulties in clinical practice because of complex sample preparation, high technical operators, and time-consuming process. Electrochemical biosensors could determine viruses with cost effectiveness, wide liner range, low limit of detection, high specificity, and the possibility of miniaturization due to the signaling mechanism between redox-active tags and electrodes. Bio-recognition probes and electrode interface nanomaterials are two basic parts of electrochemical biosensors. The detection of viruses could be enhanced by designing bio-recognition probes and developing electrode interface nanomaterials.

Keywords Electrochemical biosensors · Viruses · Bio-recognition probes · Electrode interface nanomaterials

12.1 Introduction

During the past few decades, viruses including Zika virus (ZIKV), Ebola virus (EBV), hepatitis virus, dengue virus (DENV), influenza virus, human immune deficiency virus (HIV), and recently epidemic COVID-19 et al. caused outbreaks and became the major concern of today's world [1–3]. These viruses possess a lipidic envelope and a protein capsid with a few genes and a small volume, which could be copied with millions of new particles in the host cell, leading to severe diseases [4, 5]. Worse still, viruses may induce long-term threats to public health because of their stable gene

Z. Zheng · H. Li · H. Li · F. Xia (✉)
State Key Laboratory of Biogeology and Environmental Geology, Engineering Research Center of Nano-Geomaterials of the Ministry of Education, Faculty of Materials Science and Chemistry, China University of Geosciences, Wuhan 430074, P. R. China
e-mail: xiafan@cug.edu.cn

F. Xia et al. (eds.), *Electrochemical Biosensors for Whole Blood Analysis*,
https://doi.org/10.1007/978-981-99-5644-9_12

and fast-spreading capacity. Therefore, it is crucial to detect viral diseases rapidly and accurately to prevent proliferation and take adequate treatment [6].

Conventional strategies such as reverse transcription polymerase chain reaction (RT-PCR) could accurately detect viruses [7], but encounter difficulties with the complex fabrication of samples, high requirements of technical operators, and long-time process [8–12]. Optical biosensors can detect viruses by the wavelength shift, intensity change of response signal in a label-free merit, however, the measurement requires a sophisticated light source, optical detector as well as precise optical alignments to improve the sensitivity [13]. Mechanical biosensors including micro-cantilever types, micropillar types, piezoelectric types, and photoacoustic types monitor mass or frequency change before and after virus binding [14]. However, the performance of mechanical biosensors was affected by various factors such as the mechanical and geometrical properties, cantilever dimensions, antibodies immobilization, antigen distribution, surface energy, and ambient environment, leading to low stability and repeatability [15].

For electrochemical biosensors, the immobilized biomolecules react with target analytes and produce electrons, the biochemical information is transformed into electrochemical signals such as capacitance, potential, current, and conductivity et al., electrochemical biosensors could determine viruses with high sensitivity, high specificity, and the possibility of miniaturization [16]. According to the measurement mode, electrochemical biosensors could be divided into amperometric, voltammetric, impedimetric, conductometric, polarographic, potentiometric, capacitive, and piezoelectric [17]. As shown in Fig. 12.1, virus detection based on electrochemical biosensors is widely concerned due to their signaling mechanism and easy-operated platform. In this chapter, we focus on bio-recognition probes as receptors and electrode interface nanomaterials as transducers to improve the performance of electrochemical biosensors.

12.2 Bio-recognition Probes

Bio-recognition probes, assembled by organized molecules or macromolecules, are vitally important parts of electrochemical biosensors to guarantee the recognition of virus detection specifically [18]. Nucleic acids, antibodies, and peptides are common bio-recognition probes which have been widely concerned in the detection of viruses [19].

12.2.1 Nucleic Acids

The specific binding between nucleic acids and viruses could trigger the change of electrochemical signals by modulating the transfer of electrons between electrodes and redox-active probes [20, 21]. With the merits of high specificity, stability,

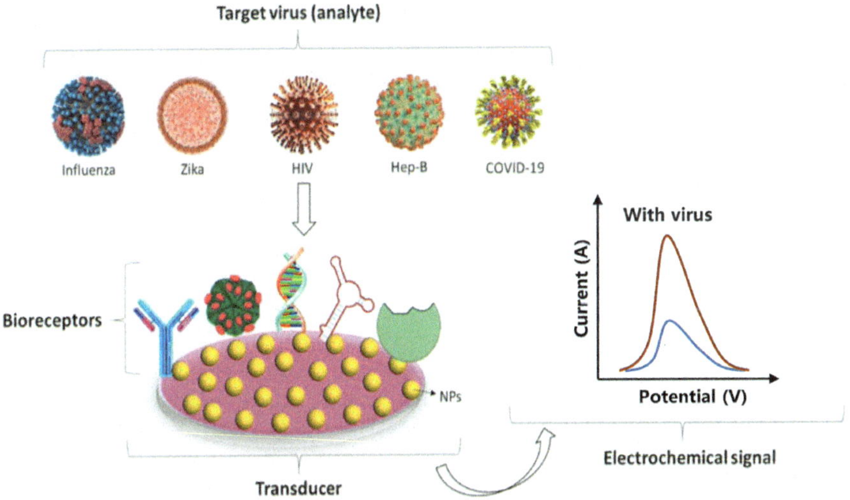

Fig. 12.1 The schematic diagram of an electrochemical biosensor platform for detecting a variety of viruses. The electrochemical biosensor platform mainly contains four parts: Target virus, bioreceptors, transducer, and electrochemical signal. The target virus binds onto the surface of electrodes by the bioreceptors, and this biochemical process could produce an obvious electrochemical signal by the transducer. (Reprinted with permission from Ref. [6]. Copyright 2020 Elsevier B.V.)

and miniaturization possibilities, nucleic acid-based electrochemical biosensors are widely applied in the fabrication of biosensors [22]. For example, Brito-Madurro et al. reported specific DNA-decorated graphite electrodes of electrochemical biosensors to detect hepatitis B virus (HBV) by monitoring oxidation peaks of ethidium bromide (EB) in the blood plasma of infected patients (Fig. 12.2a) [23]. The DNA probe was decorated on the graphite electrodes by the 4-Aminophenol (4-AMP) with two electrochemically active substituent groups (NH_2 and -OH). The hybridization process of the DNA probe and DNAgen could be detected by the electrochemical impedance spectra (EIS) at high frequencies. As the concentration of DNAgen increases, the resistivity of biosensors increases. As shown in Fig. 12.2b, the DNA probe could especially recognize the complementary region of virus, allowing the detection with a low limit of detection (35.69 ng mL^{-1}) and wide linear range (1.176–4.825 μg mL^{-1}). The biosensor could also work in the plasma with a short assay time, showing a potential application for detecting HBV.

As shown in Fig. 12.3, Liang et al. detected SARS-CoV-2 via probes of peptide-DNA (pDNA) and signal molecules of methylene blue (MB) [24]. The formation of double-stranded DNA could enhance the electrical signal by binding more MB molecules for monitoring by square wave voltammetry (SWV) measurement. The binding of SARS-CoV-2 onto the surface of electrodes induced an obvious SWV signal. This detection could be conducted in real solutions such as blood and saliva.

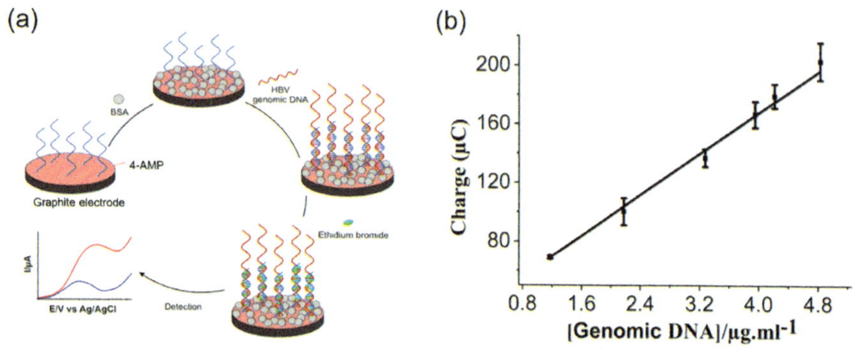

Fig. 12.2 The schematic process of constructing DNA-decorated graphite electrodes for detecting HBV based on the electrochemical biosensor. **a** The graphite electrodes were coated by 4-AMP for the decoration of HBV genomic DNA. Based on the oxidation of EB, the HBV could be detected by the electrochemical method. **b** The calibration curves of charge values at different concentrations of genomic DNA from HBV-positive patients. (Reprinted with the permission from Ref. [23]. Copyright 2023 Springer Nature Switzerland AG.)

The detection limit is low to 27.18 fM, and the linear detection range could be as wide as 1 pg mL^{-1}–10 μg mL^{-1}. Moreover, the detection of proteins shows higher SWV signals than other interferents such as albumin, demonstrating high specificity for practical application.

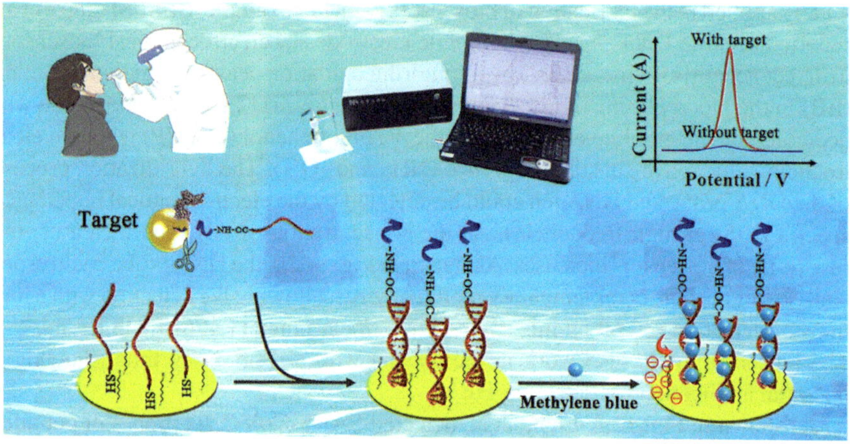

Fig. 12.3 The schematic process of detecting SARS-CoV-2 based on electrochemical biosensors. Carboxyl-modified DNA1 was cross-linked to the substrate peptide to form a composite of pDNA. Then, the pDNA bonds with Au nanoparticles (AuNPs) to form AuNP-peptide-DNA nanoprobes. In the presence of target protease, the peptide substrate was cut off, leading to the release of a DNA1 fragment. The DNA1 could hybridize with DNA2 on the surface of electrodes to form double-stranded DNA (dsDNA). Signal molecules MB could insert into the dsDNA and produce significant electrochemical signals to detect SARS-CoV-2. (Reprinted with the permission from Ref. [24]. Copyright 2022 Elsevier B.V.)

12.2.2 Antibodies

Antibodies, connected with the key hinge disulfide bridges, are two fragment-antigen binding (Fab) proteins [25]. The Fab fragments termed with disulfide (Fab') could bind with viruses through covalent interaction. Recognition molecules based on antibodies have been widely applied in diagnosing viruses due to their operational simplicity to decorate, and high specificity to bind with targets with high sensitivity [26]. The electrochemical biosensors based on antibodies could detect different types of viruses including coronaviruses, Zika, rotavirus, enterovirus, HIV, etc., which is promising to substitute traditional RT-PCR [27]. Based on the modification of alkaline phosphatase (ALP) and polyclonal antibody on bifunctional fluorescence magnetic nanospheres (FMN), the digital enzyme-linked immunosorbent assay (ELISA) could capture avian influenza virus (AIV) by controlling the proportion of magnetic nanospheres (Fig. 12.4) [28]. The higher concentration of polyclonal antibody could increase the capture efficiency of H7N9 AIV. Moreover, the biological activity of polyclonal antibodies could last about 5 weeks. In the chicken serum and liver sample, the biosensor could especially recognize H7N9 AIV and demonstrate a much higher signal than control groups from the linear sweep voltammetry. The detection limit of the biosensor for detecting H7N9 AIV is low to 7.8 fg/mL, indicating sensitive and accurate detection in early disease diagnosis. The biosensor shows a 29-fold higher response of H7N9 AIV than other interferents, indicating a high specificity and potential practical application.

Based on antibody modification to specially capture M1 protein to cause the change of EIS, Siuzdak et al. reported an electrochemical biosensor to detect several viruses per sample in saliva buffer with a low limit of detection (5×10^{-14} g/mL). This method is suitable to detect all influenza A by choosing corresponding antibodies [29]. Zourob et al. detected the SARS-CoV-2 virus by integrating collection and detection samples into screen-printed electrodes with absorbing cotton padding and immobilizing virus nucleocapsid protein antibodies. The binding of antigen by the antibody increases the intensity of the current peak of the redox couple. The electrochemical biosensors show a low LOD (0.8 pg/mL) and excellent specificity with rapid detection [30]. Hushegyi et al. designed electrochemical biosensors based on glycan for the selective and reliable sense of influenza viruses H3N2 with the visualization of atomic force microscopy [31]. The biosensors could detect influenza viruses H3N2 selectively from different subtypes based on immobilization of antibodies. The LOD of biosensors is as low as 13 viral particles μl^{-1}, and the sensitivity is 30 higher than influenza viruses H7N7.

12.2.3 Peptides

Peptides have unique properties such as simple preparation, smaller size, and diverse functionalization methods, which could be designed for a high-selectivity diagnostic

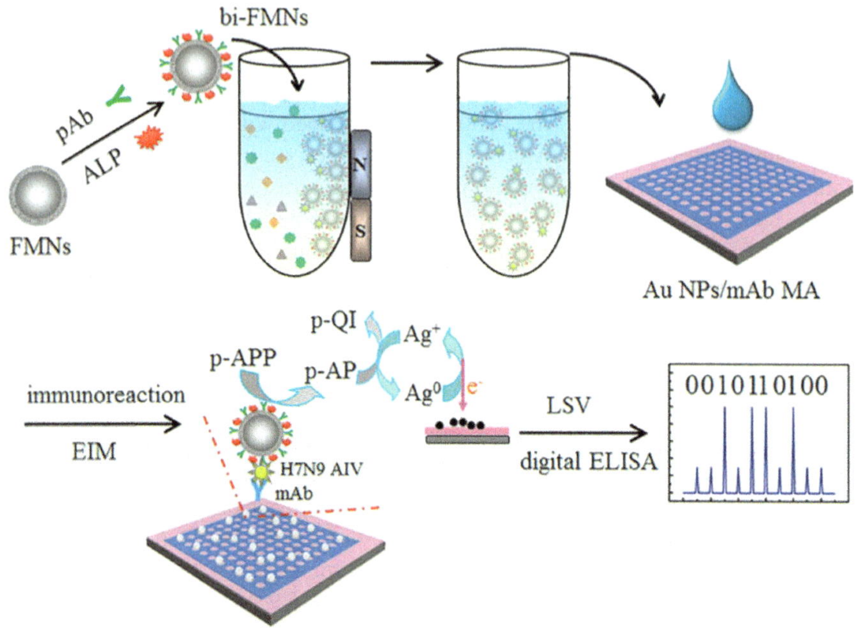

Fig. 12.4 The illustrated process of counting H7N9 AIV by the electrochemical method. Polyclonal antibodies and ALP are immobilized on FMN to separate H7N9 AIV in complex samples. ALP could catalyze the p-aminophenyl to form p-aminophenol. On the surface of the microelectrode array, the Ag^{+} is reduced to Ag^{0}, leading to the change of electrochemical signal which was counted as "0" or "1". The concentration of H7N9 AIV could be calculated through the probability of "0". (Reprinted with the permission from Ref. [28]. Copyright 2017 American Chemical Society)

of viruses [32, 33]. Moreover, peptides could provide quantitative detections of viruses as the electron transfer rate of signal markers is different before and after binding with viruses [34]. Orozco et al. detected Spike protein of SARS-CoV-2 virus resulting from COVID-19 patients based on synthetic thiolated peptide bioreceptor by EIS with $[Fe(CN)_6]^{3-/4-}/KNO_3$ as redox pair. The 23 amino acids in the synthetic peptide could specifically bind with the RBD region of Spike protein. The designed device could detect Spike protein in viral swabs with high reproducibility, sensitivity as well as specificity [35]. Reddy et al. report an ultrasensitive biosensor based on peptide probes to detect H1N1 influenza viruses and H5N2 influenza viruses, respectively [36]. The binding of H1N1 and H5N2 with peptides leads to the decreased peak current. The two recognition sequences in peptide enhance the affinity to detection for a real clinical application. The electron transfer between electrolyte solutions and peptides was accelerated by the electrostatic interactions between the redox species and N-terminus in the peptide. In this strategy, the detection limit of influenza viruses and viral proteins is low (3.63 and 2.39 nM for H1N1 and for H5N2, respectively). Moreover, the peptide-based electrochemical biosensor demonstrates specific detection of their corresponding targets. Alves et al. reported a new biosensor for detecting Breast Cancer by selecting two peptides (biotin-C3 and biotin-H2) as the biorecognition phase in serological samples [37]. Furthermore, the 3-(3-aminophenyl)propionic acid was electropolymerized during the electrochemical process. Compared to the result of ELISA, this Biotin-C3 and biotin-H2 peptide shows optimized reproducibility and stability, and a new perspective of electrochemical platform for characterization of BC. Different from traditional immobilized strategy of probes on substrates, Ito et al. dispersed peptide ligands in biological buffers to efficiently and selectively capture the virus (Fig. 12.5) [38]. The peptide ligands become electropolymerized before capturing the virus and deposit as a polymer layer on the surface of the electrode to reduce the transferred electrons between electrodes and redox molecules. The concentration of influenza virus was determined by the concentration of protein in the $K_3[Fe(CN)_6]$ solution. As the concentration of influenza virus increases, some peptide ligands cannot electropolymerize after binding with the virus, increasing the transfer of electrons to the electrode and realizing the amplified "turn-on".

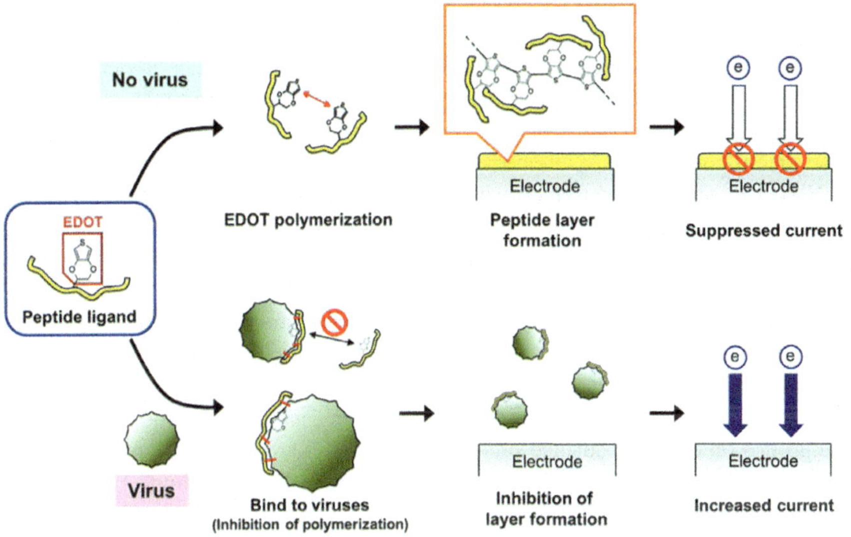

Fig. 12.5 The detection of influenza viruses based on peptide probes decorated electrochemical biosensor. The peptide ligands become electropolymerized with the interaction of 3,4-ethylenedioxythiophene (EDOT) if there is no influenza virus in the sample. The mixture of peptide ligands and EDOT is deposited as a polymer layer on the electrode surface which will decrease the transfer of electrons between electrode and redox molecules. Peptide ligands would bind with influenza viruses if they are present in the sample, inhibiting the formation of a polymer layer on the surface of the electrode, and increasing the number of transferred electrons to the electrode. (Reprinted with the permission from Ref. [38]. Copyright 2023 Royal Society of Chemistry)

12.3 Electrode Interface Nanomaterials

The nanomaterials on the working electrodes link the sensing interface and virus targets in the solution, playing an important role in fabricating high-performance electrochemical biosensors [39–41]. With modifying bio-recognition probes, the affinity of targets and sensitivity of detecting performance could be usually improved by modifying nanomaterials [42, 43]. The main reported nanomaterials and their application for detecting viruses are introduced as follows.

12.3.1 Gold-Based Nanomaterials

Gold nanoparticles (Au NPs) could link probes with high bioactivity and facilitate the transfer of electrons between the sensing interface and targets, which are widely used as electrode interface nanomaterials [44, 45]. Moreover, Au NPs could coordinate many kinds of functional groups such as –SH, forming multilayered bionanocomposite interface film [46, 47].

Palomar et al. modified Au NPs on the surface of the electrode to detect dengue toxin selectively and sensitively due to the large surface area and reactive interface for immobilizing antibodies via covalent binding interaction (Fig. 12.6) [48]. The electrochemical measurements were carried out in $K_4[Fe(CN)_6]^{4-}/K_3[Fe(CN)_6]^{3-}$ mixed solution to monitor the redox peak current intensities to investigate the electron transfer. The biosensors demonstrated a high sensitivity (- 0.44 μA/decade), a wide linear range (1×10^{-12}–1×10^{-6} g/mL), and extremely low LOD (3×10^{-13} g/mL) for detecting dengue virus 2 NS1. Moreover, the biosensor also shows excellent selectivity, reproducibility in human serum for diagnosis of dengue fever. Karakus et al. decorated Au NPs as linker molecules to immobilize spike antibody (mAb) to form Au NPs-mAb, allowing the simple and quick detection of pg/mL spike antigen in saliva [49]. During the measurement of the cathodic scan, the groups on the mAb surface such as carbonyl were reduced. The number of free groups decreases after adding an amount of spike antigen, leading to a decreased reduction signal. Moreover, the biosensor shows excellent selectivity while detecting 100 pg/mL MERS-CoV spike antigens, influenza A and Streptococcus pneumoniae with no response. This strategy is easily integrated into commercial kits with cost-efficiency and high speed, showing potential application in the diagnosis of viral diseases.

As shown in Fig. 12.7, Heo et al. modified the electrode with a flower-shaped gold nanostructure to increase its surface ratio and conductivity. The repulsive force between $[Fe(CN)_6]^{3-}$ and negative charges of SARS-CoV-2 RNA increased the resistance of electron transfer after the targets were captured, leading to the decrease of the current signal. The optimal ratio of Cas13a-crRNA could further enhance the response signal of biosensors, lowering the LOD of ORF genes of SARS-CoV-2 to 4.4×10^{-2} fg/mL and the LOD of S genes of SARS-CoV-2 to 8.1×10^{-2} fg/mL,

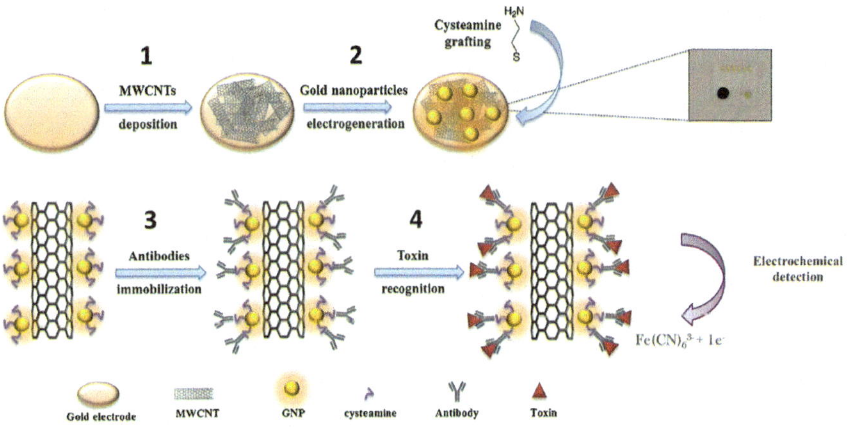

Fig. 12.6 The schematic process of detecting dengue virus with Au NPs. The Au NPs were electrodeposited onto the surface of MWCNTs. And the cysteamine was grafted on the surface of Au NPs by the Au–S binding interaction. Then the dengue antibodies were linked by cysteamine via EDC/NHS coupling for the recognition of dengue toxin. (Reprinted with the permission from Ref. [48]. Copyright 2023 Springer Nature Switzerland AG.)

respectively [50]. Hong et al. utilized the electrodeposition method to fabricate Au nanostructures on the electrode surface for the sensitive, rapid, and selective detection of norovirus based on the recognition element of concanavalin A [51]. The electrochemical measurements such as CV and EIS based on the redox probe of $Fe(CN)_6^{3-}$/ $Fe(CN)_6^{4-}$ were used to detect NoV. The binding of NoV reduced the penetration of the redox probe, leading to a low electron transfer conductance. The biosensors showed a wide linear range (10^2 and 10^6 copies/mL), and low LOD (35 copies/mL) in 1 h assay time. The biosensor also showed excellent specificity of approximately 98%.

12.3.2 *Carbon-Based Nanomaterials*

Carbon-based nanomaterials such as graphene quantum dots, 1D carbon nanotubes, or 2D graphene oxide could increase the electron transfer rates and electrical conductivity between the targets and sensing interface, resulting in high-sensitivity biosensors [52, 53]. Tancharoen et al. added graphene oxide composites into surface imprinted polymers (SIP) to increase conductivity to fabricate electrochemical biosensors for detecting ZIKV in serum with LOD (10–250 RNA copies/mL, Fig. 12.8) [54]. The detecting effect of viruses was measured by the following EIS and CV based on redox couples of $K_4Fe(CN)_6$/$K_3Fe(CN)_6$. And the impedance of the electrode decreased after increasing the concentration of ZIKV. As the concentration of ZIKV increases, only the biosensor with SIP shows the obvious response,

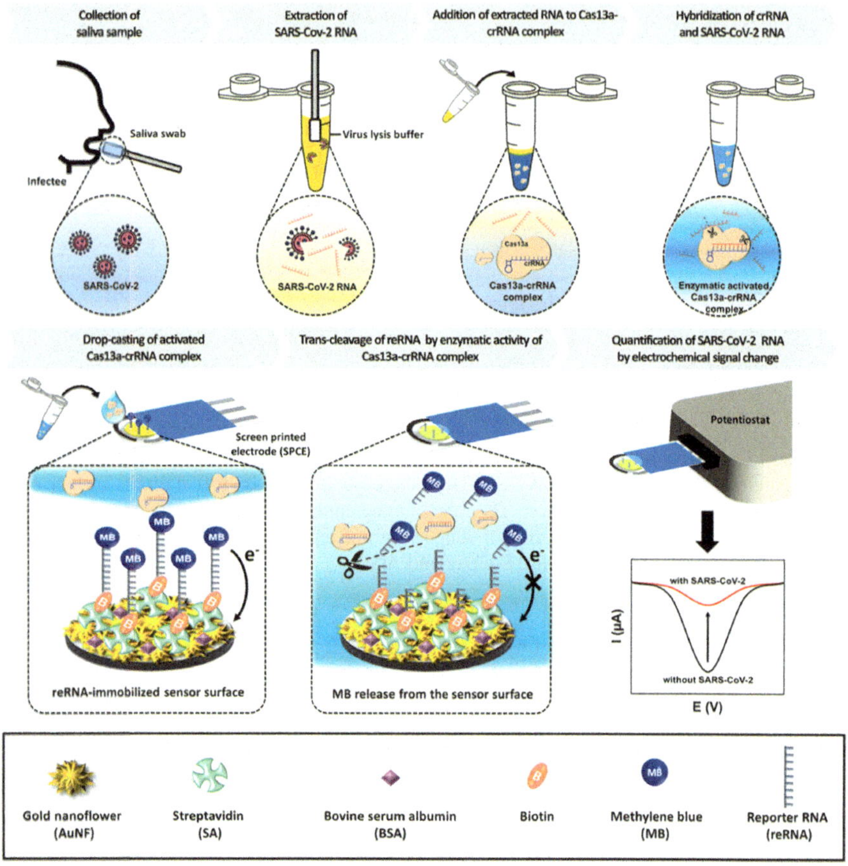

Fig. 12.7 The schematic process of detecting SARSCoV-2 viruses based on gold nanoflower. The viral RNA was extracted from infected patients in saliva with a lysis solution. Then, Cas13a-crRNA with high enzymatic activity was added to the mixture after binding with SARSCoV-2 RNA. The electrode is modified with a flower-shaped gold nanostructure followed by reporter RNA (reRNA) molecules, which were tagged with methylene blue. The activated Cas13a-crRNA complex solution was added to cleave the reRNA. The SARS-CoV-2 can be detected by quantifying the current change. (Reprinted with the permission from Ref. [50]. Copyright 2022 Elsevier B.V.)

demonstrating a high specificity. Wang et al. detected HIV based on $Ti_3C_2T_x$ modified ZIF-8 electrochemical biosensor [55]. The addition of $Ti_3C_2T_x$ could prevent the decrease of specific surface area due to the aggregation of ZIF-8, benefiting to increase the electrolyte transport and enhance the electrochemical luminescence intensity. This biosensor shows the linear response in the range of 1 fM to 1 nM and the LOD is 0.3 fM (S/N = 3) after optimization, and great recoveries in real serum samples. The response shows no obvious intensity when detecting other analogs, indicating acceptable specificity. Gogola et al. synthesize graphene quantum dots to modify electrodes to immobilize aptamer and amplify the electrochemical signal

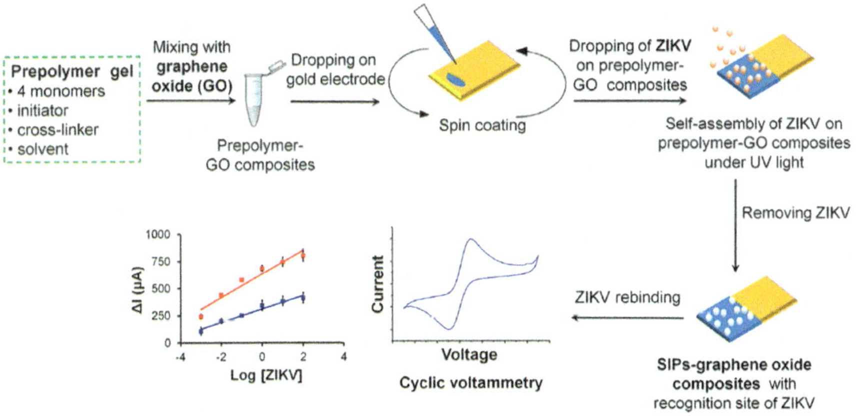

Fig. 12.8 The illustrated process of detecting Zika viruses based on decorating graphene oxides to the surface of a gold electrode. The graphene oxide was mixed with prepolymer gel to form composites, then was spin-coated to form a thin film on the surface of the gold electrode. Zika virus particles have a chance to occupy copolymers sites during the self-assembly process of prepolymer. The recognition sites of virus cavities could form after the virus was removed, which was used for detecting Zika viruses based on the electrochemical method. (Reprinted with the permission from Ref. [54]. Copyright 2018 American Chemical Society)

for detecting HIV-p24 protein with a linear response range from 0.93 ng mL^{-1} to 93 mg mL^{-1} and low LOD (51.7 pg mL^{-1}) [56]. The biosensor could identify positive and negative samples in spiked human serum for real samples. Ma et al. developed a simple method to determine HIV-p24 based on the decoration of multi-walled carbon nanotubes (MWCNTs) as electrode nanomaterials [57]. The decoration of MWCNTs was triggered by surface polymerization with acrylamide. The effect of modification was monitored by the electrochemical signal. The biosensor exhibited a wide linear range (0.1 pg cm^{-3}–2 ng cm^{-3}) and the LOD is low to 0.083 pg cm^{-3} for detecting HIV-p24. Chin et al. detected the Japanese encephalitis virus (JEV) based on carbon nanoparticles decorated screen printed carbon electrodes in human serum samples [58]. Then 3-aminopropyl triethoxysilane was functionalized for the immobilization of the antibody. The analytical performance was improved by modifying carbon nanoparticles with enhanced electron transfer and current intensity. The linear range of detecting JEV is 5–20 ng·mL^{-1} with an assay time of 20 min.

12.3.3 Magnetic Nanoparticles

Magnetic nanoparticles have a large surface area and high surface energy to allow highly efficient electron transfer. Moreover, magnetic nanoparticles could realize the separation of magnetic viruses and amplification of signals in clinical detection by external magnets [59]. For example, as shown in Fig. 12.9, Ye et al. utilized iron oxide (Fe_3O_4) magnetic nanoparticles to load with gold-silver core–shell to fabricate Fe_3O_4-Au@Ag as a label of signal DNA probe at the interface of the electrode to detect cauliflower mosaic virus 35S. The ability of the DNA probe to reduce H_2O_2 during the DPV measurement allowed for ultralow LOD (1.26×10^{-17} M) and wide linear range (1×10^{-16} to 1×10^{-10} M) by amplifying the reduction signal of hydrogen peroxide (H_2O_2) [60]. Furthermore, the response to target CaMV35S of biosensor remained 96.9% after storing at 4 °C over 15 days, indicating excellent stability. Durmus et al. developed an electrochemical biosensor with cocktail-conjugated magnetic nanoparticles to detect SARS-CoV-2 viruses in human swabs. The MNP-based biosensors showed an obvious oxidative current during the DPV measurement. After optimization, the LOD is 0.53–0.75 ng/mL. The biosensors demonstrated high discrimination of positive and negative samples. Moreover, the polyclonal antibody cocktail demonstrates better performance than commercial anti-S1 and anti-S2 methods for the detection of 10 negative patients and 40 positive patients with 100% accuracy [61]. Buyuksunetci et al. designed electrochemical biosensors with colorimetric assays for the diagnosis of SARS-CoV-2 viruses (Fig. 12.10) [62]. The magnetic γ-Fe_2O_3 nanoparticles functioned as a peroxidase-like mimic activity to monitor the oxidation–reduction reactions of 3,3′,5,5′-tetramethylbenzidine (TMB) during the reaction between S-protein and angiotensin-converting enzyme 2 receptor. The detecting results based on this colorimetric assay in the real samples are in agreement with the result of PCR. Moreover, the biosensors could detect the alpha variant of COVID-19 without changing antigens or primer reagents, showing potential application in the rapid screening test and diagnosis of COVID-19.

12.4 Conclusion

In this chapter, we reviewed the electrochemical biosensors for detecting viruses during the past few decades. For electrochemical biosensors, all kinds of bio-recognition probes including nucleic acids, antibodies, and peptides have been designed to widen the liner range, lower the limit of detection, and enhance the specificity of detecting viruses. Moreover, the development of novel electrode interface nanomaterials, including gold-based nanomaterials, carbon-based nanomaterials, and magnetic nanoparticles could provide a large surface area for functionality. Further research should concentrate on the synergetic design and modulation of bio-recognition probes and electrode interface nanomaterials to facilitate electron transfer rates and electrical conductivity between the targets and sensing interface for high performance. However, most detection of viruses based on electrochemical biosensors was conducted in the laboratory with a high-security level to prevent the infection of viruses. With efforts, it is strongly believed that the electrochemical biosensors give a fast and high-sensitivity detection of viruses, benefitting the application in the market.

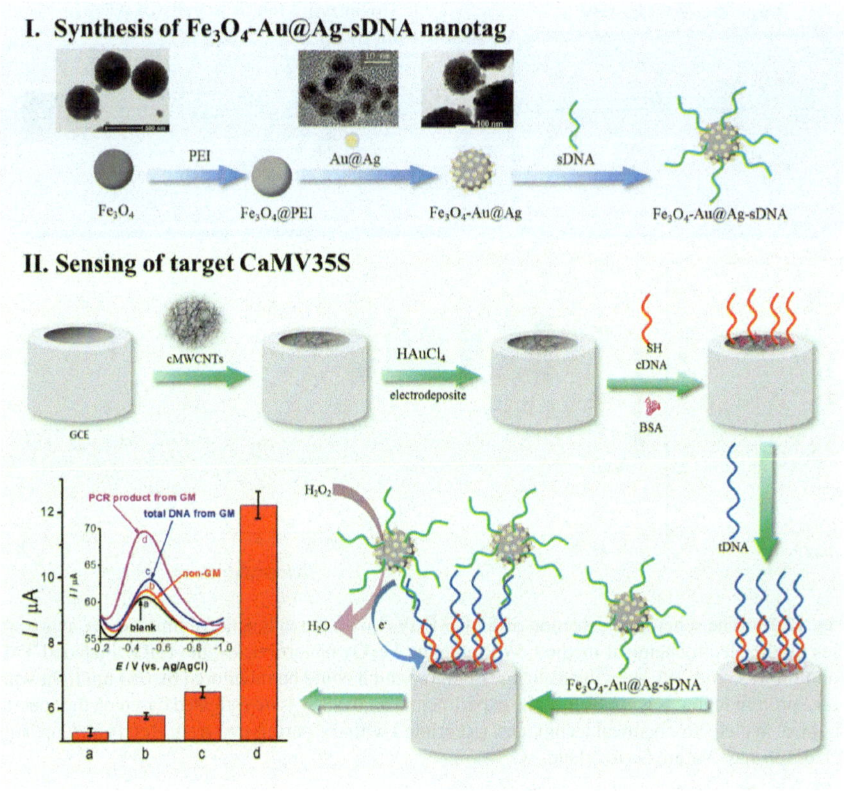

Fig. 12.9 The illustrated diagram of utilizing Fe_3O_4 magnetic particles to fabricate an electrochemical biosensor for detecting cauliflower mosaic virus 35S (CaMV35S) gene. I: Solvothermal synthesized Fe_3O_4 magnetic particles were decorated with polyethyleneimine to form Fe_3O_4-Au@Ag composites, which could further be used to decorate signal DNA (sDNA) to form Fe_3O_4-Au@Ag-sDNA probe. II: For the fabrication of the sensing platform, carboxylated multi-walled carbon nanotubes were dispersed on a glassy carbon electrode to form composites, which could be used to deposit Au nanoparticles to decorate target DNA (tDNA); Then, the tDNA was measured with incubating with Fe_3O_4-Au@Ag-sDNA solution. (Reprinted with the permission from Ref. [60]. Copyright 2019 Elsevier B.V.)

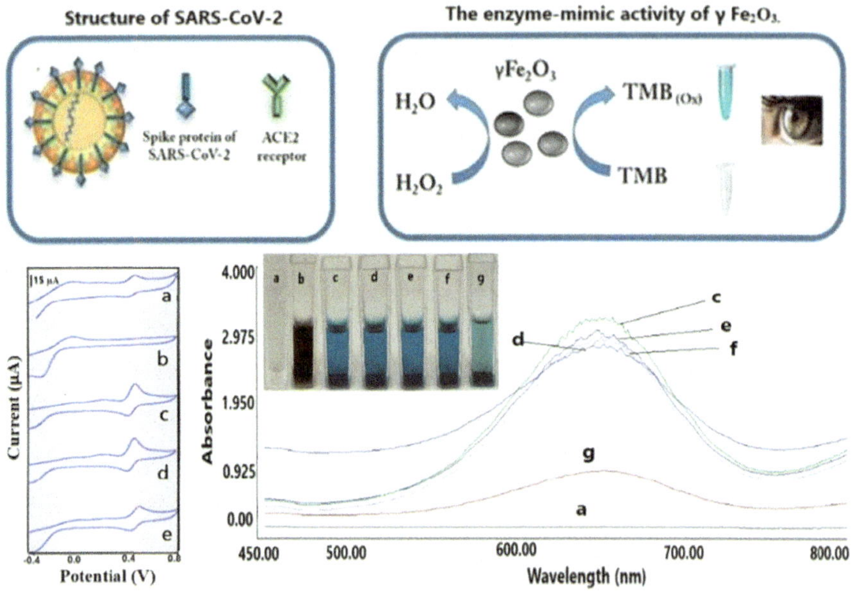

Fig. 12.10 The schematic detection of SARS-CoV-2 with the magnetic gamma Fe_2O_3 nanoparticles by the electrochemical method. With gamma Fe_2O_3 nanoparticles, the H_2O_2 included TMB solution was oxidized, showing a change of color which could be monitored by 654 nm light when spike protein interacts with ACE2. The experimental analysis was investigated based on the spectroscopical and electrochemical techniques. (Reprinted with the permission from Ref. [62]. Copyright 2023 Springer Nature Switzerland AG.)

References

1. Khater M, de la Escosura-Muniz A, Merkoçi A (2017) Biosensors for plant pathogen detection. In: Biosensors and bioelectronics, special issue selected papers from the 26th anniversary world congress on biosensors (Part II) vol 93, pp 72–86
2. Kwon PS, Ren S, Kwon SJ, Kizer ME, Kuo L, Xie M, Zhu D, Zhou F, Zhang F, Kim D, Fraser K, Kramer LD, Seeman NC, Dordick JS, Linhardt RJ, Chao J, Wang X (2020) Designer DNA architecture offers precise and multivalent spatial pattern-recognition for viral sensing and inhibition. Nat Chem 12:26–35
3. Lu M, Zhang Z, Xue M, Zhao BS, Harder O, Li A, Liang X, Gao TZ, Xu Y, Zhou J, Feng Z, Niewiesk S, Peeples ME, He C, Li J (2020) N 6-methyladenosine modification enables viral RNA to escape recognition by RNA sensor RIG-I. Nat Microbiol 5:584–598
4. Ryu W (2017) Chapter 2-Virus structure. In: W-SBT-MV of HPV Ryu. Academic Press, Boston, pp 21–29
5. Cohen FS (2016) How viruses invade cells. Biophys J 110:1028–1032
6. Khan MZH, Hasan MR, Hossain SI, Ahommed MS, Daizy M (2020) Ultrasensitive detection of pathogenic viruses with electrochemical biosensor: state of the art. Biosens Bioelectron 166:112431
7. Kwit E, Osinski Z, Rze zutka A (2019) Detection of viral DNA of myxoma virus using a validated PCR method with an internal amplification control. J Virol Methods 272:113709

8. Sakurai A, Takayama K, Nomura N, Kajiwara N, Okamatsu M, Yamamoto N, Tamura T, Yamada J, Hashimoto M, Sakoda Y, Suda Y, Kobayashi Y, Kida H, Shibasaki F (2015) Fluorescent immunochromatography for rapid and sensitive typing of seasonal influenza viruses. PLoS ONE 10:e0116715
9. Tani H, Kimura M, Tan L, Yoshida Y, Ozawa T, Kishi H, Fukushi S, Saijo M, Sano K, Suzuki T, Kawasuji H, Ueno A, Miyajima Y, Fukui Y, Sakamaki I, Yamamoto Y, Morinaga Y (2021) Evaluation of SARS-CoV-2 neutralizing antibodies using a vesicular stomatitis virus possessing SARS-CoV-2 spike protein. Virol J 18:16
10. Zhao Z, Huang CF, Huang ZY, Lin FJ, He QL, Tao D, Jaffrezic-Renault N, Guo ZZ (2021) Advancements in electrochemical biosensing for respiratory virus detection: a review. Trac-Trends Ana Chem 139:116253
11. Long QX, Liu BZ, Deng HJ, Wu GC, Deng K, Chen YK, Liao P, Qiu JF, Lin Y, Cai XF, Wang DQ, Hu Y, Ren JH, Tang N, Xu YY, Yu LH, Mo Z, Gong F, Zhang XL, Tian WG, Hu L, Zhang XX, Xiang JL, Du HX, Liu HW, Lang CH, Luo XH, Wu SB, Cui XP, Zhou Z, Zhu MM, Wang J, Xue CJ, Li XF, Wang L, Li ZJ, Wang K, Niu CC, Yang QJ, Tang XJ, Zhang Y, Liu XM, Li JJ, Zhang DC, Zhang F, Liu P, Yuan J, Li Q, Hu JL, Chen J, Huang AL (2020) Antibody responses to SARS-CoV-2 in patients with COVID-19. Nat Med 26:845–848
12. Liu J, Wang RK, Zhou H, Mathesh M, Dubey M, Zhang WG, Wang B, Yang WR (2022) Nucleic acid isothermal amplification-based soft nanoarchitectonics as an emerging electrochemical biosensing platform. Nanoscale 14:10286
13. Zheng YQ, Song XX, Fredj Z, Bian S, Sawan M (2023) Challenges and perspectives of multi-virus biosensing techniques: a review. Anal Chim Acta 1244:340860
14. SoltanRezaee M, Bodaghi M (2020) Simulation of an electrically actuated cantilever as a novel biosensor. Sci Rep 10:3385
15. Rosario MR, Mutharasan R (2014) Nucleic acid electrochemical and electromechanical biosensors: a review of techniques and developments. Rev Anal Chem 33:213–230
16. Presnova G, Presnov D, Krupenin V, Grigorenko V, Trifonov A, Andreeva I, Ignatenko O, Egorov A, Rubtsova M (2017) Biosensor based on a silicon nanowire field-effect transistor functionalized by gold nanoparticles for the highly sensitive determination of prostate specific antigen. Biosens Bioelectron 88:283–289
17. Silva MMS, Dias ACMS, Cordeiro MT, Marques E, Goulart MOF, Dutra RF (2014) A thiophene-modified screen printed electrode for detection of dengue virus NS1 protein. Talanta 128:505–510
18. Victorious A, Zhang ZJ, Chang DR, Maclachlan R, Pandey R, Xia JR, Gu J, Hoare T, Soleymani L, Li YF (2022) A DNA barcode-based aptasensor enables rapid testing of porcine epidemic diarrhea viruses in swine saliva using electrochemical readout. Angew Chem Int Edit 61:e202204252
19. Chaibun T, Thanasapburachot P, Chatchawal P, Yin LS, Jiaranuchart S, Jearanaikoon P, Promptmas C, Buajeeb W, Lertanantawong B (2022) A multianalyte electrochemical genosensor for the detection of high-risk HPV genotypes in oral and cervical cancers. Biosensors-Basel 12:290
20. Zhang QX, Zhang MR, Guo ZQ, Li JJ, Zhu ZX, Wang Y, Liu S, Huang JD, Yu JH (2023) DNA tetrahedron-besieged primer and DNAzyme-activated programmatic RCA for low-background electrochemical detection of ochratoxin A. Anal Chim Acta 1242:340782
21. Yan BY, Li M, Luo F, Jin XY, Qiu B, Lin ZY (2022) Photothermal biosensor for HPV16 based on strand-displacement amplification and gold nanoparticles using a thermometer as readout. Microchim Acta 189:437
22. Saylan Y, Erdem O, Ünal S, Denizli A (2019) An alternative medical diagnosis method: biosensors for virus detection. Biosensors 9:65
23. De Castro ACH, Kochi LT, Flauzino JMR, Soares MMCN,·Alves VA, Da Silva LA, Madurro JM, Brito-Madurro AG (2022) Electrochemical biosensor for sensitive detection of hepatitis B in human plasma. Appl Biochem Biotechnol 194:2604–2619
24. Liang QZ, Huang Y, Wang MH, Kuang DQ, Yang JH, Yi YX, Shi H, Li JL, Yang J, Li GX (2022) An electrochemical biosensor for SARS-CoV-2 detection via its papain-like cysteine protease and the protease inhibitor screening. Chem Eng J 452:139646

25. Crivianu-Gaita V, Thompson M (2016) Aptamers, antibody scFv, and antibody Fab' fragments: an overview and comparison of three of the most versatile biosensor biorecognition elements. Biosens Bioelectron 85:32–45
26. Da Silva ETSG, Souto DEP, Barragan JTC, De J, Giarola F, De Moraes ACM, Kubota LT (2017) Electrochemical biosensors in point-of-care devices: recent advances and future trends. ChemElectroChem 4:778–794
27. Saylan Y, Erdem O, Unal S, Denizli A (2021) An alternative medical diagnosis method: biosensors for virus detection. Biosensors-Basel 9:65
28. Wu Z, Guo WJ, Bai YY, Zhang L, Hu J, Pang DW, Zhang ZL (2018) Digital single virus electrochemical enzyme-linked immunoassay for ultrasensitive H7N9 avian influenza virus counting. Anal Chem 90:1683–1690
29. Siuzdak K, Niedziałkowski P, Sobaszek M, Łęga T, Sawczak M, Czaczyk E, Dziąbowska K, Ossowski T, Nidzworski D (2019) Biomolecular influenza virus detection based on the electrochemical impedance spectroscopy using the nanocrystalline boron-doped diamond electrodes with covalently bound antibodies. Sens Actuat B-Chem 280:263–271
30. Eissa S, Zourob M (2021) Development of a low-cost cotton-tipped electrochemical immunosensor for the detection of SARS-CoV-2. Anal Chem 93:1826–1833
31. Hushegyi A, Pihíková D, Bertok T, Adam V, Kizek R, Tkac J (2016) Ultrasensitive detection of influenza viruses with a glycan-based impedimetric biosensor. Biosens Bioelectron 79:644–649
32. Kim JH, Cho CH, Shin JH, Hyun MS, Hwang E, Park TJ, Park JP (2021) Biomimetic isolation of affinity peptides for electrochemical detection of influenza virus antigen. Sens Actuator B-Chem 343:130161
33. Matsubara T, Ujie M, Yamamoto T, Einaga Y, Daidoji T, Nakaya T, Sato T (2020) Avian influenza virus detection by optimized peptide termination on a boron-doped diamond electrode. ACS Sens 5:431–439
34. Negahdary M, Angnes L (2022) Electrochemical nanobiosensors equipped with peptides: a review. Microchim Acta 189:94
35. Soto D, Orozco J (2022) Peptide-based simple detection of SARS-CoV-2 with electrochemical readout. Anal Chim Acta 1205:339739
36. Reddy YVM, Shin JH, Hwang J, Kweon DH, Choi CH, Park K, Kim SK, Madhavi G, Yi HYM, Park JP (2022) Fine-tuning of MXene-nickel oxide-reduced graphene oxide nanocomposite bioelectrode: sensor for the detection of influenza virus and viral protein. Biosens Bioelectron 214:114511
37. Alves RD, Martins IC, Franco DL, Silva AD, Santos PD, Goulart LR, Maia YCD, Silva ATF, Araujo TG, Sotomayor MDT (2022) A novel peptide-based electrochemical biosensor for breast cancer characterization over a poly 3-(3-aminophenyl) propionic acid matrix. Biosens Bioelectron 205:114081
38. Bahadur KCT, Tada S, Zhu LP, Uzawa T, Minagawa N, Luo SC, Zhao HC, Yu HH, Aigaki T, Ito Y (2018) In vitro selection of electrochemical peptide probes using bioorthogonal tRNA for influenza virus detection. Chem Commun 54:5201
39. Allegra A, Petrarca C, Di Gioacchino M, Mirabile G, Gangemi S (2023) Electrochemical biosensors in the diagnosis of acute and chronic leukemias. Cancers 146:15
40. Vermisoglou E, Panacek D, Jayaramulu K, Pykal M, Frebort I, Kolar M, Hajduch M, Zboril R, Otyepka M (2020) Human virus detection with graphene-based materials. Biosens Bioelectron 166:112436
41. Hasanjani HRA, Zarei K (2021) DNA/Au-Pt bimetallic nanoparticles/graphene oxide-chitosan composites modified pencil graphite electrode used as an electrochemical biosensor for sub-picomolar detection of anti-HIV drug zidovudine. Microchem J 164:106005
42. Hwang HS, Jeong JW, Kim YA, Chang M (2020) Carbon nanomaterials as versatile platforms for biosensing applications. Micromachines 11:814
43. Wang Y, Hu YQ, He QY, Yan JH, Xiong HJ, Wen NC, Cai SD, Peng DM, Liu YF, Liu ZB (2020) Metal-organic frameworks for virus detection. Biosens Bioelectron 169:112604
44. Nasrin F, Khoris IM, Chowdhury AD, Boonyakida J, Park EY (2022) Impedimetric biosensor of Norovirus with low variance using simple bioconjugation on conductive polymer-Au nanocomposite. Sens Actuat B-Chem 369:132390

45. Silva LRG, Stefano JS, Orzari LO, Brazaca LC, Carrilho E, Marcolino LH, Bergamini MF, Munoz RAA, Janegitz BC (2022) Electrochemical biosensor for SARS-CoV-2 cDNA detection using AuPs-modified 3D-printed graphene electrodes. Biosens Basel 12:622
46. Qian L, Durairaj S, Prins S, Chen A (2020) Nanomaterial-based electrochemical sensors and biosensors for the detection of pharmaceutical compounds. Biosens Bioelectron 175:112836
47. Hatamluyi B, Rezayi M, Jamehdar SA, Rizi KS, Mojarrad M, Meshkat Z, Choobin H, Soleimanpour S, Boroushaki MT (2022) Sensitive and specific clinically diagnosis of SARS-CoV-2 employing a novel biosensor based on boron nitride quantum dots/flower-like gold nanostructures signal amplification. Biosens Bioelectron 207:114209
48. Palomar Q, Xu XX, Gondran C, Holzinger M, Cosnier S, Zhang Z (2020) Voltammetric sensing of recombinant viral dengue virus 2 NS1 based on Au nanoparticle-decorated multiwalled carbon nanotube composites. Microchim Acta 187:363
49. Karakus E, Erdemir E, Demirbilek N, Liv L (2021) Colorimetric and electrochemical detection of SARS-CoV-2 spike antigen with a gold nanoparticle-based biosensor. Anal Chim Acta 1182:338939
50. Heo W, Lee K, Park S, Hyun KA (2022) Electrochemical biosensor for nucleic acid amplification-free and sensitive detection of severe acute respiratory syndrome coronavirus 2 (SARS-CoV-2) RNA via CRISPR/Cas13a trans-cleavage reaction. Biosens Bioelectron 201:113960
51. Hong SA, Kwon J, Kim D, Yang S (2015) A rapid, sensitive and selective electrochemical biosensor with concanavalin A for the preemptive detection of norovirus. Biosens Bioelectron 64:338–344
52. Wu K, Saha R, Su D, Krishna VD, Liu J, Cheeran MCJ, Wang JP (2020) Magnetic-nanosensor-based virus and pathogen detection strategies before and during COVID-19. ACS Appl Nano Mater 3:9560
53. Li CM, Che B, Deng LH (2023) Electrochemical biosensors based on carbon nanomaterials for diagnosis of human respiratory diseases. Biosensors-Basel 13:12
54. Tancharoen C, Sukjee W, Thepparit C, Jaimipuk T, Auewarakul P, Thitithanyanont A, Sangma C (2019) Electrochemical biosensor based on surface imprinting for Zika virus detection in serum. ACS Sens 4:69–75
55. Wang YF, Sun WJ, Li YX, Zhuang XM, Tian CY, Luan F, Fu XL (2021) Imidazole metal-organic frameworks embedded in layered $Ti_3C_2T_x$ Mxene as a high-performance electrochemiluminescence biosensor for sensitive detection of HIV-1 protein. Microchem J 167:106332
56. Gogola JL, Martins G, Gevaerd A, Blanes L, Cardoso J, Marchini FK, Banks CE, Bergamini MF, Marcolino-Junior LH (2021) Label-free aptasensor for p24-HIV protein detection based on graphene quantum dots as an electrochemical signal amplifier. Anal Chim Acta 1166:338548
57. Ma Y, Shen XL, Zeng Q, Wang HS, Wang LS (2017) A multi-walled carbon nanotubes based molecularly imprinted polymers electrochemical sensor for the sensitive determination of HIV-p24. Talanta 164:121–127
58. Chin SF, Lim LS, Pang SC, Sum MSH, Perera D (2017) Carbon nanoparticle modified screen printed carbon electrode as a disposable electrochemical immunosensor strip for the detection of Japanese encephalitis virus. Microchim Acta 184:491–497
59. Li S, Tian T, Zhang T, Cai X, Lin Y (2019) Advances in biological applications of self-assembled DNA tetrahedral nanostructures. Mater Today 24:57–68
60. Ye YK, Mao S, He SD, Xu X, Cao XD, Wei ZJ, Gunasekaran S (2020) Ultrasensitive electrochemical genosensor for detection of CaMV35S gene with Fe_3O_4-Au@Ag nanoprobe. Talanta 206:120205
61. Durmus C, Hanoglu SB, Harmanci D, Moulahoum H, Tok K, Ghorbanizamani F, Sanli S, Zihnioglu F, Evran S, Cicek C, Sertoz R, Arda B, Goksel T, Turhan K, Timur S (2022) Indiscriminate SARS-CoV-2 multivariant detection using magnetic nanoparticle-based electrochemical immunosensing. Talanta 243:123356
62. Buyuksunetci YT, Citil BE, Tapan U, Anik U (2021) Development and application of a SARS-CoV-2 colorimetric biosensor based on the peroxidase-mimic activity of gamma-Fe_2O_3 nanoparticles. Microchim Acta 188:335

Chapter 13
Electrochemical Biosensors for Cells and Cell Exosomes Detection

Zhi Zheng, Haiyang Li, Hui Li, and Fan Xia

Abstract Cancer has been the major disease of human death worldwide because of the deterioration of the living environment and lack of advanced early diagnosis techniques. Generally, cancer is induced by the unlimited proliferation, division, and differentiation of cancer cells. Cell exosomes are cell-derived vesicles with a lipid bilayer, containing nucleic acid information and protein of the parent cells. The diagnosis of cancer depends on the rapid and accurate detection of biomarkers including cells and cell exosomes. However, the quantity of cancer cells and cell exosomes is very rare in the early stage and it is a big challenge to detect them in complex blood samples. Conventional strategies such as the optical methods depend on the advanced instruments with high costs and requirements of technical operation. Electrochemical biosensors based on the design of bio-recognition probes such as nucleic acids, antibodies as well as peptides could improve the performance with the promotion of target hybridization and electron transport, and further modification of electrode interface nanomaterials such as metal nanomaterials, carbon nanomaterials, magnetic nanoparticles and metal–organic framework nanomaterials could increase the electrical conductivity and improve the capture efficiency. The optimization of electrochemical biosensors could quantify cells and cell exosomes with low detection limit, a wide linear range, providing promising strategies for cancer diagnosis.

Keywords Electrochemical biosensors · Cells · Cell exosomes · Bio-recognition elements · Electrode interface nanomaterials

Z. Zheng · H. Li · H. Li · F. Xia (✉)
State Key Laboratory of Biogeology and Environmental Geology, Engineering Research Center of Nano-Geomaterials of the Ministry of Education, Faculty of Materials Science and Chemistry, China University of Geosciences, Wuhan 430074, P. R. China
e-mail: xiafan@cug.edu.cn

F. Xia et al. (eds.), *Electrochemical Biosensors for Whole Blood Analysis*,
https://doi.org/10.1007/978-981-99-5644-9_13

13.1 Introduction

Cancer has been the major disease of human death worldwide and threatens public health because of the deterioration of the living environment and the lack of advanced early diagnosis and accurate treatment [1, 2]. The conventional clinical prognosis strategies such as computerized tomography (CT) could not effectively distinguish tumor from inflammation, magnetic resonance imaging (MRI) requires complex operation and high cost, and histological biopsy faces the difficulty of time loss and early metastasis screening [3, 4].

Liquid biopsy possesses merits of rapid detection, low cost, good reproducibility, and noninvasiveness, and has emerged as a promising candidate for conventional cancer diagnosis [5, 6]. Cancer cells, including circulating tumor cells (CTCs), are the most stable and reported biomarkers for preliminary examination of cancers [7, 8]. Recent reports demonstrate that various cancer cells could secrete a large number of exosomes which play an important role in tumor oncogenesis, tumor metastasis, and immunoregulation [7, 8]. However, the quantity of cancer cells and cell exosomes are very rare in the early stage and it is a big challenge to detect them in a complex blood sample. Until now, many strategies including enzyme-linked immunosorbent assay (ELISA), fluorescence [9], colorimetric [10], flow cytometry [11], microfluidic [12] and surface-enhanced Raman scattering (SERS) [13] are widely developed for the determination of cells and cell exosomes [14, 15]. Therein, optical methods such as fluorescence and colorimetry approaches are widely applied in the fields of point-of-care (POC) diagnosis, having significant importance for home healthcare [16, 17]. But the quantitative and sensitive diagnosis based on optical strategies is dependent on the advanced instruments.

Electrochemical biosensors based on quantitative analysis of signal change between biochemical molecules and electrochemical transducers were easily integrated for POC devices with high performance [18, 19]. For detection in living cells, the process of electrochemical detection is more economical and effective than conventional methods [20, 21]. Generally, researchers design highly sensitive bio-recognition probes and electrode interface nanomaterials to increase the signal sensitivity. Design probes could recognize more variety of signal molecules, increasing the electrochemical sensitivity by accelerating the electron transport. Moreover, the decoration of nanomaterials on the surface of the electrode interface with new physical and chemical properties could benefit to the discovery of novel mechanisms and improve the detecting performance of biosensors.

13.2 Bio-recognition Probes

The species and structures of bio-recognition probes including nucleic acids, antibodies, and peptides are crucial to the detecting performance of cells and cell exosomes [22]. The design and immobilization of bio-recognition probes on the surface of electrodes could improve the electron transport and activity of redox tags, thus increasing the detection sensitivity of electrochemical biosensors [23, 24].

13.2.1 Nucleic Acids

With the merits of facile preparation, easy modification, high temperature, pH stability, and excellent stability, electrochemical biosensors based on nucleic acids probes are reported with lower detection limits and could be large-scale fabricated at a lower price [25, 26]. The 3D folded structure through the systematic screening process could recognize various cells and cell exosomes [27]. For the capture of targets, Dong et al. detected tumor exosomes of LNCaP cells based on prostate-specific membrane antigen (PSMA) aptamer instead of capturing targets directly on the surface of the electrode to enhance the response [28]. The messenger DNAs (mDNAs) contain two complementary sequences for the linking of PSMA aptamer and P1 at the same time. In the experiment, PSMA aptamers hybridized by three mDNAs could capture one tumor exosome. Then the mDNAs were released in the supernatant. The number of target exosomes is proportional to the number of released mDNAs. Moreover, the released mDNAs could reduce an obvious electrochemical signal based on cyclic enzymatic amplification. The limit of detection (LOD) is low to 70 particles/μL for detecting exosomes. The biosensor also shows high sensitivity which has a much lower response to HeLa cell-derived exosomes and MCF cell-derived exosomes. Therefore, this biosensor shows high performance in complex biosamples and potential application in clinical samples. As shown in Fig. 13.1, Guo et al. designed dual-recognition proximity probes with an on–off-on strategy based on the binding-induced DNA walker [29]. First, two probes, one probe contains a DNAzyme tail sequence which is Pb^{2+}-dependent (cholesterol probe), and another probe contains identifying units, both were integrated onto the surface of the exosome (aptamer probe). Then, the conformational change of hairpin DNA (H) was observed by the binding of dual recognition proximity. This binding process could result in the "off" and "on" states and allow for the detection of exosomes. This dual-recognition strategy demonstrates accurate identification and satisfactory reproducibility, the linear range is from 5.0×10^4 to 1.0×10^8 particles/mL, and the LOD is 1.6×10^4 particles/mL. Hallaj et al. utilized RNA-cleavage DNAzymes to detect breast cancer cells [30]. The RNA-cleavage DNAzymes contain a substrate strand and enzyme strand which were linked by the thionine-modified gold nanoparticles and magnetic nanoparticles, respectively. The thionine caused the electrochemical signal and magnetic nanoparticles induced the movement of DNAzyme to the electrode's surface. The transfer of electrons between the electrode and solution before and after adding with target protein was evaluated by the differential pulse response. The protein of breast cancer cells reacted with the enzyme sequence and induced the cleavage of gold nanoparticle-thionine labels to release into the solution, decreasing the current. After optimization, the linear detection was in the range of 1×10^{-6} to 10 pg/ml, and the LOD was low to 1×10^{-7} pg/ml. This DNAzyme-based biosensor

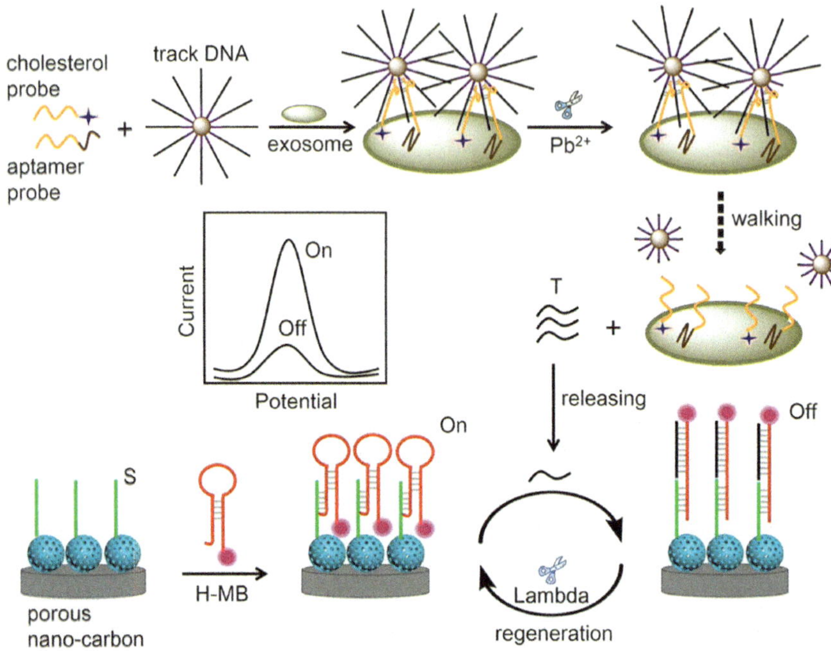

Fig. 13.1 The working principle of detecting exosomes based on DNA probes. Cholesterol probes and CD63 aptamer probes with 5′ terminals could bind with the lipid bilayer and CD63 protein of exosomes surface, respectively. And these two probes had Pb^{2+}-dependent DNA-zyme tail sequences in 3′ terminals. With the existence of Pb^{2+}, the DNA-zyme would quickly cleave the track DNA, which could further activate the walking of track DNA. The separation of track DNA from probes induced a large number of generation of T strands, which could open the stem-loop structure of hairpin DNA and methylene blue mixtures (H-MB). The opened H-MB could benefit from the formation of a double-stranded configuration, leading to the separation of MB from the electrode. This separation could decrease the electron transfer and current response ("off" status). T strands would degrade from 5′ to 3′when adding of Lambda exonuclease, leading to a shorter distance between MB and electrodes, exhibiting a higher current response ("on" status). (Reprinted with the permission from Ref. [29]. Copyright 2021 Elsevier B.V.)

also shows high selectivity of detecting MDA-MB-231, and is stable for about two months.

13.2.2 Antibodies

Antibodies could bind with exosomes in a strong affinity and high specificity. Due to the derivative relationships, the exosomes inherited much structured information from the original cells [31]. The immobilization of antibody probes is important for constructing high-performance biosensors. Liu et al. constructed novel electrochemical biosensors for detecting exosomes based on decorating the epidermal growth

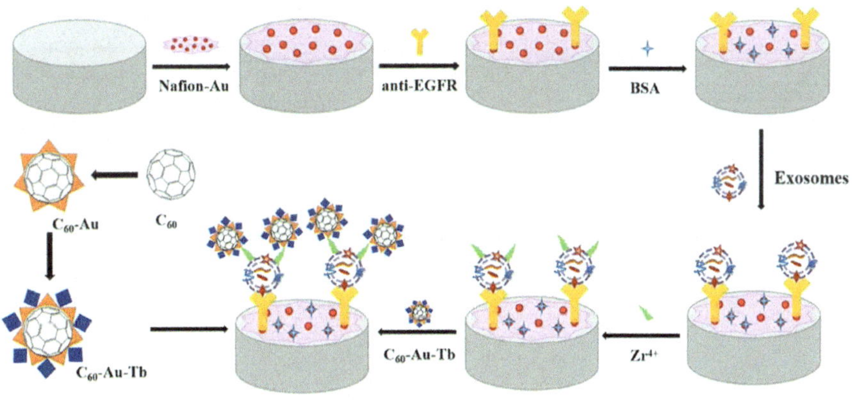

Fig. 13.2 The schematic detection of exosomes based on antibody probes. Firstly, the EGFR antibody decorated on the Au NPs surface through the binding interaction of Au-NH_2. And the and unbinding Au sites were blocked by BSA. Captured by EGFR antibody, the exosomes adsorbed Zr^{4+} and connected C_{60}-Au-Tb composites through "carboxyl-Zr^{4+}-phosphate" binding interaction, which could produce a current signal. (Reprinted with the permission from Ref. [32]. Copyright 2021 Elsevier B.V.)

factor receptor (EGFR) antibody with Au nanoparticles (NPs) substrate with high specificity, good repeatability, and stability (Fig. 13.2) [32]. The results showed that the appropriate concentration of EGFR antibody was favorable to the response of exosomes. The square wave voltammetry (SWV) signal increases as the concentration of exosomes increases. After optimization, the linear range of detecting exosomes could reach 5×10^4–5×10^9 exosomes/mL, and the LOD was low to 2.67×10^4 exosomes/mL. The constructed biosensors also showed good selectivity, reproducibility, and stability, demonstrating potential clinical application.

As shown in Fig. 13.3, Zhou et al. proposed a dual-recognition electrochemical biosensors for detecting CTCs by decorating anti-EpCAM onto the surface of Ce organometallic frame with gold nanoparticles (CeMOF-AuNPs). The mesoporous nanospheres of PdPtCuRu were modified with mucin 1 aptamer [33]. The detection of A549 cells was realized by catalyzing H_2O_2 with PdPtCuRu MNSs during the differential pulse voltammetry (DPV) measurement. The biosensors show a LOD of detecting A549 cells with less than 10 cells mL^{-1}, and excellent specificity in spiked serum samples. Diaz-Fernandez et al. functionalized CD34 antibodies on the surface of screen-printed electrodes for specifically detecting CD34 T-cells by a self-assembled monolayer thin film [34]. The detection of CD34 was achieved by monitoring the R_{ct} change in the electrochemical impedance spectroscopy (EIS), which shows an increase as the concentration increases. In Faradaic conditions, the biosensor shows a linear work range in 230–1.4×10^3 cell mL^{-1} cell mL^{-1} for detecting CD34 T-cells. This linear range could expand to 50–1×10^5 cell mL^{-1} without Faradaic condition. The non-Faradaic impedance of the biosensor was continuously measured in a microfluidic system, showing a comparable work range and LOD (27 cells mL^{-1}).

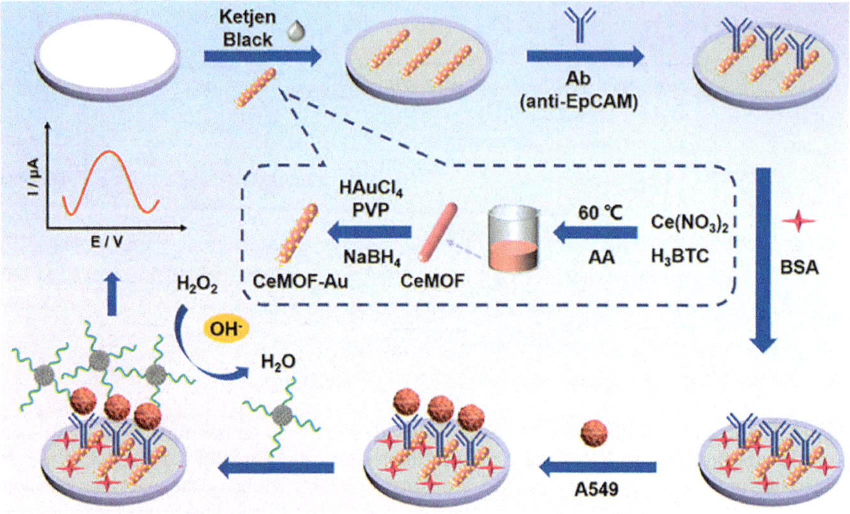

Fig. 13.3 The schematic detection of A549 cells based on antibody probes. The electrode surface was decorated by CeMOF-AuNPs. Then anti-EpCAM was immobilized on the surface of CeMOF-AuNPs through Au-NH_2 bonds. The anti-EpCAM of biosensors could capture A549 cells after blocking the inactive sites by BSA. The electrochemical signals were monitored through the catalysis of NH_2-MUC1 aptamer decorated PdPtCuRu mesoporous nanospheres on H_2O_2. (Reprinted with the permission from Ref. [33]. Copyright 2023 Elsevier B.V.)

13.2.3 Peptides

With the merits of easy synthesis and functionalization, simple chemical structure as well as high specificity and strong binding affinity, peptides-based biosensors could recognize cell and cell exosomes with high excellent stability and reliability in biological samples, which have attracted wide attention [35]. As shown in Fig. 13.4, Wang et al. designed a dual-trapping peptide to recognize aggressive BC cells by detecting HER-2 and MMP-2, and combined with AuNPs@HRP@ZIF-8 composites to enhance the performance [36]. One end of peptides is immobilized on the active sites of AuNPs@HRP@ZIF-8, and the other end is linked on the ITO electrode. The HER-2 protein could bind with the recognition sequence of peptides, and then be detected by the DPV. On the other hand, the MMP-2 could cut the sequence site of peptides and release AuNPs@HRP@ZIF-8, making it possible to detect MMP-2. The designed electrochemical biosensors could linearly detect HER-2 in the range of 50 fg mL^{-1}–50 ng mL^{-1}, and detect MMP-2 in the range of 10 fg mL^{-1}–10 ng mL^{-1}, respectively.

Liu et al. constructed an electrochemiluminescence biosensor with a specific binding peptide (FNFRLKAGAKIRFGRGC) for highly sensitive detection of exosomes derived from ovarian tumor cells, allowing for quantitative and accurate detection in biological samples (Fig. 13.5) [35]. One end of the peptide not only could be linked on the Au surface, but also could be linked with g-C3N4 nanosheets,

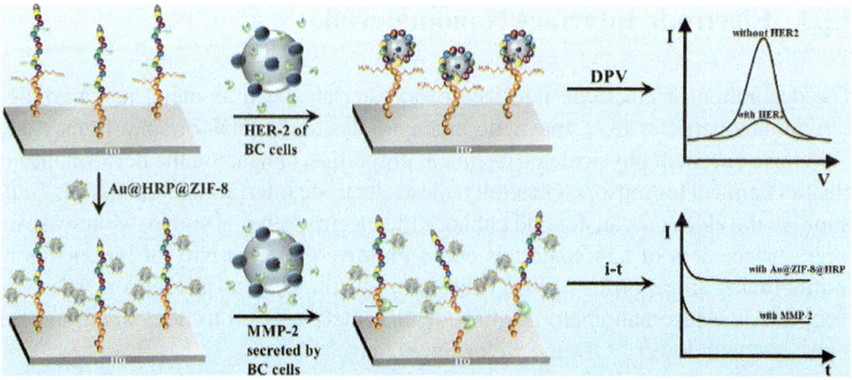

Fig. 13.4 The schematic detection of BC cells based on peptide probes. The peptide probes with two sequences could bind with HER-2 of BC by the interaction of Au–S, and bind with MMP-2 of BC cells by the interaction of Au–N. At the same time, AuNPs@HRP@ZIF-8 composites could bind with peptides and produce electrochemical responses by HRP-contained ZIF-8. The electrochemical biosensors could recognize different BC cells by utilizing the outputs of two signals. (Reprinted with the permission from Ref. [36]. Copyright 2023 The Royal Society of Chemistry)

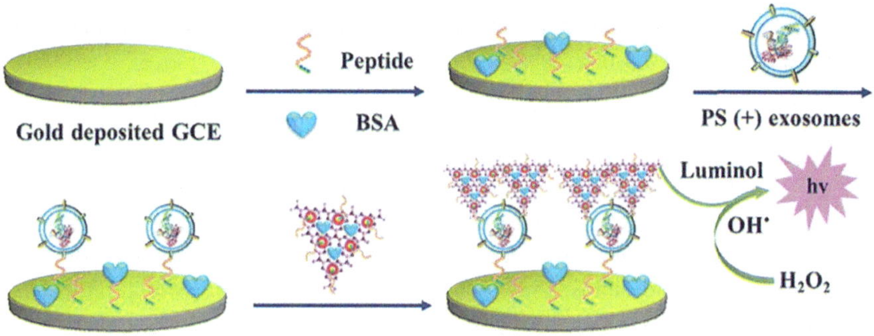

Fig. 13.5 The schematic exosomes detection based on electrochemical biosensors with peptide probes. For the modification of peptide probes on glassy carbon electrode, Au nanoflowers were decorated. The g-C_3N_4 nanosheets could immobilize exosomes and catalyze H_2O_2 to decomposition. The sandwich electrochemiluminescence biosensor based on peptide probes was designed for determining exosomes. (Reprinted with the permission from Ref. [35]. Copyright 2020 Elsevier B.V.)

the other end of the peptide could be used for capturing exosomes. The H_2O_2 was catalyzed by g-C3N4 nanosheets to amplify the electrochemiluminescence signal. Compared with conventional methods, the biosensors showed 5 order magnitude of linear range, and the LOD is lower to 39 particles μL^{-1}. This biosensor based on peptide recognition for signal amplification shows a great potential application for detecting exosomes in clinical medicine.

13.3 Electrode Interface Nanomaterials

The decoration of electrode interface nanomaterials, such as metal nanoparticles, carbon nanoparticles [37], magnetic nanoparticles, and metal–organic frameworks [38], have excellent physical and chemical properties, enhancing the performance of electrochemical biosensors. Generally, these electrode interface nanomaterials could improve the electron transfer and enhance the electrochemical signal. Moreover, the large surface area of nanomaterials could improve the sensitivity of biosensors by immobilizing more probes [39]. Therefore, electrochemical biosensors based on electrode interface nanomaterials have been widely applied to detect cell and cell exosomes with lower LOD and wider linear range.

13.3.1 Metal Nanomaterials

As shown in Fig. 13.6, precious metal materials, such as gold nanoparticles, silver nanoparticles, and platinum nanoparticles, are crucial in electrochemical biosensors due to their excellent biocompatibility, electrical and electrochemical properties [40, 41]. The decoration of metal nanoparticles could increase the active area for capturing more targets, improving the sensitivity [42, 43].

Wang et al. decorated gold nanostars on the surface of glassy carbon electrodes as plasmon to enhance the detection of circulating tumor cells (CTCs) [44]. The CTCs were selectively captured in samples by immobilizing aptamer probes on the surface of gold nanostars. The gold nanostars could induce the effect of localized surface plasmon resonance to increase the electrochemical signal by enhancing the transport of hot electrons from gold nanostars to circuits. The capture of CTCs on the surface of gold nanostars decreased the response current which reduced the transport efficiency of hot electrons while utilizing ascorbic acid as a redox couple. Furthermore, the biosensor could selectively detect CTCs in the mixture of other cells in the blood. This plasmon-enhanced method could detect CTCs with a low LOD (5 cells/mL), and the LOD of detecting MCF-7 cells is low to 10 cells/mL, providing a promising application in biological samples. Wu et al. deposited nickel (Ni) micropillars on the surface of indium tin oxide to increase electrical conductivity and increase the cells' capture efficiency. The enhanced local topographic interactions between 3D Ni micropillars and PLGA nanofibers could improve the capture yields of cells. Then quantum dots decorated anti-EpCAM antibodies were functionalized after electrospinning nanofibers on the surface of Ni micropillars. The released Cd^{2+} from bioprobes could change the current during the detecting process. The constructed biosensors demonstrated a wide range of 10^1–10^5 cells mL^{-1} and a low LOD of 8 cells mL^{-1} [45]. Moreover, these biosensors also showed high detection sensitivity and selectivity of CTC among a series of human plasma samples.

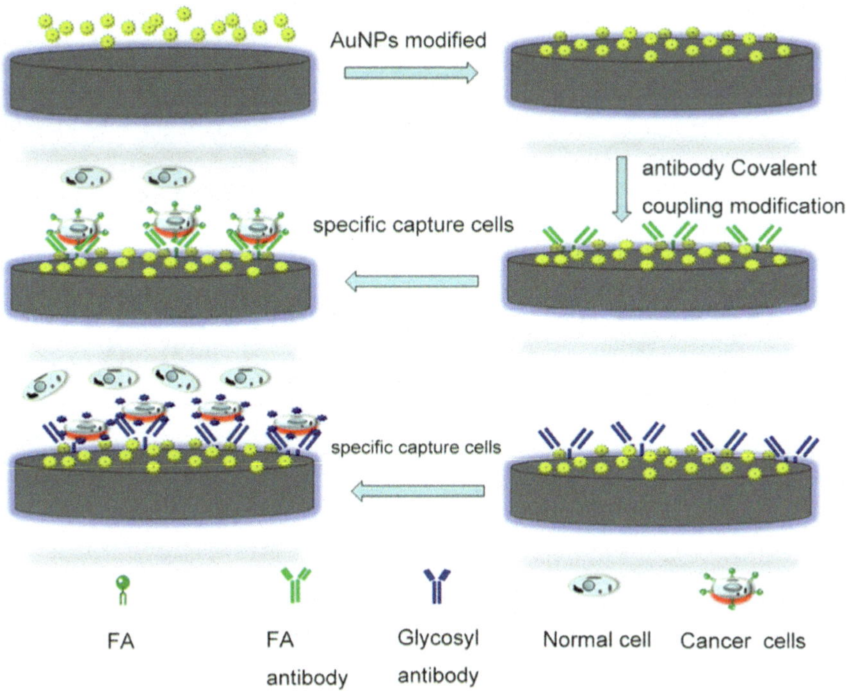

Fig. 13.6 The schematic cancer cell detection based on AuNPs. AuNPs were decorated on the surface of electrodes, and then folic acid (FA) antibodies were decorated on the surface of AuNPs to capture cancer cells by recognizing FA. The glycosyl antibodies were also decorated on the surface of AuNPs to specifically capture normal cells. (Reprinted with the permission from Ref. [41]. Copyright 2020 John Wiley & Sons, Inc.)

13.3.2 Carbon Nanomaterials

Carbon-based materials such as fullerenes, carbon nanotubes (CNT), graphene, and carbon dots could satisfy most demand for biosensors [46, 47]. For example, one-dimensional (1D) CNT possesses a larger aspect ratio, excellent conductivity, and chemical stability, and have been widely used as an immobilization layer and redox facilitator [48].

Liu et al. designed a sandwich-type electrochemical biosensor for quantitatively detecting CD44-overexpressing HeLa cells based on the modification of multi-walled carbon nanotubes (MWCNT) on the surface of indium tin oxide electrodes (Fig. 13.7) [49]. This 1D structure of MWCNT had high chemical reactivities, provided a conjugation interface for hyaluronic acid and ethylenediamine, and improved the electronic electron transfer rate. The DPV and chronocoulometry were used as electrochemical measurements by oxidation–reduction of $[Fe(CN)_6]^{3-/4-}$. The fabricated biosensors demonstrated a high sensitivity of detecting HeLa cells when using fluorescent probes as tracing labels. The capture of HeLa cells shows a linear decrease of current in the

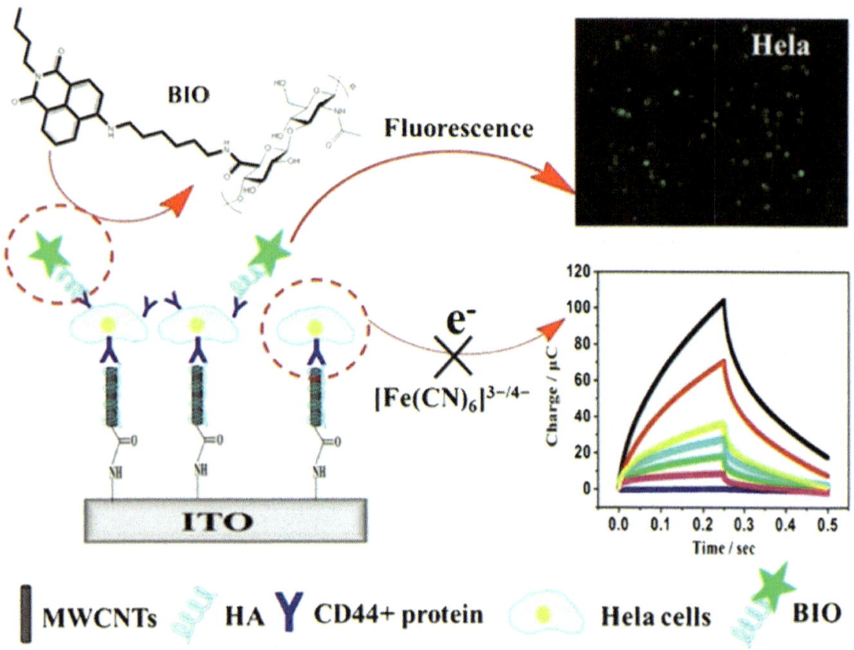

Fig. 13.7 The schematic process of detecting HeLa cells based on carbon nanomaterials. Hyaluronic acid (HA) decorated 3D-MWCNT structure were used for ultrasensitive determination of CD44 HeLa cells, which were recognized by the fluorescent BIO. The binding of CD44 HeLa cells decreased the chronocoulometric response by decreasing the electron transfer of the redox probe ($[Fe(CN)_6]^{3-}$/$[Fe(CN)_6]^{4-}$) (Reprinted with the permission from Ref. [49]. Copyright 2023 Springer Nature Switzerland AG.)

concentration range of 2.1×10^2–2.1×10^7 HeLa cells·mL^{-1}, and the LOD is low to 70 cells·mL^{-1}. This biosensor demonstrates high specificity in detecting HeLa cells with five times higher chronocoulometric signal from CD44, ROS 1728, BXPC-3, Chang liver, and MCF-7. Yazdanparast et al. utilized MWCNT nanocomposites on the surface of electrodes to detect human breast cancer cells [50]. They construct a sandwich to selectively recognize MCF-7 and amplify the detecting signal based on the DPV method. The optimized linear range of detecting MCF-7 cells could reach 1.0×10^2 to 1.0×10^7 cells·mL^{-1}, and the LOD is low to 25 cells mL^{-1}. The biosensor nearly shows no obvious current change for the control cells including HeGp2, HEK, SKBR3, and Hela cells. Moreover, the biosensor also has excellent long-time stability when measured for more than half a month.

13.3.3 Magnetic Nanoparticles

The size of magnetic nanoparticles (MNPs) ranges from nanoscale to microscale, matching the requirements of immobilizing biomolecules for biosensors. MNPs could also provide a modifiable surface for recognizing and capturing different target molecules [51]. Moreover, MNPs could preconcentrate magnetic target biomolecules onto the surface of electrodes by the magnetic field. Thus, MNPs have significant potential for detecting cells and cell exosomes [52]. Boriachek et al. directly isolated tumor-derived exosomes by designing nanoporous ferric oxide nanocubes with gold nanoparticles (AuNPs-Fe_2O_3NC) which were decorated by anti-CD63 in the magnetic field [53]. The nanoporous ferric oxide nanocubes with gold nanoparticles and exosomes were transferred onto the electrodes surface with secondary antibody decoration. The magnetic ferric oxide nanocubes possess catalytic activity of tetramethylbenzidine (TMB) for electrochemical detection. The electrochemical detection based on nanoporous ferric oxide nanocubes exhibits a wide linear range from 10^3 to 10^7 exosomes/mL and an LOD of 10^3 exosomes/mL (Fig. 13.8).

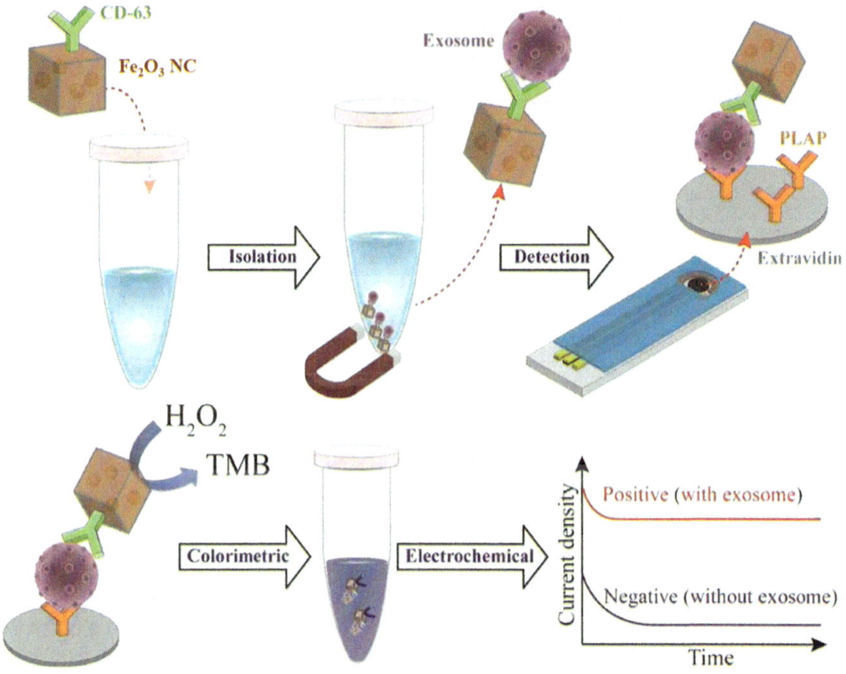

Fig. 13.8 The schematic process of detecting exosomes based on magnetic nanoparticles. The AuNPs-Fe_2O_3NC were decorated by antibody CD63 and captured exosomes in sample fluids. Then, this AuNPs-Fe_2O_3NC was transferred to the electrodes surface, which was decorated by the placenta alkaline phosphatase. The catalytic activity of AuNPs-Fe_2O_3NC enhanced the naked-eye detection and quantification of exosomes. (Reprinted with the permission from Ref. [53]. Copyright 2023 American Chemical Society)

As shown in Fig. 13.9, with the separation of magnetic nanospheres (about 70 nm), Shen et al. detected CTCs in whole blood. MNs were decorated on the surface of electrodes and isolate CTCs from the samples with anti-EpCAM antibodies [54]. The reaction between the phosphate in the DNA backbone and Na_2MoO_4 generates an electrochemical current. The biosensor could detect MCF-7 in blood samples with a low LOD when utilizing the rolling circle amplification (RCA) signal amplification strategy, indicating potential clinical applications. The biosensor shows a weak response to 1000 cells mL^{-1} HeLa cells during the SWV measurement, and the same response to the MCF-7 sample with different concentrations of HeLa cells.

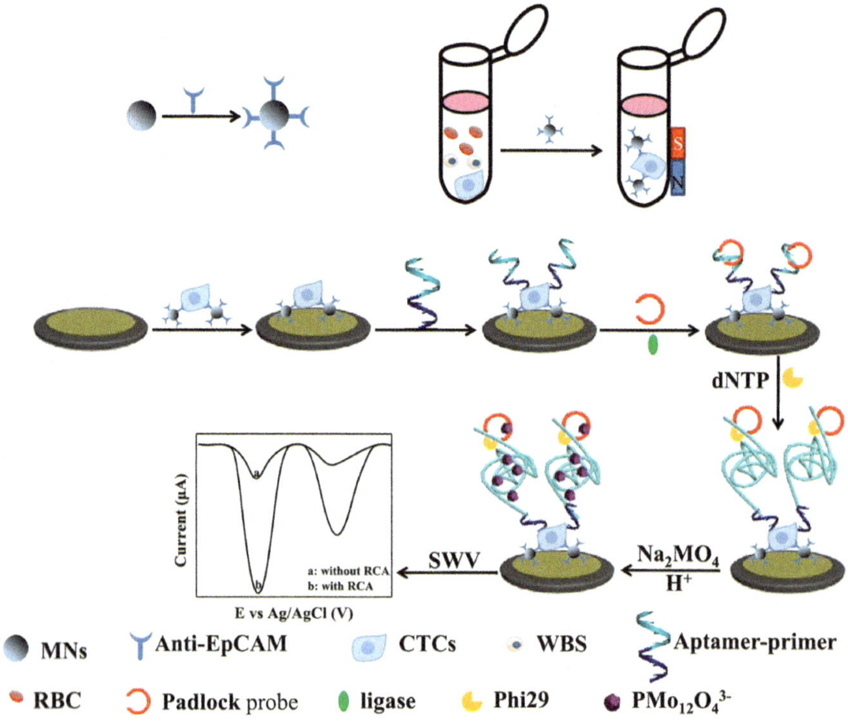

Fig. 13.9 The schematic process of detecting CTCs based on MNs. The MNs could enrich and isolate MCF-7 cells, and were decorated on the surface of electrodes. The reaction between the phosphate in the DNA backbone and Na_2MoO_4 generates an electrochemical current. RCA was performed to amplify the detection and enhance the immobilization of DNA molecules on the surface of the electrode to increase the current intensity. (Reprinted with the permission from Ref. [54]. Copyright 2019 American Chemical Society)

13.3.4 Metal–Organic Framework Nanomaterials

Due to its porous structure, metal–organic framework (MOF) possesses large surface areas and contains a large number of metal clusters and organic linkers, which is widely applied in biosensors [55, 56]. MOF not only provides excellent decoration and encapsulated room of aptamers, antibodies, ions, and enzymes because of the conjugated π-electron interaction, but also modulates the distribution and improves the stability of modified biomolecules, greatly enhances the performance of biosensors [57, 58]. As shown in Fig. 13.10, Liu et al. [59] developed an enzyme-free and label free POC biosensor for quantitative and high-performance detection of exosomes based on the recognition system of Zr-MOFs and aptamers, and signal amplification of hybridization chain reaction (HCR) and DNAzyme-mediated catalysis. Exosomes were adsorbed on the Zr-MOFs surface via the Zr-O-P binding interaction. The designed hairpin probes (HP) with single-stranded regions could bind with CD63 in exosomes. Then large numbers of G-quadruplexes were formed, displaying high catalytic activity of 3, 3′,5,5′-TMB. The electrochemical biosensors constructed on paper have a low LOD (5×10^3 particles/mL), and are easily operated and affordable for POC diagnosis. Jiang et al. [58] reported electrochemical biosensors with silver nanoparticles (AgNPs) decorated MOFs to detect living cells (U87) based on the non-enzymatic superoxide anion radical ($O_2^{\bullet-}$). The electrochemical measurement with $[Fe(CN)_6]^{3-/4-}$ redox couple in KNO_3 solution. The biosensors demonstrated 8 orders magnitude linear range, and a low LOD of 0.0564 pM due to the effective release of $O_2^{\bullet-}$ from living cells. Moreover, the biosensor shows high selectivity even though the concentration of interferent is 20 times higher and high stability (94.12%) after 50 segments of cyclic voltammetry.

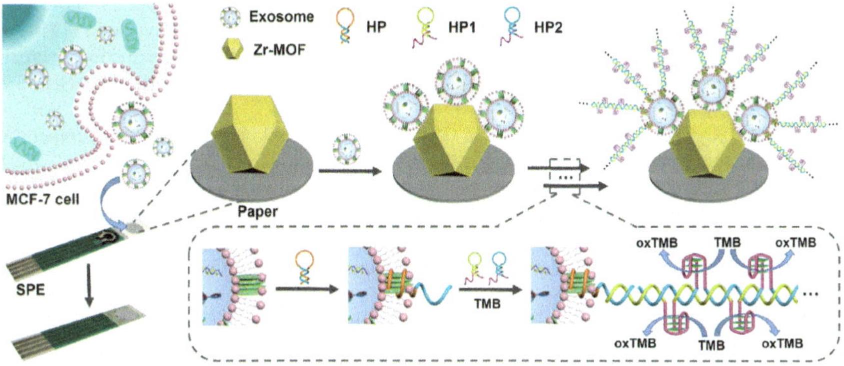

Fig. 13.10 The schematic process of detecting exosomes based on functionalization of MOF. With the recognition of DNA aptamer decorated Zr-MOFs, DNAzyme atalysis, and HCR for signal amplification, a paper-based point-of-care (POC) biosensor was fabricated. (Reprinted with the permission from Ref. [59]. Copyright 2021 American Chemical Society)

13.4 Conclusion

In this chapter, the recent progress of electrochemical biosensors to detect cells and cell exosomes for cancer diagnosis was summarized. The advantages of electrochemical biosensors result from the designable bio-recognition probes and electrode interface nanomaterials for selective recognization of cells and cell exosomes with wide linear range as well as low LOD. On one side, designing probes for electrochemical biosensors based on nucleic acids, antibodies as well as peptides could expand the range of detecting cells and cell exosomes, and improve the sensitivity with the promotion of target hybridization and electron transport. On the other side, modification of nanomaterials on the surface of the electrode interface for electrochemical biosensors such as metal nanomaterials, carbon nanomaterials, magnetic nanoparticles, and metal–organic framework nanomaterials could increase the electrical conductivity and improve the capture efficiency. With the development of the microelectronic technique, the main challenge is to construct a miniaturized biosensor device for POC detection of live-state activity of cells and cell exosomes in real samples. Also, further effort should concentrate on the integration of device arrays to joint detection of different cells or cell exosomes to improve the detecting accuracy.

References

1. Feng RM, Zong YN, Cao SM, Xu RH (2019) Current cancer situation in China: good or bad news from the 2018 global cancer statistics? Cancer Commun (Lond Engl) 39:22
2. He Z, Chen Z, Tan M, Elingarami S, Liu Y, Li T, Deng Y, He N, Li S, Fu J, Li W (2020) A review on methods for diagnosis of breast cancer cells and tissues. Cell Prolif 53:e12822
3. Cortez E, Gladh H, Braun S, Bocci M, Cordero E, Björkström NK, Miyazaki H, Michael IP, Eriksson U, Folestad E (2016) Functional malignant cell heterogeneity in pancreatic neuroendocrine tumors revealed by targeting of PDGF-DD. Proc Natl Acad Sci USA 113:E864–E873
4. Kleeff J, Korc M, Apte M, La Vecchia C, Johnson CD, Biankin AV, Neale RE, Tempero M, Tuveson DA, Hruban RH, Neoptolemos JP (2016) Pancreatic cancer. Nat Rev Dis Prim 2:16022
5. Oshi M, Murthy V, Takahashi H, Huyser M, Okano M, Tokumaru Y, Rashid OM, Matsuyama R, Endo I, Takabe K (2021) Urine as a source of liquid biopsy for cancer. Cancers 13:2652
6. Palmirotta R, Lovero D, Cafforio P, Felici C, Mannavola F, PellèE QD, Tucci M, Silvestris F (2018) Liquid biopsy of cancer: a multimodal diagnostic tool in clinical oncology. Ther Adv Med Oncol 10:1758835918794630
7. Liu L, Yang K, Gao H, Li X, Chen Y, Zhang L, Peng X, Zhang Y (2019) Artificial antibody with site-enhanced multivalent aptamers for specific capture of circulating tumor cells. Anal Chem 91:2591–2594
8. Negishi R, Takai K, Tanaka T, Matsunaga T, Yoshino T (2018) High-throughput manipulation of circulating tumor cells using a multiple single-cell encapsulation system with a digital micromirror device. Anal Chem 90:9734–9741
9. Wang Y, Luo D, Fang Y, Wu W, Wang Y, Xia Y, Wu F, Li C, Lan J, Chen J (2019) An aptasensor based on upconversion nanoparticles as LRET donors for the detection of exosomes. Sensor Actuat B Chem 298:126900

10. Xia Y, Liu M, Wang L, Yan A, He W, Chen M, Lan J, Xu J, Guan L, Chen J (2017) A visible and colorimetric aptasensor based on DNA-capped single-walled carbon nanotubes for detection of exosomes. Biosens Bioelectron 92:8–15
11. Tian Y, Ma L, Gong M, Su G, Zhu S, Zhang W, Wang S, Li Z, Chen C, Li L, Wu L, Yan X (2018) Protein profiling and sizing of extracellular vesicles from colorectal cancer patients via flow cytometry. ACS Nano 12:671–680
12. Zhang P, Zhou X, He M, Shang Y, Tetlow AL, Godwin AK, Zeng Y (2019) Ultrasensitive detection of circulating exosomes with a 3D-nanopatterned microfluidic chip. Nat Biomed Eng 3:438–451
13. Avella-Oliver M, Puchades R, Wachsmann-Hogiu S, Maquieira A (2017) Label-free SERS analysis of proteins and exosomes with large-scale substrates from recordable compact disks. Sensor Actuat B Chem 252:657–662
14. Zhou YG, Mohamadi RM, Poudineh M, Kermanshah L, Ahmed S, Safaei TS, Stojcic J, Nam RK, Sargent EH, Kelley SO (2016) Interrogating circulating microsomes and exosomes using metal nanoparticles. Small 12:727–732
15. Wang S, Zhang L, Wan S, Cansiz S, Cui C, Liu Y, Cai R, Hong C, Teng IT, Shi M (2017) Aptasensor with expanded nucleotide using DNA nanotetrahedra for electrochemical detection of cancerous exosomes. ACS Nano 11:3943–3949
16. Miao P, Yin J, Ning L, Li X (2014) Peptide-based electrochemical approach for apoptosis evaluation. Biosens Bioelectron 62:97101
17. Liu SG, Han L, Li N, Fan YZ, Yang YZ, Li NB, Luo HQ (2019) A ratiometric fluorescent strategy for alkaline phosphatase activity assay based on g-C_3N_4/CoOOH nanohybrid via target-triggered competitive redox reaction. Sensor Actuat B Chem 283:515–523
18. Wang Y, Sun H, Xu G, Guan M, Zhang Q, Wang Z, Dong Z, Chen W, Yang X, Qiao A (2022) A multiplexed electrochemical quantitative polymerase chain reaction platform for single-base mutation analysis. Biosens Bioelectron 214:114496
19. Wang X, Su J, Zeng D, Liu G, Liu L, Xu Y, Wang C, Liu X, Wang L, Mi X (2019) Gold nano-flowers (Au NFs) modified screen-printed carbon electrode electrochemical biosensor for label-free and quantitative detection of glycated hemoglobin. Talanta 201:119–125
20. Ruan X, Wang Y, Kwon EY, Wang L, Cheng N, Niu X, Ding S, Van Wie BJ, Lin Y, Du D (2021) Nanomaterial-enhanced 3D-printed sensor platform for simultaneous detection of atrazine and acetochlor. Biosens Bioelectron 184:113238
21. Su S, Wu Y, Zhu D, Chao J, Liu X, Wan Y, Su Y, Zuo X, Fan C, Wang L (2016) On-electrode synthesis of shape-controlled hierarchical flower-like gold nanostructures for efficient interfacial DNA assembly and sensitive electrochemical sensing of microRNA. Small 12:3794–3801
22. Filippova TA, Masamrekh RA, Shumyantseva VV, Latsis IA, Farafonova TE, Ilina IY, Kanashenko SL, Moshkovskii SA, Kuzikov AV (2023) Electrochemical biosensor for trypsin activity assay based on cleavage of immobilized tyrosine-containing peptide. Talanta 257:124341
23. Shahbazi-Derakhshi P, Mahmoudi E, Majidi MM, Sohrabi H, Amini M, Majidi MR, Niaei A, Shaykh-Baygloo N, Mokhtarzadeh A (2023) An ultrasensitive miRNA-based genosensor for detection of microRNA 21 in gastric cancer cells based on functional signal amplifier and synthesized perovskite-graphene oxide and AuNPs. Biosensors-Basel 13:172
24. Zhang XH, Jiang XJ, Wang W, Luo SH, Guan SJ, Li WB, Situ B, Li B, Zhang Y, Zheng L (2023) A simple and sensitive electrochemical biosensor for circulating tumor cell determination based on dual-toehold accelerated catalytic hairpin assembly. Microchim Acta 190:65
25. Qi X, Yan X, Zhao Y, Li L, Wang S (2020) Highly sensitive and specific detection of small molecules using advanced aptasensors based on split aptamers: a review. TrAC Trends Anal Chem 133:116069
26. Ali MR, Bacchu MS, Das S, Akter S, Rahman MM, Aly MAS, Khan MZH (2023) Label free flexible electrochemical DNA biosensor for selective detection of Shigella flexneri in real food samples. Talanta 253:123909

27. Zhang XN, Wang ZZ, Li X, Xiao W, Zou XB, Huang QL, Zhou LL (2021) Competitive electrochemical sensing for cancer cell evaluation based on thionine-interlinked signal probes. Analyst 148:912
28. Dong H, Chen H, Jiang J, Zhang H, Cai C, Shen Q (2018) Highly sensitive electrochemical detection of tumor exosomes based on aptamer recognition-induced multi-DNA release and cyclic enzymatic amplification. Anal Chem 90:4507–4513
29. Guo YH, Liu SH, Yang HL, Wang P, Feng QM (2021) Regenerable electrochemical biosensor for exosomes detection based on the dual-recognition proximity binding-induced DNA walker. Sens Actuat B-Chem 349:130765
30. Hallaj R, Ghafary Z, Mohammed OK, Shakeri R (2023) Induced ultrasensitive electrochemical biosensor for target MDA-MB-231 cell cytoplasmic protein detection based on RNA-cleavage DNAzyme catalytic reaction. Biosens Bioelectron 227:115168
31. Shiohara A, Wojnilowicz M, Lyu QX, Pei Y, Easton CD, Chen Y, White JF, McAuley A, Prieto-Simon B, Thissen H, Voelcker NH (2022) SARS-CoV-2 virus detection via a polymeric nanochannel-based electrochemical biosensor. Small 18:2205281
32. Liu ZH, Wang H, Li J, Wang ML, Yang HX, Si FC, Kong JM (2021) Detection of exosomes via an electrochemical biosensor based on C60-Au-Tb composite. Microchem J 170:106772
33. Zhou X, Bai D, Fu YX, Song L, Wu Y, Chen KA, Li JJ, Yang YJ, Chen HJ, Wang ZZ, Xie GM (2023) Detection of rare CTCs by electrochemical biosensor built on quaternary PdPtCuRu nanospheres with mesoporous architectures. Talanta 25:123955
34. Diaz-Fernandez A, Bernalte E, Fernandez-Ramos C, Moise S, Estrela P, Di Lorenzo M (2022) An impedimetric immunosensor for the selective detection of CD34(+) T-cells in human serum. Sens Actuator B-Chem 356:131306
35. Liu XJ, Wang QE, Chen J, Chen X, Yang WS (2021) Ultrasensitive electrochemiluminescence biosensor for the detection of tumor exosomes based on peptide recognition and luminol-AuNPs@g-C_3N_4 nanoprobe signal amplification. Talanta 221:121379
36. Wang L, Xie HJ, Zhou XY, Lin YX, Qin YJ, Yang J, Zhao J, Li GX (2023) An electrochemical biosensor to identify the phenotype of aggressive breast cancer cells. Chem Commun 59:3890–3893
37. Ramadan S, Lobo R, Zhang Y, Xu L, Shaforost O, Kwong Hong Tsang D, Feng J, Yin T, Qiao M, Rajeshirke A (2021) Carbon-dot-enhanced graphene field-effect transistors for ultrasensitive detection of exosomes. ACS Appl Mater Interfaces 13:7854–7864
38. Wang Y, Mao Z, Chen Q, Koh K, Hu X, Chen H (2022) Rapid and sensitive detection of PD-L1 exosomes using Cu-TCPP 2D MOF as a SPR sensitizer. Biosens Bioelectron 201:113954
39. Holzinger M, Le Goff A, Cosnier S (2014) Nanomaterials for biosensing applications: a review. Front Chem 2:63
40. Wu DH, He YH, Tong LJ, Wang JX, Liu L, Yi XY, Hu SQ (2021) Electrochemical determination of caspase-3 using signal amplification by HeLa cells modified with silver nanoparticles. Microchim Acta 188:110
41. Zhang ZH, Li QC, Du X, Liu M (2020) Application of electrochemical biosensors in tumor cell detection. Talanta 11:840–850
42. Lee M, Kim J, Jang M, Park C, Lee JH, Lee T (2022) Introduction of nanomaterials to biosensors for exosome detection: case study for cancer analysis. Biosensors 2:648
43. Zhao Y, Yao J, Wu ZL, Liu H, Zheng WW (2021) AuPt NPs with enhanced electrochemical oxidization activity for ratiometric electrochemical aptasensor. Biosens Bioelectron 196:113733
44. Wang SS, Zhao XP, Liu FF, Younis MR, Xia XH, Wang C (2019) Direct plasmon-enhanced electrochemistry for enabling ultrasensitive and label-free detection of circulating tumor cells in blood. Anal Chem 91:4413–4420
45. Wu X, Xiao TY, Luo Z, He RX, Cao YP, Guo ZZ, Zhang WY, Chen Y (2018) A micro-/nanochip and quantum dots-based 3D cytosensor for quantitative analysis of circulating tumor cells. J Nanobiotechnol 16:65
46. Hu K, Cheng JM, Wang KB, Zhao YQ, Liu YJ, Yang HX, Zhang ZQ (2021) Carbon nanotube based biosensors. Talanta 238:122987

47. Wei SG, Chen XJ, Zhang XY, Chen L (2021) Recent development of graphene based electrochemical sensor for detecting hematological malignancies-associated biomarkers: a mini-review. Front Chem 9:735668
48. Yang N, Chen X, Ren T, Zhang P, Yang D (2014) Carbon nanotube based biosensors. Sens Actuat B Chem 207:690–715
49. Liu MH, Xu YN, Huang CS, Jia T, Zhang XY, Yang DP, Jia NQ (2018) Hyaluronic acid-grafted three-dimensional MWCNT array as biosensing interface for chronocoulometric detection and fluorometric imaging of CD44-overexpressing cancer cells. Microchim Acta 185:338
50. Yazdanparast S, Benvidi A, Banaei M, Nikukar H, Tezerjani MD, Azimzadeh M (2018) Dual-aptamer based electrochemical sandwich biosensor for MCF-7 human breast cancer cells using silver nanoparticle labels and a poly(glutamic acid)/MWNT nanocomposite. Mikrochim Acta 185:405
51. Vajhadin F, Mazloum-Ardakani M, Shahidi M, Moshtaghioun SM, Haghiralsadat F (2021) Carbon nanotube based biosensors. Biosens Bioelectron 195:113626
52. Xia N, Wu DH, Yu HQ, Sun WW, Yi XY, Liu L (2020) Magnetic bead-based electrochemical and colorimetric assays of circulating tumor cells with boronic acid derivatives as the recognition elements and signal probes. Talanta 221:121640
53. Boriachek K, Masud MK, Palma C, Phan HP, Yamauchi Y, Hossain MSA, Nguyen NT, Salomon C, Shiddiky MJA (2019) Avoiding pre-isolation step in exosome analysis: direct isolation and sensitive detection of exosomes using gold-loaded nanoporous ferric oxide nanozymes. Anal Chem 91:3827–3834
54. Shen CC, Liu SP, Li XQ, Yang MH (2023) Electrochemical detection of circulating tumor cells based on DNA generated electrochemical current and rolling circle amplification. Anal Chem 91:11614–11619
55. Ke XY, Zhao Z, Huang JY, Liu C, Huang GS, Tan J, Zhu HQ, Xiao ZJ, Liu XY, Mei YF, Chu JH (2019) Growth control of metal-organic framework films on marine biological carbon and their potential-dependent dopamine sensing. ACS Appl Mater Interfaces 15:12005–12016
56. Mo GC, Qin DM, Jiang XH, Zheng XF, Mo WM, Deng BY (2020) A sensitive electrochemiluminescence biosensor based on metal-organic framework and imprinted polymer for squamous cell carcinoma antigen detection. Sens Actuat B-Chem 310:127852
57. Li XY, Li XM, Li DD, Zhao M, Wu HP, Shen B, Liu P, Ding SJ (2020) Electrochemical biosensor for ultrasensitive exosomal miRNA analysis by cascade primer exchange reaction and MOF@Pt@MOF. Biosens Bioelectron 168:112554
58. Jiang XC, Liu XH, Wu TD, Li L, Zhang RJ, Lu XQ (2019) Metal-organic framework derived carbon-based sensor for monitoring of the oxidative stress of living cell and assessment of antioxidant activity of food extracts. Talanta 194:591–597
59. Liu X, Gao X, Yang L, Zhao Y, Li F (2021) Metal-organic framework-functionalized paper-based electrochemical biosensor for ultrasensitive exosome assay. Anal Chem 93:11614–11619

Chapter 14
Electrochemical Biosensor with Machine Learning

Yunliang Chen, Xiaohui Huang, Ao Long, and Wei Han

Abstract The advancement of electrochemical biosensors poses new challenges for researchers to analyze data with interference and overlap. Machine learning shows its powerful ability in processing tasks such as classification, regression, etc., through the analysis of electrical signal data generated by electrochemical biosensors in recent years. The reason is that machine learning methods overcome subjectivity, which is usually the limitation of traditional methods. This chapter first introduces the background of electrochemical biosensor data analysis, followed by the detailed basics of machine learning techniques. Finally, it summarizes partial representative works on coupling machine learning techniques with electrochemical biosensor data analysis.

Keywords Biosensors · Electrochemical biosensor · Data analysis · Machine learning · Supervised learning · Unsupervised learning · Reinforcement learning · Active learning

14.1 Introduction

An electrochemical biosensor is an integrated device that transforms biological or chemical reaction signals into a detectable electrical signal in real time [1]. Electrochemical biosensors are sensitive, robust, and selective [2]. Therefore, they are now widely applied in applications such as DNA detection [3], early diagnosis of breast cancer [4], monitoring of human fitness [5], and foodborne pathogens detection [6]. Figure 14.1 depicts the schematic diagram of an electrochemical biosensor composed

Y. Chen · X. Huang (✉) · A. Long · W. Han
School of Computer Science, China University of Geosciences, Wuhan, Hubei, P. R. China
e-mail: xhhuang@cug.edu.cn

A. Long
e-mail: long.ao@cug.edu.cn

W. Han
e-mail: weihan@cug.edu.cn

F. Xia et al. (eds.), *Electrochemical Biosensors for Whole Blood Analysis*,
https://doi.org/10.1007/978-981-99-5644-9_14

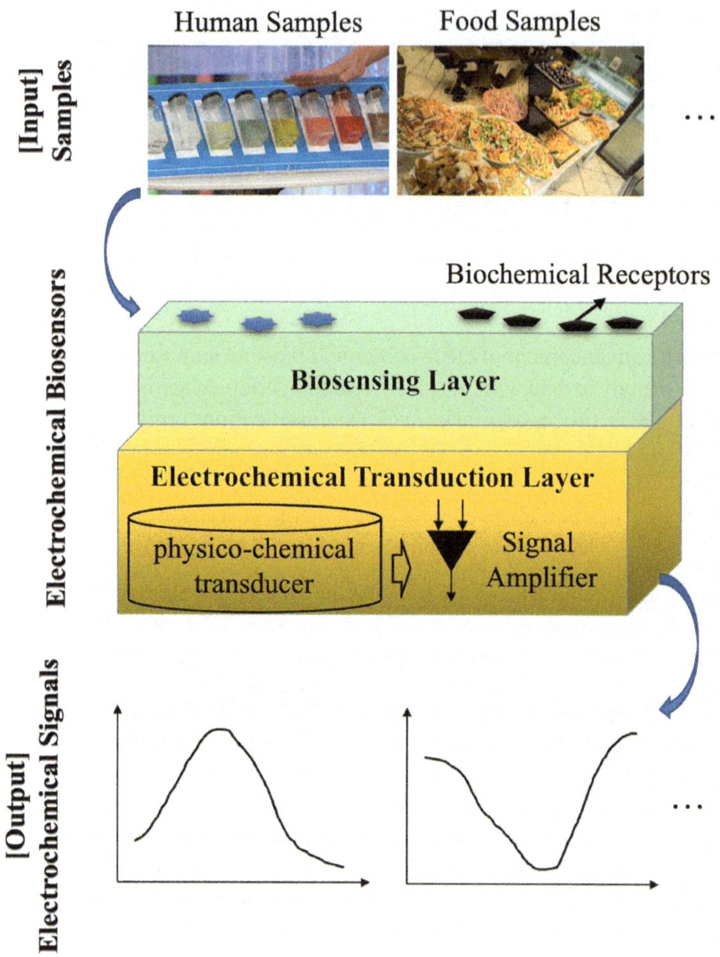

Fig. 14.1 The schematic diagram of an electrochemical biosensor

of two layers. The biosensing layer monitors the biological or chemical reaction with diverse biochemical receptors and then directly contacts the electrochemical transduction layer. The biological or chemical reaction signal is transformed into a detectable electrical signal, which is then enhanced with a signal amplifier as the final output.

The remarkable achievements of biosensors make it more convenient to detect or identify diverse analytes, thus remaining multiple signal data to be analyzed. In recent years, machine learning has been gradually applied in analyzing electrochemical biosensor data for its ability to learn from data automatically and to discover hidden relations of a sheer amount of data [7]. Machine learning benefits electrochemical biosensors because it can solve challenges caused by the interference and overlap in electrochemical biosensor data through the analysis of electrical signal data. For example, Orlandic et al. [1] propose a measure using non-invasive, wearable electrochemical biosensors and machine learning to predict an individual's passage of time perception. The goal is to provide real-time insights into an individual's mental state for continuous mental health monitoring. In this study, a machine learning optimization procedure is implemented to select the best machine learning model and hyperparameters for predicting POTP from the biomarkers that are derived from wearable electrochemical biosensors. Experimental results indicate that the model prediction accuracy is significantly improved by comparing to results without machine learning methods. Investigating how machine learning becomes an emerging opportunity for electrochemical biosensor data analysis is vital for researchers in electrochemical fields. Therefore, this chapter presents the basics of machine learning techniques and summarizes state-of-the-art works on applying machine learning techniques for electrochemical biosensors data analysis to convey insights into developing machine learning-based data analysis for electrochemical biosensors.

14.2 Backgrounds of Basic Machine Learning Techniques

The concept of machine learning is proposed to describe a set of algorithms (models) that can learn from sufficient datasets automatically to find patterns during the data analytical process [8]. Once a machine learning algorithm is built from the parts of a dataset (e.g., the training set), it can be adopted to infer the rest of the dataset (e.g., the testing set) on specific tasks according to the patterns it has learned from the dataset. Building a machine learning-based application consists of four steps [9], i.e., data collection, data preprocessing, model training and validation, and model evaluation, shown in Fig. 14.2. The data collection step collects historical data from diverse data sources such as sensors, databases, and logs. In this way, we have a sheer amount of data at the disposal of researchers. Later, the historical dataset is processed in the data preprocessing step. This step aims to conduct operations, e.g., data cleaning, de-duplication, and filling missing values, to solve data's heterogeneities. Therefore, data quality and integrity are ensured before feeding into the next step. After that, preprocessed data is divided into the training data and the testing data in a random way. The volume of training data usually accounts for a majority of the total preprocessed dataset. Users choose a desired machine learning algorithm to train and tune with the training data. Once the algorithm is ready, its efficiencies and effectiveness are validated using the validation data and are evaluated using the testing data. Lastly, the well-prepared machine learning algorithm is deployed in real applications to perform tasks such as prediction, classification, etc. Years have witnessed the great strides of machine learning techniques, which have become an enabling technology in diverse applications, i.e., natural language processing, traffic prediction, medical diagnosis, and analysis of biosensor data, through the past years [10, 11].

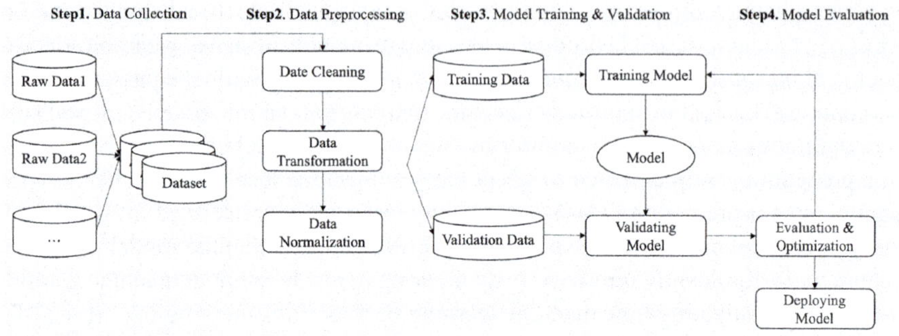

Fig. 14.2 A general overview of a machine learning process

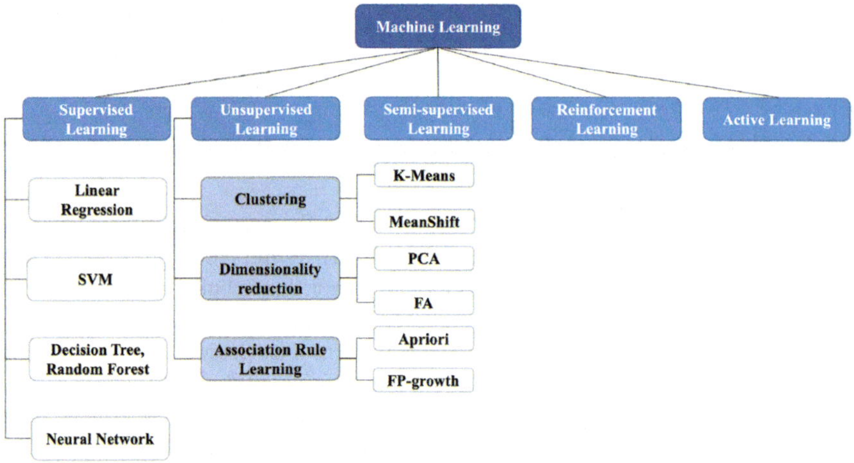

Fig. 14.3 The classification of machine learning techniques

In general, machine learning techniques can be categorized into five classes, as depicted in Fig. 14.3, namely unsupervised learning, semi-supervised learning, supervised learning, reinforcement learning, and active learning, respectively [12]. Each class can be performed on at least one of five major tasks including classification, clustering, association, regression and control [13]. In the rest of this section, each class of machine learning techniques is introduced respectively.

14.2.1 Supervised Learning

As a class of widely used machine learning techniques, supervised learning trains a model with a labeled dataset that is composed of data objects and their labels.

Supervised learning aims to find a function that maps the input data to the desired data, i.e., labels. It can be applied to new, unseen input data. Linear regression, support vector machine, decision tree and random forest, and neural networks are commonly used supervised learning methods.

14.2.1.1 Linear Regression

Linear regression is a statistical method and is used to establish the relationship between two continuous variables, where one variable, a.k.a. the dependent variable, is predicted based on the values of another variable, a.k.a. the independent variable, [14]. In this way, it can be used to make predictions about the value of one variable when giving the value of another variable. The basic idea of linear regression is to fit a straight line through a number of data points, where the line represents the best approximation of the relationship between the two variables. The line is determined by finding the values of the slope and intercept that minimize the sum of the squared differences between the observed data points and the predicted values on the line. This method is also known as the method of least squares.

The linear regression method is easy to understand and has strong interpretability, wide applicability, fast calculation speed, etc. Therefore, it is widely used in diverse fields, e.g., the analysis of proportions in finance with self-selection [15], etc. However, it is sensitive to outliers. In addition, if independent variables have a strong correlation, the model's coefficient could be unstable. Besides, linear regression can only deal with continuous data, which is unsuitable for classification and nonlinear problems.

14.2.1.2 Support Vector Machine

Support Vector Machine (SVM) is generally adopted in classification and regression analysis. The basic idea of SVM is to find a hyperplane that maximizes the margin between the classes, where the margin is the distance between the hyperplane and the closest data points of each class. Figure 14.4 depicts the basic idea of SVM in two-dimensional space. The dataset is divided into two classes, i.e., *Class* 1 and *Class* 2 by the hyperplane represented in the full line. SVM can work with both linear and non-linear data by using different kernels to transform the data into higher dimensions.

SVM has the ability to process high-dimensional data. It also has a strong generalization ability and can solve nonlinear problems. However, SVM still suffers from the huge computation overhead and thus needs a long training time when dealing with a sheer volume of the dataset. Besides, SVM shows insufficient classification effects when processing noisy, overlapping, or unbalanced datasets. For multi-class classification problems, extra techniques such as one-to-many and one-to-one strategies are generally adopted to ensure the effectiveness of SVM.

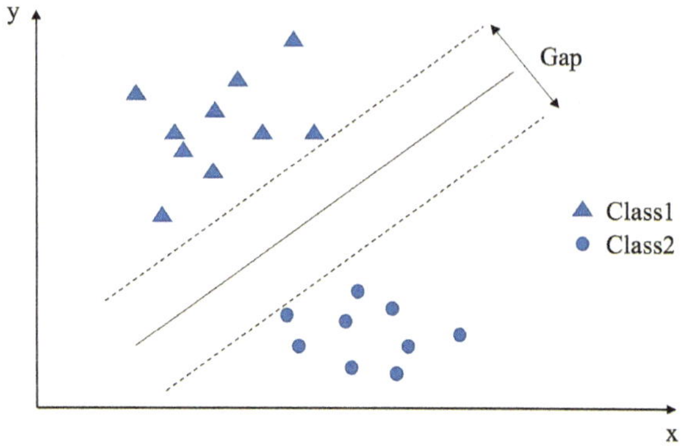

Fig. 14.4 An example of a SVM in two-dimensional space

14.2.1.3 Decision Tree and Random Forest

Decision tree is a tree structure-based supervised learning algorithm applied to solve classification and regression problems. It divides the dataset into many small decision units, each unit represents an interval of attribute values, and the last decision unit is a leaf node representing the final classification or regression results [16]. The generation process of the decision tree is a recursive process. Firstly, an attribute is selected as the root node, and then other attributes are divided to generate sub-nodes. The division process is repeated for each sub-node until the leaf node is generated. The generation of the decision tree needs to select the optimal attribute as the basis for division. Generally, the optimal attribute is selected using techniques including information gain, information gain ratio, Gini index, etc. Random forest is an integrated learning method aiming to build a classifier based on the results of a set of decision trees. It constructs multiple decision trees by selecting training samples and feature subsets in a random way and synthesizes the results of multiple decision trees by voting to carry out the final classification results.

Decision tree is explainable because it can present the decision process in a visual way. However, it is easy to over-fit. Random forest reduces the over-fitting risk of a single decision tree and improves the generalization ability of the model through random sampling, but it becomes difficult to visualize the decision-making process of the whole model. Moreover, the training process of random forest is time-consuming, which requires us to build multiple decision trees and select features on each decision tree. Decision trees and random forests are suitable for classification and regression tasks. They can handle a large number of features and samples. The difference between decision trees and random forests is that decision trees are suitable for scenarios that need to explain the decision-making process of the model, while random forests are suitable for scenarios that need high precision and generalization ability.

14.2.1.4 Neural Networks

Neural networks are a machine learning algorithm inspired by the human brain's structure and function [17]. A neural network is a set of connected nodes, namely neurons, each of which works together with other neurons to solve complex problems. A neural network typically consists of an input layer, at least one hidden layer, and an output layer. Figure 14.5 shows an example of a neural network, which is composed of an input layer, multiple hidden layers, and an output layer. The network takes a n-dimensional input and outputs a m-dimensional result. Each layer is a collection of multiple neurons, each of which is connected to neurons in the adjacent layers. The neurons in each layer process the input data, perform mathematical operations, and pass the output to the next layer. The connections between the neurons are weighted, and these weights are adjusted during the training process to improve the accuracy of the model. The training process involves feeding the network with input data and comparing the output with the actual output to calculate the error. The weights are then adjusted to minimize the error, and the process is repeated until the model achieves a stable and converged loss and accuracy.

As a powerful machine learning method, neural networks have the following advantages: they can process a large number of complex nonlinear data following the end-to-end learning pattern. That is to say, neural networks learn features directly from the original data without manual feature extraction. In addition, they can perform parallel computation and thus is suitable for analyzing large-scale data sets. Besides, even if the input data has a certain noise or error, the neural networks can automatically adjust the parameters to adapt to the data as much as possible, which means they are fault-tolerant. Therefore, neural networks are widely applied in various applications such as natural language processing [18], and image processing [19].

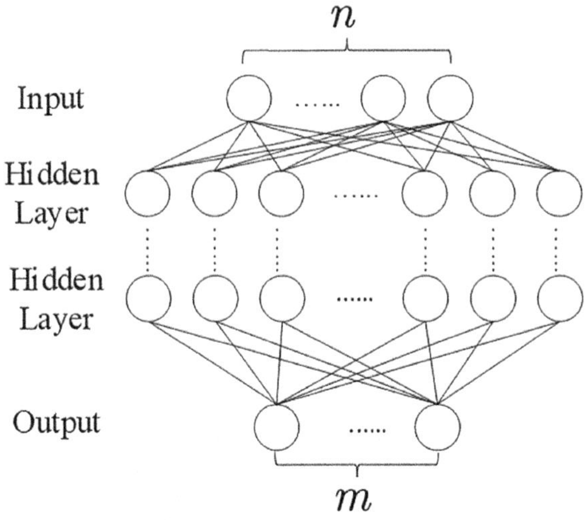

Fig. 14.5 An example of a neural network consisting of multiple hidden layers

However, the neural network also still suffers from the following limitations: firstly, training a neural network is time-consuming and requires a lot of computing resources to carry out the back-propagation algorithm, aiming to calculate the gradient and update the weights. Secondly, sometimes the trained neural network may be over-fitting. That is to say, the model performs well in training data but not well in testing data. Thirdly, neural networks become "black boxes" that are difficult for us to understand when the number of hidden layers is increased.

14.2.2 Unsupervised Learning

Unsupervised learning finds patterns from unlabeled data, which is different from supervised learning. It has the ability to process unlabeled data, aiming to discover hidden patterns and relationships. Unsupervised learning methods, including clustering, dimensionality reduction, and association rule learning, focus on finding patterns or relationships within the dataset that are not explicitly labeled [20].

Clustering [21] is used to divide a group of data samples into several independent subsets, each of which is called a cluster. The goal of the clustering algorithm is to make the data samples in the same cluster as similar as possible while the samples in different clusters are as different as possible. Generally speaking, the input of the clustering algorithm is a group of unlabeled data, while the output is a group of labeled clusters. The process of a clustering algorithm usually includes the following steps:

Step 1: Determining appropriate distance measurement methods, such as Euclidean distance, and Manhattan distance.

Step 2: Choosing appropriate clustering algorithms, such as *k*-Means and hierarchical clustering.

Step 3: Initializing a group of cluster center points according to the selected clustering algorithm.

Step 4: Allocating data samples to the nearest cluster center point.

Step 5: Updating the center point of each cluster.

Step 6: Repeating Step 4 and Step 5 until the stop condition is met. For example, the change of the cluster center point is less than a certain threshold or reaches the specified number of iterations.

Dimension reduction [22] is a data processing technology used to convert high-dimensional data into low-dimensional data. The purpose of dimensionality reduction is to reduce the dimension of the data while preserving vital features of the data as many as possible, aiming to better analyze and visualize the data. Dataset with very high dimensions is often encountered and may contain a lot of redundant information or noise, which will lead to difficulties in data analysis and modeling. The dimensionality reduction algorithm can eliminate redundant information and noise to improve the efficiency and accuracy of data processing. Dimension reduction algorithms can be divided into two categories, i.e., linear dimension reduction and nonlinear dimension reduction. Linear dimension reduction reduces the dimension of data through linear transformation, such as principal component analysis (PCA), and factor analysis (FA). Nonlinear dimensionality reduction reduces the dimension of data through nonlinear transformation, such as local linear embedding (LLE), and isometric mapping (Isomap).

Association rule learning [23] is mainly used to discover the relationships and dependencies between items in a dataset, aiming to make decisions and predict behaviors. Many algorithms belong to learning association rules., e.g., the widely used the priori algorithm, which is divided into two stages. In the first stage, the a priori algorithm generates all frequent item sets. Note that the a priori algorithm gradually generates a larger itemset from the monomial set until it cannot be regenerated. Then, the a priori algorithm generates association rules based on frequent item sets and calculates the support and confidence of association rules in the second stage.

14.2.3 Semi-supervised Learning

Semi-supervised learning trains a model on both labeled and unlabeled data. This type of technique combines the advantages of both supervised and unsupervised learning [24]. In semi-supervised learning, the model is initially trained on a small amount of labeled data at first. Then unsupervised learning techniques are adopted to analyze and extract patterns from the remaining unlabeled data. The patterns learned from the unlabeled data are used to improve the accuracy and generalization of the model on the labeled data. Semi-supervised learning is usually divided into two categories, i.e., generation-based methods and graph-based methods [25, 26]. The generation-based methods use existing labeled data to train a model, which is then applied to predict the unlabeled data and obtain their Pseudo-labels. The graph-based model constructs a graph of labeled data and unlabeled data and then uses the information of nodes and edges to predict labels of the above unlabeled data in the graph.

14.2.4 Reinforcement Learning

Reinforcement learning [27] trains a model to make decisions based on feedback from the environment. It is inspired by the concept of trial-and-error learning, where an agent learns to make decisions by taking actions in an environment and receiving feedback in the form of rewards or penalties. In general, reinforcement learning consists of five elements, i.e., state, action, reward, policy, and value function. A state refers to the environmental state of the agent. It is used to describe the observation of the agent and the characteristics of the environment. An action taken by an agent has an impact on the environment. A reward is calculated once an action is performed by an agent in a specific state, the reward is then used to guide the agent's behaviors. The policy is usually presented with probability distributions and is applied to describe the way an agent takes action. A value function measures the long-term reward value of an agent in a specific state or action, which is used to guide the decision of the agent. Reinforcement learning enables agents to learn the best strategy through interaction with the environment to achieve autonomous learning. It can also learn the best strategy in different environments and can adapt to different environments through continuous trial and error.

14.2.5 Active Learning

Active learning is a machine learning technique that involves iteratively selecting the most informative samples or instances for labeling by a human expert, aiming to improve the accuracy and efficiency of the learning process [28]. It is useful when labeled data is scarce or expensive to obtain, and when the cost of labeling data can be reduced by selecting the most informative data points. The process of how active learning works is illustrated in Fig. 14.6. The main idea behind active learning is to let the model play an active role in selecting which data to label and use for training. That is to say, the model selects the most valuable data samples from the unlabeled data, and then manually marks data samples to retrain the model. Hence, the accuracy and generalization ability of the model can be incrementally improved.

Active learning is widely applied in machine learning tasks such as classification, regression, and clustering. The advantage of active learning is that it can maximize

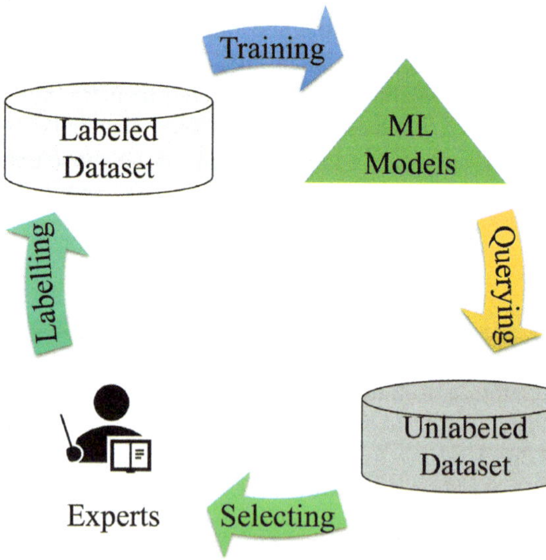

Fig. 14.6 The process of how active learning works

the use of data and improve the learning efficiency and performance of the model, especially in the case of the volume of data sets is limited. However, active learning requires labeling the selected data manual. Therefore, domain experts with experience and expertise are required to carry out the labeling work. In addition, Selection bias needs to be taken into consideration in active learning because it selects data samples to be labeled based on the model's own selection strategy.

14.3 Machine-Learning-Driven Electrochemical Biosensor Data Analysis

Machine learning has been gradually used in the field of electrochemical biosensor data analysis in recent years. The reason is that machine learning provides new visions to overcome the challenges faced when analyzing data generated by electrochemical biosensors. Electrochemical biosensors are used as a kind of detection or diagnostic equipment and inevitably have some irregular signal noise. What's more, some electrochemical biosensors rely heavily on antibodies or aptamers as biological receptors, which limits the short shelf life and poor stability of biosensors. At present, these issues limit their commercialization. At the same time, machine learning provides the ability to process signal data for complex matrices effectively. Besides, machine learning makes it possible to obtain analytical results from noisy and signal data that may be overlapped. In addition, proper deployment of machine learning techniques can discover hidden relations between sample parameters and signal data. Currently, machine learning is used to analyze the raw data generated by electrochemical biosensors in several ways, including categorization, anomaly detection, etc. [7].

Multiple researchers have applied machine learning algorithms to electrochemical biosensors. We reviewed the representative works in machine-learning-driven electrochemical biosensors and showed the result as shown in Table 14.1. It can be observed from Table 14.1 that although these papers involve different machine learning tasks and use different machine learning methods, the motivation of all the surveyed works for electrochemical biosensors is to boost the performance of the biosensors. For example, Xu et al. [29] demonstrate a machine-learning-based EIS biosensor for E.coli detection with improved accuracy, Puthongkham et al. [30] analyze complicated electrochemical data to improve calibration and analyte classification, etc.

In the perspective of machine learning tasks, machine learning algorithms are generally used to solve classification and regression for data analysis of electrochemical biosensors. For example, the authors in [32] identify and predict the concentration of glucose and lactic acid by using a neural network from a classification view. Authors in [31] focus on Glucose-Oxidase Biosensor (GOB) modeling from a regression perspective. The main reason why clustering problems are rare is that most data sets made from the raw data generated by electrochemical biosensors are usually

Table 14.1 Summarization of recent works on electrochemical biosensors data analysis with machine learning

References	ML Task [a] & its Description	ML Techniques [b]
Xu et al. [29]	(Regression) Finding the relationship between multiple impedimetric parameters and bacterial concentrations	SVM and PCA
Gonzalez-Navarro et al. [31]	(Regression) Modeling GOB through statistical learning methods from a regression work perspective	SVM and Neural Networks
Zhou et al. [32]	(Classification) Predicting the concentration of glucose and lactic acid	Neural Networks
Amethiya et al. [33]	(Classification) Detecting breast cancer quickly and effectively with various machine learning algorithms	SVM, Random Forest and Neural Networks
Puthongkham et al. [30]	(Regression and Classification) Performing electrochemical analysis by extracting relationships between chemical structures	Linear Regression, Neural Networks, and SVM
Zhu et al. [34]	(Regression) Analyzing carbendazim (CBZ) residues in tea and rice samples	Neural Networks
Du et al. [35]	Regression) Using a microbial electrochemical sensor to quantify multiple toxicants simultaneously, aiming to enable multiple-parameter biotoxicity detection for environmental monitoring	SVM and Neural Networks

[a]ML: Machine Learning
[b]SVM: Support Vector Machine, PCA: Principal Component Analysis

labeled. So lots of issues belong to the category of supervised learning in which classification and regression are the core tasks.

In the view of machine learning methods, SVM and neural networks are the most used algorithms. This is because SVM can process high-dimensional data, reduce the amount of computation, and have strong generalization ability. Xu et al. [29] first uses PCA to reduce dimensions and then applies SVM to regression for E. coli detection. SVM is not suitable for processing large-scale datasets, has insufficient classification effect for noisy data and overlapping data, and needs to use some special skills for multi-class classification problems. But neural network solves these problems exactly. Zhu et al. [34] finds neural network outperforms traditional regression when analyzing CBZ. In summary, SVM and neural network algorithms complement each other and can solve most classification and regression problems in electrochemical biosensors. Puthongkham et al. [30], Du et al. [35] use both SVM and neural network to extract complex relationships and analyze complicated electrochemical data.

By integrating machine learning techniques, electrochemical biosensors become more intelligent and overcome the limitation of low selectivity and detect with improved accuracy compared to traditional biosensors, even some biosensors have the ability to predict the type or concentration of analytes. Such as authors in

[33] point machine learning makes it possible to detect breast cancer quickly and effectively.

14.4 Conclusion

Machine learning techniques can process electrical signal data with interference and overlap, thus making electrochemical biosensors intelligent. This chapter presents the adoption of machine learning techniques in electrochemical biosensor data analysis.

Both the efficiencies and effectiveness of electrochemical biosensor data analytical tasks are improved after machine learning techniques such as SVM and neural networks are adopted. However, machine learning techniques are still limited due to the lack of standard datasets and evaluation metrics in electrochemical biosensor data analysis.

From linear calibrations and nonlinear fittings to advanced machine learning methods for the classification, regression, and clustering of complex biological samples, chemometrics provides robust mathematical tools for biosensor data interpretation. Machine learning methods, including deep learning, are becoming popular in electrochemical biosensor data analysis in recent years. Besides, machine learning is explainable instead of a black box. That is to say, the machine's decision is understandable to chemical professionals and decision-makers. At the same time, machine learning techniques, especially deep learning need a sheer amount of data to train a well-performed model. To break through the data bottleneck for connecting the machine learning and electrochemical biosensors, some federal agencies need to make elements of the biosensor data available to trusted research teams.

Acknowledgements This work is supported by the National Natural Science Foundation of China (No. 62076224) and the Open Research Project of The Hubei Key Laboratory of Intelligent Geo-Information Processing (KLIGIP-2022-B16).

References

1. Ronkainen NJ, Halsall HB, Heineman WR (2010) Electrochemical biosensors. Chem Soc Rev 39(5):1747–1763. https://doi.org/10.1039/B714449K. https://pubs.rsc.org/en/content/articlelanding/2010/cs/b714449k
2. Moon JM, Thapliyal N, Hussain KK, Goyal RN, Shim YB (2018) Conducting polymer-based electrochemical biosen- sors for neurotransmitters: a review. Biosens Bioelectron 102:540–552. https://doi.org/10.1016/j.bios.2017.11.069. https://www.sciencedirect.com/science/article/abs/pii/S0956566317307984
3. Wang YH, Huang KJ, Wu X (2017) Recent advances in transition- metal dichalcogenides based electrochemical biosensors: a review. Biosens Bioelectron 97:305–316. https://doi.org/10.1016/j.bios.2017.06.011. https://www.sciencedirect.com/science/article/abs/pii/S0956566317303871

4. Jing L, Xie C, Li Q, Yang M, Li S, Li H, Xia F (2021) Electrochemical biosensors for the analysis of breast cancer biomarkers: from design to application. Anal Chem 94(1):269–296. https://doi.org/10.1021/acs.analchem.1c04475
5. Min J, Sempionatto JR, Teymourian H, Wang J, Gao W (2021) Wearable electrochemical biosensors in north america. Biosens Bioelectron 172:112750. https://doi.org/10.1016/j.bios.2020.112750. https://www.sciencedirect.com/science/article/abs/pii/S0956566320307387
6. Mei Y, He C, Zeng W, Luo Y, Liu C, Yang M, Kuang Y, Lin X, Huang Q (2022) Electrochemical biosensors for foodborne pathogens detection based on carbon nanomaterials: recent advances and challenges. Food Bioproc Technol 15(3):498–513. https://doi.org/10.1007/s11947-022-02759-7
7. Cui F, Yue Y, Zhang Y, Zhang Z, Zhou HS (2020) Advancing biosensors with machine learning. ACS sensors 5(11):3346–3364. https://doi.org/10.1021/acssensors.0c01424. https://pubmed.ncbi.nlm.nih.gov/33185417/
8. Zhong Z, Xu M, Rodriguez MA, Xu C, Buyya R (2022) Machine learning-based orchestration of containers: a taxonomy and future directions. ACM Comp Surv 54(10s):1–35. https://doi.org/10.1145/3510415
9. Zhou ZH (2021) Machine learning. Springer Nature
10. Jordan MI, Mitchell TM (2015) Machine learning: trends, perspectives, and prospects. Science 349(6245):255–260. https://doi.org/10.1126/science.aaa8415
11. Ashmore R, Calinescu R, Paterson C (2021) Assuring the machine learning life- cycle: desiderata, methods, and challenges. ACM Comp Surv 54(5):1–39. https://doi.org/10.1145/3453444
12. Wu N, Xie Y (2022) A survey of machine learning for computer architecture and systems. ACM Comp Surv (CSUR) 55(3):1–39. https://doi.org/10.1145/3494523
13. Parmar J, Chouhan S, Raychoudhury V, Rathore S (2023) Open- world machine learning: applications, challenges, and opportunities. ACM Comp Surv 55(10):1–37. https://doi.org/10.1145/3561381
14. Maulud D, Abdulazeez AM (2020) A review on linear regression comprehensive in machine learning. J Appl Sci Technol Trends 1(4):140–147. https://doi.org/10.38094/jastt1457
15. Cook DO, Kieschnick R, McCullough BD (2008) Regression analysis of proportions in finance with self selection. J Empir Finance 15(5):860–867. https://doi.org/10.1016/j.jempfin.2008.02.001. https://www.sciencedirect.com/science/article/abs/pii/S0927539808000145
16. Qi Y (2012) Random forest for bioinformatics. In: Ensemble machine learning: methods and applications. Springer, pp 307–323
17. Yuste R (2015) From the neuron doctrine to neural networks. Nat Rev Neurosci 16(8):487–497. https://doi.org/10.1038/nrn3962. https://www.nature.com/articles/nrn3962
18. Goldberg Y (2016) A primer on neural network models for natural language processing. J Artif Intell Res 57:345–420. https://doi.org/10.1613/jair.4992. https://www.jair.org/index.php/jair/article/view/11030
19. Szegedy C, Ioffe S, Vanhoucke V, Alemi A (2017) Inception-v4, inception resnet and the impact of residual connections on learning. In: Proceedings of the AAAI conference on artificial intelligence, vol 31. https://doi.org/10.1609/aaai.v31i1.11231. https://ojs.aaai.org/index.php/aaai/article/view/11231
20. Alayrac JB, Bojanowski P, Agrawal N, Sivic J, Laptev I, Lacoste-Julien S (2016) Unsupervised learning from narrated instruction videos. In: Proceedings of the IEEE conference on computer vision and pattern recognition, pp 4575–4583
21. Al-Omary AY, Jamil MS (2006) A new approach of clustering based machine-learning algorithm. Knowledge-Based Syst 19(4):248–258. https://doi.org/10.1016/j.knosys.2005.10.011. https://www.sciencedirect.com/science/article/abs/pii/S0950705106000189
22. Sorzano COS, Vargas J, Montano AP (2014) A survey of dimensionality reduction techniques. arXiv preprint https://arxiv.org/abs/1403.2877
23. Zheng Z, Kohavi R, Mason L (2001) Real world performance of association rule algorithms. In: Proceedings of the seventh ACM SIGKDD international conference on Knowledge discovery and data mining, pp 401–406. https://doi.org/10.1145/502512.502572

24. Van Engelen JE, Hoos HH (2020) A survey on semi-supervised learning. Mach Learn 109(2):373–440. https://doi.org/10.1007/s10994-019-05855-6
25. Yang X, Song Z, King I, Xu Z (2022) A survey on deep semi-supervised learning. IEEE Trans Knowl Data Eng https://doi.org/10.1109/TKDE.2022.3220219. https://ieeexplore.ieee.org/document/9941371
26. Chong Y, Ding Y, Yan Q, Pan S (2020) Graph- based semi-supervised learning: a review. Neurocomputing 408:216–230. https://doi.org/10.1016/j.neucom.2019.12.130. https://www.sciencedirect.com/science/article/abs/pii/S0925231220304938
27. LP Kaelbling ML Littman AW Moore 1996 Reinforcement learning: a survey J Artif Intell Res 4:237–285. https://doi.org/10.1613/jair.301. https://www.jair.org/index.php/jair/article/view/10166
28. Ren P, Xiao Y, Chang X, Huang PY, Li Z, Gupta BB, Chen X, Wang X (2021) A survey of deep active learning. ACM Comput Surv 54(9):1–40. https://doi.org/10.1145/3472291
29. Xu Y, Li C, Jiang Y, Guo M, Yang Y, Yang Y, Yu H (2020) Electrochemical impedance spectroscopic detection of e. coli with machine learning. J Electrochem Soc 167(4):047508. https://doi.org/10.1149/1945-7111/ab732f
30. Puthongkham P, Wirojsaengthong S, Suea-Ngam A (2021) Ma- chine learning and chemometrics for electrochemical sensors: moving forward to the future of analytical chemistry. Analyst 146(21):6351–6364. https://doi.org/10.1039/D1AN01148K. https://pubs.rsc.org/en/content/articlelanding/2021/an/d1an01148k
31. Gonzalez-Navarro FF, Stilianova-Stoytcheva M, Renteria-Gutierrez L, Belanche-Muñoz LA, Flores-Rios BL, Ibarra-Esquer JE (2016) Glucose oxidase biosensor modeling and predictors optimization by machine learning methods. Sensors 16(11):1483. https://doi.org/10.3390/s16111483. https://www.mdpi.com/1424-8220/16/11/1483
32. Zhou Z, Wang L, Wang J, Liu C, Xu T, Zhang X (2022) Machine learning with neural networks to enhance selectivity of nonenzymatic electrochemical biosensors in multianalyte mixtures. ACS Appl Mater Interf 14(47):52684–52690. https://doi.org/10.1021/acsami.2c17593
33. Amethiya Y, Pipariya P, Patel S, Shah M (2022) Comparative analysis of breast cancer detection using machine learning and biosensors. Intell Med 2(2):69–81. https://doi.org/10.1016/j.imed.2021.08.004. https://www.sciencedirect.com/science/article/pii/S2667102621000887
34. Zhu X, Liu P, Ge Y, Wu R, Xue T, Sheng Y, Ai S, Tang K, Wen Y (2020) Mos2/mwcnts porous nanohybrid network with oxidase-like characteristic as electrochemical nanozyme sensor coupled with machine learning for intelligent analysis of carbendazim. J Electroanalyt Chem 862:113940. https://doi.org/10.1016/j.jelechem.2020.113940. https://www.sciencedirect.com/science/article/abs/pii/S1572665720301235
35. Du L, Yan Y, Li T, Liu H, Li N, Wang X (2021) Machine learning enables quantification of multiple toxicants with microbial electrochemical sensors. ACS ES&T Eng 2(1):92–100. https://doi.org/10.1021/acsestengg.1c00287

The manufacturer's authorised representative in the EU is Springer Nature Customer Service Centre GmbH, Europaplatz 3, 69115 Heidelberg, Germany. If you have any concerns regarding our products, please contact ProductSafety@springernature.com

Printed and bound by CPI Group (UK) Ltd, Croydon, CR0 4YY
01/10/2026
02225867-0006